For Mary

ὅ τι καλὸν φίλον ἀεί

Preface

This book is intended for use by scientists and engineers in their research or in their professional practice, and for use as a textbook in one-semester or one-quarter courses on soil chemistry or biogeochemistry. A background in basic soil science as found, for example, in *Introduction to the Principles and Practice of Soil Science* by R. E. White, *Soils* by William Dubbin, or *Soils: Genesis and Geomorphology* by Randall Schaetzl and Sharon Anderson is assumed on the part of the reader. An understanding of chemistry and calculus at an elementary level also is necessary, although the latter topic is in fact applied very sparingly throughout the text. Familiarity with statistical analysis is of considerable benefit while solving some of the problems accompanying the text.

The general plan of the book is to introduce the principal reactive components of soils in the first four chapters, then to describe important soil chemical processes in the next six. One hopes that the reader will notice that a conscious effort has been made throughout the text to blur somewhat the distinction between soil chemistry and soil microbiology. The final pair of chapters discusses applications of soil chemistry to the two most important issues attending the maintenance of soil quality for agriculture: soil acidity and soil salinity. These two chapters are not intended to be comprehensive reviews, but instead to serve as guides to the soil chemistry underlying the topics discussed in more specialized courses or books on soil quality management.

A brief appendix on *le Système International d'Unités* (SI units) and physical quantities used in soil chemistry is provided at the end of the book. Readers are advised to review this appendix and work the problems in it before beginning to read the book itself, not only as a prequel to the terminology appearing in the text, but also as an aid to evaluating the status of their understanding of introductory chemistry. The 180 problems following the chapters in this book have been designed to reinforce or extend the main points discussed and thus are regarded as an integral part of the text. No reader should be satisfied with her or his understanding of soil chemistry without undertaking at least a substantial portion of these problems. In addition to the problems, an annotated reading list at the end of each chapter is offered to those who wish to explore in greater depth the subject matter discussed. Both the problems and the reading lists should figure importantly in any course of university lectures based on this book.

Acknowledgments

I must thank Angela Zabel for her excellent preparation of the typescript of this book and Cynthia Borcena for her most creative preparation of its figures. I must also express gratitude to Stephen Judge for initially suggesting that I write a soil chemistry textbook, to Harvey Doner and John Hsu for providing lengthy commentary on and corrections to its first edition (now nearly 20 years old), and to Kirk Nordstrom for his generosity in sharing material related to aluminum geochemistry that strongly influenced the writing of Chapter 11. Thanks also to Kideok Kwon, Sung-Ho Park, and Rebecca Sutton for providing original artwork for some of the figures, and to Teri Van Dorston for her able assistance in providing an image of the painting that adorns the front cover of this book. Finally, I must express my great indebtedness to two of my Berkeley colleagues—Bob Hass and Tony Long—humanists whose example and perspectives have influenced my writing in ways I could scarcely have imagined two decades ago.

Berkeley, California
May 2007

Cover art: Grant Wood, *Young Corn*, 1931. Oil on Masonite panel, 24 × 29 7/8 in. Collection of the Cedar Rapids Community School District, on loan to the Cedar Rapids Museum of Art. Used with permission of the owner.

The Chemistry of Soils

Second Edition

Garrison Sposito

2008

OXFORD
UNIVERSITY PRESS

Oxford University Press, Inc., publishes works that further
Oxford University's objective of excellence
in research, scholarship, and education.

Oxford New York
Auckland Cape Town Dar es Salaam Hong Kong Karachi
Kuala Lumpur Madrid Melbourne Mexico City Nairobi
New Delhi Shanghai Taipei Toronto

With offices in
Argentina Austria Brazil Chile Czech Republic France Greece
Guatemala Hungary Italy Japan Poland Portugal Singapore
South Korea Switzerland Thailand Turkey Ukraine Vietnam

Published by Oxford University Press, Inc.
198 Madison Avenue, New York, New York 10016

www.oup.com

Oxford is a registered trademark of Oxford University Press.

Library of Congress Cataloging-in-Publication Data
Sposito, Garrison, 1939–
The chemistry of soils / Garrison Sposito.
p. cm.
Includes bibliographical references and index.
ISBN 978-0-19-531369-7
1. Soil chemistry. I. Title.
S592.5.S656 2008
631.4′1—dc22 2007028057

1 3 5 7 9 8 6 4 2

Printed in the United States of America
on acid-free paper

Contents

The Chemistry of Soils

Right thinking is the greatest excellence,
and wisdom is to speak the truth
and act in accordance with Nature,
while paying attention to it.

—HERACLITUS OF EPHESUS

Now we give place to the genius of soils,
the strength of each, its hue,
its native power for bearing.

—VERGIL, *GEORGICS II*

1

The Chemical Composition of Soils

1.1 Elemental Composition

Soils are porous media created at the land surface through weathering processes mediated by biological, geological, and hydrological phenomena. Soils differ from mere weathered rock, however, because they show an approximately vertical stratification (the *soil horizons*) that has been produced by the continual influence of percolating water and living organisms. From the point of view of chemistry, soils are open, multicomponent, biogeochemical systems containing solids, liquids, and gases. That they are open systems means soils exchange both matter and energy with the surrounding atmosphere, biosphere, and hydrosphere. These flows of matter and energy to or from soils are highly variable in time and space, but they are the essential fluxes that cause the development of soil profiles and govern the patterns of soil quality.

The role of soil as a dynamic reservoir in the cycling of chemical elements can be appreciated by examining tables 1.1 and 1.2, which list average mass concentrations of important nonmetal, metal, and metalloid chemical elements in continental crustal rocks and soils. The rock concentrations take into account both crustal stratification and the relative abundance of sedimentary, magmatic, and metamorphic subunits worldwide. The soil concentrations refer to samples taken approximately 0.2 m beneath the land surface from uncontaminated mineral soils in the conterminous United States. These latter concentration data are quite comparable with those for soils sampled worldwide. The average values listed have large standard deviations, however, because of spatial heterogeneity on all scales.

Table 1.1
Mean content (measured in milligrams per kilogram) of nonmetal elements in crustal rocks and United States soils.

Element	Crust[a]	Soil[b]	Element	Crust[a]	Soil[b]
B	17	26	P	757	260
C	1990	16,000	S	697	1200
N	60	2000	Cl	472	100
O	472,000	490,000	Se	0.12	0.26

[a]Wedepohl, K. H. (1995) The composition of the continental crust. *Geochim. Cosmochim. Acta* 59: 1217.
[b]Schacklette, H. T., and J. G. Boerngen. (1984) *Element concentrations in soils and other surficial materials of the conterminous United States.* U.S. Geological Survey Professional Paper 1270.

Table 1.2
Mean content (measured in milligrams per kilogram) of metal and metalloid elements and their anthropogenic mobilization factors (AMFs).

Element	Crust[a]	Soil[b]	AMF[c]	Element	Crust[a]	Soil[b]	AMF[c]
Li	18	20	3	Cu	25	17	632
Be	2.4	0.6	2	Zn	65	48	115
Na	23600	5900	2	As	1.7	5.2	27
Mg	22000	4400	<1	Sr	333	120	3
Al	79600	47000	<1	Zr	203	180	4
Si	288000	310000	<1	Mo	1.1	0.6	80
K	21400	15000	<1	Ag	0.07	0.05	185
Ca	38500	9200	2	Cd	0.1	0.2	112
Ti	4010	2400	1	Sn	23	0.9	65
V	98	58	14	Sb	0.3	0.5	246
Cr	126	37	273	Cs	3.4	4.0	12
Mn	716	330	10	Ba	584	440	4
Fe	43200	18000	16	Hg	0.04	0.06	342
Co	24	7	4	Pb	14.8	16	127
Ni	56	13	56	U	1.7	2.3	12

[a]Wedepohl, K. H. (1995) The composition of the continental crust. *Geochim. Cosmochim. Acta* 59:1217.
[b]Schacklette, H. T., and J. G. Boerngen. (1984) *Element concentrations in soils and other surficial materials of the conterminous United States.* U.S. Geological Survey Professional Paper 1270.
[c]AMF = mass extracted annually by mining and fossil fuel production ÷ mass released annually by crustal weathering and volcanic activity. Data from Klee, R. J., and T. E. Graedel. (2004) Elemental cycles: A status report on human or natural dominance. *Annu. Rev. Environ. Resour.* 29:69.

The *major elements* in soils are those with concentrations that exceed 100 mg kg^{-1}, all others being termed *trace elements.* According to the data in tables 1.1 and 1.2, the major elements include O, Si, Al, Fe, C, K, Ca, Na, Mg, Ti, N, S, Ba, Mn, P, and perhaps Sr and Zr, in decreasing order of concentration. Notable among the major elements is the strong enrichment of C and N in soils relative to crustal rocks (Table 1.1), whereas Ca, Na, and Mg show significant depletion (Table 1.2). The strong enrichment of C and N is a result of the principal chemical forms these elements assume in soils—namely, those associated with organic matter. The average C-to-N, C-to-P, and C-to-S ratios (8, 61, and 13 respectively) in soils, indicated by the data in Table 1.1, are very low and, therefore, are conducive to microbial mineralization processes, further reflecting the active biological milieu that distinguishes soil from crustal rock.

The major elements C, N, P, and S also are *macronutrients,* meaning they are essential to the life cycles of organisms and are absorbed by them in significant amounts. The global biogeochemical cycles of these elements are therefore of major interest, especially because of the large anthropogenic influence they experience. Mining operations and fossil fuel production, for example, combine to release annually more than 1000 times as much C and N, 100 times as much S, and 10 times as much P as is released annually worldwide from crustal weathering processes. In soils, these four elements undergo biological and chemical transformations that release them to the vicinal atmosphere, biosphere, and hydrosphere, as illustrated in Figure 1.1, a flow diagram that applies to natural soils at spatial scales ranging from pedon to landscape. The two storage components in Figure 1.1 respectively depict the litter layer and *humus,* the organic matter not identifiable as unaltered or partially altered biomass. The microbial transformation of litter to humus is termed *humification.* The content of humus in soils worldwide varies systematically with climate, with accumulation being favored by low temperature and high precipitation. For example, the average humus content in desert soils increases by about one order of magnitude as the mean annual surface temperature drops fivefold. The average humus content of tropical forest soils increases approximately threefold as the mean annual precipitation increases about eightfold. In most soils, the microbial degradation of litter and humus is the process through which C, N, S, and P are released to the contiguous aqueous phase (the *soil solution*) as inorganic ions susceptible to uptake by the biota or loss by the three processes indicated in Figure 1.1 by arrows outgoing from the humus storage component.

Important losses of C from soils occur as a result of leaching, erosion, and runoff, but most quantitative studies have focused on emissions to the atmosphere in the form of either CO_2 or CH_4 produced by respiring microorganisms. The CO_2 emissions do not arise uniformly from soil humus, but instead are ascribed conventionally to three humus "pools": an *active pool,* with C residence times up to a year; a *slow pool,* with residence times up to a century, and a *passive pool,* with residence times up to a millennium. Natural

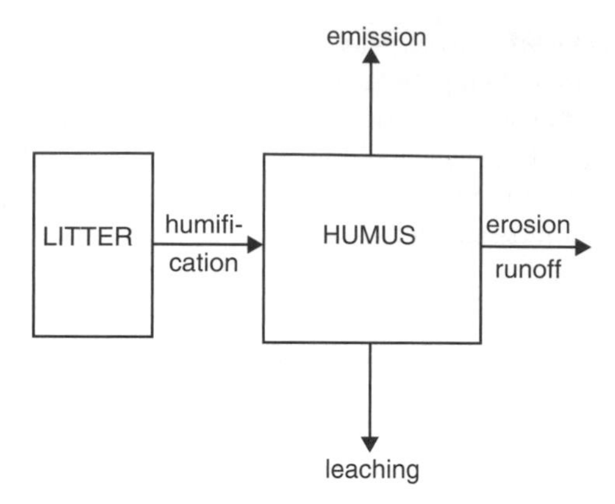

Figure 1.1. Flow diagram showing storage components (boxes) and transfers (arrows) in the soil biogeochemical cycling of C, N, P, and S.

soils can continue to accumulate C for several millennia, only to lose it over decades when placed under cultivation. The importance of this loss can be appreciated in light of the fact that soils are the largest repository of nonfossil fuel organic C on the planet, storing about four times the amount of C contained in the terrestrial biosphere.

The picture for soil N flows is similar to that for soil C, in that humus is the dominant storage component and emissions to the atmosphere are an important pathway of loss. The emissions send mainly N_2 along with N_2O and NH_3 to the atmosphere. The N_2O, like CO_2 and CH_4, is of environmental concern because of its very strong absorption of terrestrial infrared radiation (*greenhouse gas*). Unlike the case of CO_2 and CH_4, however, the source of these gases is dissolved *inorganic* N, the transformation of which is termed *denitrification* when N_2 and N_2O are the products, and *ammonia volatilization* when NH_3 is the product. Denitrification is typically mediated by respiring microorganisms, whereas ammonia volatilization results from the deprotonation of aqueous NH_4^+ (which itself may be bacterially produced) under alkaline conditions. Dissolved inorganic N comprises the highly soluble, "free-ion" chemical species, NO_3^-, NO_2^-, and NH_4^+, which can transform among themselves by electron transfer processes (*redox reactions*), be complexed by other dissolved solutes, react with particle surfaces, or be absorbed by living organisms, as illustrated in the competition diagram shown in Figure 1.2, which pertains to soils at the ped spatial scale. Natural soils tend to cycle N without significant loss through leaching (as NO_3^-), but denitrification losses can be large if soluble humus, which is readily decomposed by microorganisms, is abundant and flooding induces anaerobic conditions, thereby eliminating O as a competitor with N for the electrons made available when humus is degraded. Cultivated soils, on the other hand, often show excessive leaching and runoff losses of N, as well as significant emissions—both of which are of major environmental concern—because of high inputs of nitrate or ammonium fertilizers that artificially and suddenly increase inorganic N content. A similar problem occurs when organic wastes with low C-to-N ratios are applied to these soils as fertilizers, because rapid microbial mineralization of such materials is favored.

Sulfur flows in soils that form outside arid regions or tidal zones can be described as shown in Figure 1.1, with humus as the dominant reservoir and losses through leaching, runoff, and emission processes. Mineralization of organic S in humus usually produces SO_4^{2-}, which can be leached, react with particle surfaces, or be absorbed by living organisms (Fig. 1.2). In flooded soils, soluble H_2S and other potentially volatile sulfides are produced under microbial mediation from the degradation of humus or the reduction of sulfate (electron transfer to sulfate to produce sulfide). They can be lost by emission to the atmosphere or by precipitation along with ferrous iron or trace metals as solid-phase sulfides. The competition for aqueous SO_4^{2-} in soil peds thus follows the paradigm in Figure 1.2, with the main differences from NO_3^- being the much stronger reactions between sulfate and particle surfaces and the possibility of precipitation as a solid-phase sulfide, as well as emission to the atmosphere, under flooded conditions.

Phosphorus flows in soils follow the diagram in Figure 1.1 with the important caveat that inorganic P reservoirs—phosphate on particle surfaces and in solid phases—can sometimes be as large as or larger than that afforded by humus, depending on precipitation. Leaching losses of soil P are minimal, and gaseous P emissions to the atmosphere essentially do not occur from natural

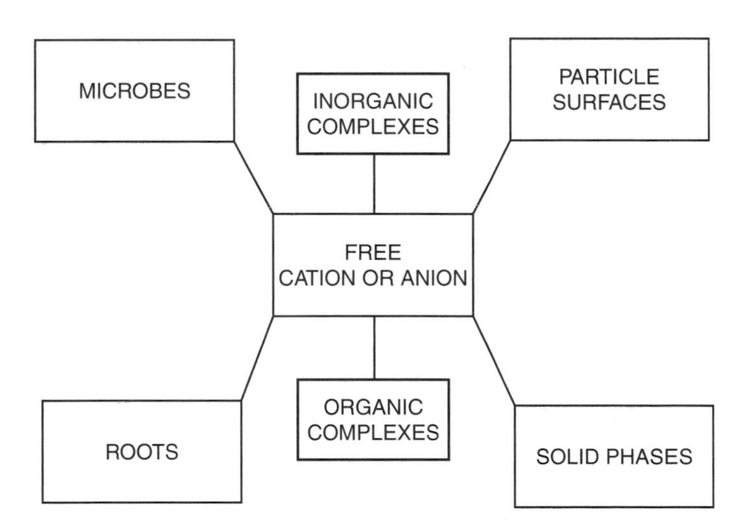

Figure 1.2. Competition diagram showing biotic and abiotic sources/sinks for aqueous species (inner three boxes) in a soil ped. Coupling among the four sources/sinks is mediated by the free ionic species of an element.

soils. Mineralization of humus and dissolution of P from solid phases both produce aqueous PO_4^{3-} or its proton complexes (e.g., $H_2PO_4^-$), depending on pH, and these dissolved species can be absorbed by living organisms or lost to particle surfaces through adsorption reactions, which are yet stronger than those of sulfate, and through precipitation, along with Ca, Al, or Fe, as a solid-phase phosphate, again depending on pH. As is the case with N, fertilizer additions and organic waste applications to soils can lead to P losses, mainly by erosion, that pose environmental hazards.

Even this brief summary of the soil cycles of C, N, S, and P can serve to illustrate their biogeochemical similarities in the setting provided by Figures 1.1 and 1.2. Humus is their principal reservoir (with P sometimes as an exception), and all four elements become *oxyanions* (CO_3^{2-}, NO_3^-, SO_4^{2-}, PO_4^{3-}, and their proton complexes) when humus is mineralized by microorganisms under aerobic conditions at circumneutral pH. The affinity of these oxyanions for particle surfaces, as well as their susceptibility to precipitation with metals, has been observed often to increase in the order $NO_3^- < SO_4^{2-} \leq CO_3^{2-} \ll PO_4^{3-}$. This ordering is accordingly reversed for their potential to be lost from soils by leaching or runoff processes, whereas it remains the same for their potential to be lost by erosion processes.

1.2 Metal Elements in Soils

Table 1.2 lists average crustal and soil concentrations of 27 metals and three metalloids (Si, As, and Sb) along with their *anthropogenic mobilization factors* (AMFs). The value of AMF is calculated as the mass of an element extracted annually, through mining operations and fossil fuel production, divided by the mass released annually through crustal weathering processes and volcanic activity, with both figures being based on data obtained worldwide. If AMF is well above 10, an element is said to have significant anthropogenic perturbation of its global biogeochemical cycle. A glance along the fourth and eighth columns in Table 1.2 reveals that, according to this criterion, the transition metals Cr, Ni, Cu, Zn, Mo, and Sn; the "heavy metals" Ag, Cd, Hg, and Pb; and the metalloids As and Sb have significantly perturbed biogeochemical cycles. Not surprisingly, these 12 elements also figure importantly in environmental regulations.

Metal elements are classified according to two important characteristics with respect to their biogeochemical behavior in soils and aquatic systems. The first of these is the *ionic potential* (IP), which is the valence of a metal cation divided by its ionic radius in nanometers. Metal cations with IP $< 30\ nm^{-1}$ tend to be found in circumneutral aqueous solutions as solvated chemical species (*free cations*); those with $30 <$ IP $< 100\ nm^{-1}$ tend to hydrolyze readily in circumneutral waters; and those with IP $> 100\ nm^{-1}$ tend to be found as oxyanions. As examples of these three classes, consider Na^+ (IP $= 9.8\ nm^{-1}$), Al^{3+}(IP $= 56\ nm^{-1}$), and Cr^{6+}(IP $= 231\ nm^{-1}$). (See Table 2.1 for a listing of ionic radii used to calculate IP.) If a metal element has different valence states, it may fall into different classes: Cr^{3+}(IP = 49 nm^{-1}) hydrolyzes, whereas it has just been shown that hexavalent Cr forms an oxyanion species in aqueous solution. The physical basis for this classification can be understood in terms of coulomb repulsion between the metal cation and a solvating water molecule that binds to it in aqueous solution through ion–dipole interactions. If IP is low, so is the positive coulomb field acting on and repelling the protons in the solvating water molecule; but, as IP becomes larger, the repulsive coulomb field becomes strong enough to cause one of the water protons to dissociate, thus forming a hydroxide ion. If IP is very large, the coulomb field then becomes strong enough to dissociate both water protons, and an oxyanion forms instead.

Evidently any monovalent cation with an ionic radius larger than 0.033 nm will be a solvated species in aqueous solution, whereas any bivalent cation will require an ionic radius larger than 0.067 nm to be a solvated species. The alkali metal in Table 1.2 with the smallest cation is Li (ionic radius, 0.076 nm) and the alkaline earth metal with the smallest cation is Be (ionic radius, 0.027 nm), followed by Mg (ionic radius, 0.072 nm). Thus alkali and alkaline earth metals, with the notable exception of Be, will be free cations in circumneutral aqueous solutions. The same is true for the monovalent heavy metals (e.g., Ag^{+}) and the bivalent transition metals and heavy metals (e.g., $Mn^{2+}$and Hg^{2+}), although the bivalent transition metals come perilously close to the IP hydrolysis threshold. Trivalent metals, on the other hand, tend always to be hydrolyzed [e.g., Al^{3+}, Cr^{3+}, and Mn^{3+} (IP = 46 nm^{-1})], and quadrivalent or higher valent metals tend to be oxyanions. The soluble metal species in circumneutral waters are either free cations or free oxyanions, whereas hydrolyzing metals tend to precipitate as insoluble oxides or hydroxides. Thus, falling into the middle IP range (30–100 nm^{-1}) is the signature of metal elements that are not expected to be soluble at circumneutral pH.

The second important characteristic of metal elements is their *Class A* or *Class B* behavior. A metal cation is Class A if (1) it has low polarizability (a measure of the ease with which the electrons in an ion can be drawn away from its nucleus) and (2) it tends to form stronger complexes with O-containing ligands [e.g., carboxylate (COO^{-}), phosphate, or a water molecule] than with N- or S-containing ligands. A metal is Class B if it has the opposite characteristics. If a metal is neither Class A nor Class B, it is termed *borderline.* The Class B metals in Table 1.2 are the heavy metals Ag, Cd, Hg, and Pb, whereas the borderline metals are the transition metals Ti to Zn, along with Zr, Mo, and Sn, each of which can behave as Class A or Class B, depending on their valence and local bonding environment (*stereochemistry*). All the other metals in Table 1.2 are Class A. We note in passing that Class A metals tend to form strong *hydrophilic* (water-loving) complexes with ligands in aqueous solution through ionic or even electrostatic bonding, whereas Class B metals tend to form strong *lipophilic* (fat-loving) complexes with ligands in aqueous solution through more covalent bonding. Hydrophilic complexes seek polar molecular environments (e.g., cell surfaces), whereas lipophilic complexes seek nonpolar environments (e.g., cell membranes). These tendencies are a direct result of (1) the polarizability of a metal cation (with high polarizability implying a labile "electron cloud," one that can be attracted toward and shared with a ligand) and (2) the less polar nature of N- or S-containing ligands, which makes them less hydrophilic than O-containing ligands.

The description of metals according to these two characteristics can be applied not only to understand the behavior of metals in terms of solubility and complex formation, but also to predict their status as plant (and microbial) *toxicants* (see the flow diagram in Fig. 1.3). For a given metal cation, if IP $< 30\ nm^{-1}$ and the metal is Class A, then it is unlikely to be toxic (e.g., Ca^{2+}), except possibly at very high concentrations (e.g., Li^+, Na^+). Moving toward the right in Figure 1.3, we see that if IP $> 100\ nm^{-1}$, or if IP $< 30\ nm^{-1}$ and the metal is borderline, then it is quite possibly toxic, examples being Cr^{6+} in the first case and bivalent transition metal cations in the second case. If, instead, $30 <$ IP $< 100\ nm^{-1}$, or the metal cation is Class B, then it is *very* likely to be toxic, examples being Be^{2+} and Al^{3+} in the first case; and Ag^+, Hg^+, along with the bivalent heavy metals in the second case. The chemistry underlying these conclusions is simple: If a metal tends to hydrolyze in aqueous solution or has covalent binding characteristics, it is very likely to be toxic, whereas if it tends to be solvated in aqueous solution and has ionic

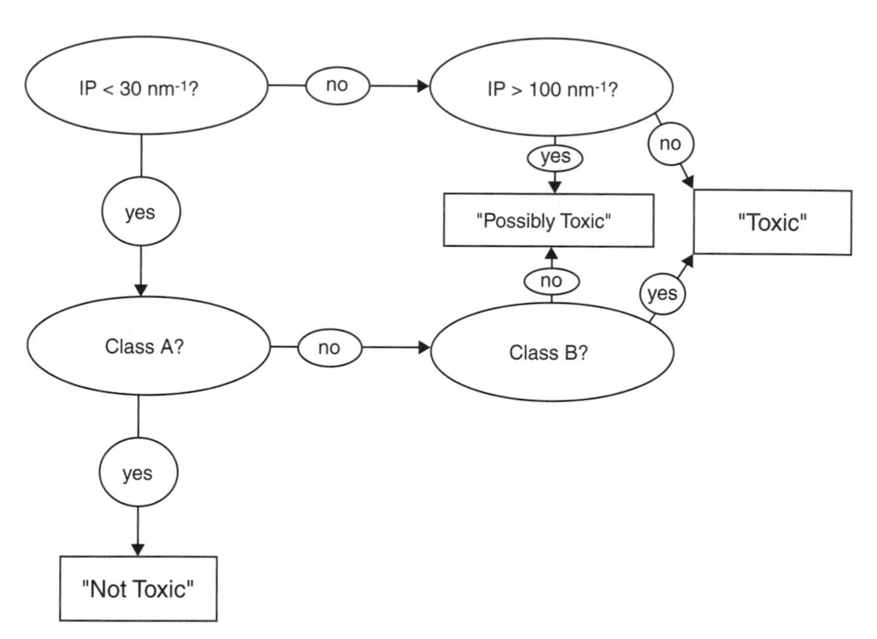

Figure 1.3. Flow diagram (beginning at upper left corner) for the toxicological classification of a metal cation at circumneutral pH using the criteria of ionic potential (IP) and Class A or B character.

or electrostatic binding characteristics, it is not as likely to be toxic. Toxicity is thus associated with insoluble metal cations and with those that tend to form covalent bonds in complexes with ligands. The first property evidently reflects low abundance in aquatic systems and, therefore, the nonavailability of a metal element as life evolved, whereas the second property is inimical to the relatively labile metal cation binding that characterizes most biochemical processes. Indeed, borderline metals become toxicants when they displace Class A metals from essential binding sites in biomolecules, bonding to these sites more strongly, and Class B metals are always toxicants, simply because they can displace either borderline metals (which often serve as cofactors in enzymes) or Class A metals from essential binding sites through much more tenacious bonding mechanisms. Note that the large AMF values in Table 1.2 are associated with borderline and Class B metals, implying, unfortunately, that human perturbations of metal biogeochemical cycles have enhanced the concentrations of toxicant metals in soil and water environments.

1.3 Solid Phases in Soils

About one half to two thirds of the soil volume is made up of solid matter. Of this material, typically more than 90% represents inorganic compounds, except for Histosols (peat and muck soils), wherein organic material accounts for more than 50% of the solid matter. The inorganic solid phases in soils often do not have a simple stoichiometry (i.e., they do not exhibit molar ratios of one element to another which are rational fractions), because they are actually in a metastable state of transition from an inhomogeneous, irregular structure to a more homogeneous, regular structure as a result of weathering processes. Nonetheless, a number of solid phases of relatively uniform composition (*minerals*) has been identified in soils worldwide. Table 1.3 lists the most common soil minerals along with their chemical formulas. Details of the atomic structures of these minerals are given in Chapter 2.

According to Table 1.1, the two most abundant elements in soils are oxygen and silicon, and these two elements combine chemically to form the 15 silicates listed in Table 1.3. The first nine silicates in the table are termed *primary minerals* because they are typically inherited from parent material, particularly crustal rock, as opposed to being precipitated through weathering processes. The key structural entity in these minerals is the Si–O bond, which is a more covalent (and, therefore, stronger) bond than typical metal–oxygen bonds (see Section 2.1). The relative resistance of any one of the minerals to decomposition by weathering is correlated positively with the Si-to-O molar ratio of its fundamental silicate structural unit, as a larger Si-to-O ratio means a lesser need to incorporate metal cations into the mineral structure to neutralize the oxygen anion charge. To the extent that metal cations are so excluded, the degree of covalency in the overall bonding arrangement will be greater and the mineral will be more resistant to decomposition in the soil environment.

Table 1.3
Common soil minerals.

Name	Chemical formula	Importance
Quartz	SiO_2	Abundant in sand and silt
Feldspar	$(Na, K)AlO_2[SiO_2]_3$ $CaAl_2O_4[SiO_2]_2$	Abundant in soil that is not leached extensively
Mica	$K_2Al_2O_5[Si_2O_5]_3Al_4(OH)_4$ $K_2Al_2O_5[Si_2O_5]_3(Mg, Fe)_6(OH)_4$	Source of K in most temperate-zone soils
Amphibole	$(Ca, Na, K)_{2,3}(Mg, Fe, Al)_5(OH)_2[(Si, Al)_4O_{11}]_2$	Easily weathered to clay minerals and oxides
Pyroxene	$(Ca, Mg, Fe, Ti, Al)_2(Si, Al)_2O_6$	Easily weathered
Olivine	$(Mg, Fe)_2SiO_4$	Easily weathered
Epidote	$Ca_2(Al,Fe)Al_2(OH)Si_3O_{12}$	Highly resistant to chemical weathering (Epidote, Tourmaline, Zircon, Rutile)
Tourmaline	$NaMg_3Al_6B_3Si_6O_{27}(OH, F)_4$	
Zircon	$ZrSiO_4$	
Rutile	TiO_2	
Kaolinite	$Si_4Al_4O_{10}(OH)_8$	Abundant in soil clay fractions as products of weathering (Kaolinite through Chlorite)
Smectite, Illite, Vermiculite, Chlorite	$M_x(Si, Al)_8(Al, Fe, Mg)_4O_{20}(OH)_4$ M = interlayer cation $0.4 \leq x \leq 2.0$ = layer charge	
Allophane	$Si_yAl_4O_{6+2y} \cdot nH_2O$, $1.6 \leq y \leq 4, n \geq 5$	Abundant in soils derived from volcanic ash deposits
Imogolite	$Si_2Al_4O_{10} \cdot 5\ H_2O$	
Gibbsite	$Al(OH)_3$	Abundant in leached soils
Goethite	FeOOH	Abundant Fe oxide in temperate soils
Hematite	Fe_2O_3	Abundant Fe oxide in aerobic soils
Ferrihydrite	$Fe_{10}O_{15} \cdot 9\ H_2O$	Abundant in seasonally wet soils
Birnessite	$M_xMn(IV)_aMn(III)_b\blacktriangle_cO_2$ M = interlayer cation, $x = b + 4c$ = layer charge, $a + b + c = 1$	Most abundant Mn oxide
Lithiophorite	$LiAl_2(OH)_6Mn(IV)_2Mn(III)O_6$	Found in acidic soils
Calcite	$CaCO_3$	Most abundant carbonate
Gypsum	$CaSO_4 \cdot 2H_2O$	Most abundant sulfate

For the first six silicates listed in Table 1.3, the Si-to-O molar ratios of their fundamental structural units are as follows: 0.50 (quartz and feldspar, SiO_2), 0.40 (mica, Si_2O_5), 0.36 (amphibole, Si_4O_{11}), 0.33 (pyroxene, SiO_3), and 0.25 (olivine, SiO_4). The decreasing order of the Si-to-O molar ratio is the same as the observed decreasing order of resistance of these minerals to weathering in soils (see Section 2.2).

The minerals epidote, tourmaline, zircon, and rutile, listed in the middle of Table 1.3, are highly resistant to weathering in the soil environment. Under the assumption of uniform parent material, measured variation in the relative number of single-crystal grains of these minerals in the fine sand or coarse silt fractions of a soil profile can serve as a quantitative indicator of mass changes in soil horizons produced by chemical weathering.

The minerals listed from kaolinite to gypsum in Table 1.3 are termed *secondary minerals* because they nearly always result from the weathering transformations of primary silicates. Often these secondary minerals are of clay size and many exhibit a relatively poorly ordered atomic structure. Variability in their composition through the substitution of ions into their structure (*isomorphic substitution*) also is noted frequently. The layer-type aluminosilicates, smectite, illite, vermiculite, and chlorite, bear a net charge on their surfaces (*layer charge*) principally because of this variability in composition, as shown in Section 2.3. Kaolinite and the secondary metal oxides below it in the list—with the important exception of birnessite—also bear a net surface charge, but because of proton adsorption and desorption reactions, not isomorphic substitutions. Birnessite, a layer-type Mn oxide, also bears a surface charge, mainly because of vacancies in its structure (quantified by an x subscript in the chemical formula) where Mn^{4+} cations should reside. Secondary metal oxides like gibbsite and goethite tend to persist in the soil environment longer than secondary silicates because Si is more readily leached than Al, Fe, or Mn, unless significant amounts of soluble organic matter are present to render these latter metals more soluble.

Organic matter is, of course, an important constituent of the solid phase in soils. The structural complexity of soil humus has thus far precluded the making of a simple list of component solids like that in Table 1.3, but something can be said about the overall composition of *humic substances*—the dark, microbially transformed organic materials that persist in soils (slow and passive humus pools) throughout profile development. The two most investigated humic substances are *humic acid* and *fulvic acid*. Their chemical behavior is discussed in Section 3.2. Worldwide, the average chemical formula for these two substances in soil is $C_{185}H_{191}O_{90}N_{10}S$ (humic acid) and $C_{186}H_{245}O_{142}N_9S_2$ (fulvic acid). These two average chemical formulas can be compared with the average C-to-N-to-S *molar* ratio of bulk soil humus, which is 140:10:1.3, and with the average chemical formula for terrestrial plants, which is $C_{146}H_{227}O_{123}N_{10}$. Relative to soil humus as a whole, humic and fulvic acids are depleted in N. Their C-to-N molar ratio is 30% to 50% larger than that of soil humus, indicating their greater resistance to net microbial mineralization. Relative to terrestrial plants, humic and fulvic acids are enriched in C but depleted in H. The depletion of H, from roughly a 1.5:1 H-to-C molar ratio in plant material to roughly 1.3 in fulvic acid and 1.0 in humic acid, suggests a greater degree of aromaticity (e.g., H-to-C ratio is 1.0 in benzene, the prototypical aromatic organic molecule) in the latter materials, which is consistent with their resistance to microbial attack.

The 21 trace elements listed in Tables 1.1 and 1.2 seldom occur in soils as separate mineral phases, but instead are found as constituents of host minerals and humus. The principal ways in which important trace elements occur in primary and secondary soil minerals are summarized in tables 1.4 and 1.5. Table 1.5 also indicates the trace elements found typically in association with soil humus. The chemical process underlying these trace element occurrences is called *coprecipitation.* Coprecipitation is the simultaneous precipitation of a chemical element with other elements by any mechanism and at any rate. The three broad types of coprecipitation are *inclusion, adsorption,* and *solid-solution formation.*

Table 1.4
Occurrence of trace elements in primary minerals.

Element	Principal modes of occurrence in primary minerals
B	Tourmaline, borate minerals; isomorphic substitution for Si in micas
Ti	Rutile and ilmenite ($FeTiO_3$); oxide inclusions in silicates
V	Isomorphic substitution for Fe in pyroxenes and amphiboles, and for Al in micas; substitution for Fe in oxides
Cr	Chromite ($FeCr_2O_4$); isomorphic substitution for Fe or Al in other minerals of the spinel group
Co	Isomorphic substitution for Mn in oxides and for Fe in pyroxenes, amphiboles, and micas
Ni	Sulfide inclusions in silicates; isomorphic substitution for Fe in olivines, pyroxenes, amphiboles, micas, and spinels
Cu	Sulfide inclusions in silicates; isomorphic substitution for Fe and Mg in olivines, pyroxenes, amphiboles, and micas; and for Ca, K, or Na in feldspars
Zn	Sulfide inclusions in silicates; isomorphic substitution for Mg and Fe in olivines, pyroxenes, and amphiboles; and for Fe or Mn in oxides
As	Arsenopyrite (FeAsS) and other arsenate minerals
Se	Selenide minerals; isomorphic substitution for S in sulfides; iron selenite
Mo	Molybdenite (MoS_2); isomorphic substitution for Fe in oxides
Cd	Sulfide inclusions and isomorphic substitution for Cu, Zn, Hg, and Pb in sulfides
Pb	Sulfide, phosphate, and carbonate inclusions; isomorphic substitution for K in feldspars and micas; for Ca in feldspars, pyroxenes, and phosphates; and for Fe and Mn in oxides

Table 1.5
Trace elements coprecipitated with secondary soil minerals and soil humus.

Solid	Coprecipitated trace elements
Fe and Al oxides	B, P, Ti, V, Cr, Mn, Co, Ni, Cu, Zn, Mo, As, Se, Cd, Pb
Mn oxides	P, Fe, Co, Ni, Cu, Zn, Mo, As, Se, Cd, Pb
Ca carbonates	P, V, Mn, Fe, Co, Cd, Pb
Illites	B, V, Ni, Co, Cr, Cu, Zn, Mo, As, Se, Pb
Smectites	B, Ti, V, Cr, Mn, Fe, Co, Ni, Cu, Zn, Pb
Vermiculites	Ti, Mn, Fe
Humus	B, Al, V, Cr, Mn, Fe, Ni, Cu, Zn, Se, Cd, Pb

If a solid phase formed by a trace element has a very different atomic structure from that of the host mineral, then it is likely that the host mineral and the trace element will occur together only as morphologically distinct phases. This kind of association is termed *inclusion* with respect to the trace element. For example, CuS occurs as an inclusion—a small, separate phase—in primary silicates (Table 1.4).

If there is only limited structural compatibility between a trace element and a corresponding major element in a host mineral, then coprecipitation produces a mixture of the two elements restricted to the host mineral–soil solution interface. This mechanism is termed *adsorption* because the mixed solid phase that forms is restricted to the interfacial region (including the interlayer region of layer-type minerals). Well-known examples of adsorption are the incorporation of metals like Pb and oxyanions like arsenate into secondary metal oxides (Table 1.5).

Finally, if structural compatibility is high and diffusion of a trace element within the host mineral is possible, a major element in the host mineral can be replaced sparingly but uniformly throughout by the trace element. This kind of homogeneous coprecipitation is *solid-solution formation*. It is enhanced if the size and valence of the substituting element are comparable with those of the element replaced. Examples of solid-solution formation occur when precipitating secondary aluminosilicates incorporate metals like Fe to replace Al in their structures (Table 1.5) or when calcium carbonate precipitates with Cd replacing Ca in the structure. Soil solid phases bearing trace elements serve as reservoirs, releasing the trace elements slowly into the soil solution as weathering continues. If a trace element is also a nutrient, then the rate of weathering becomes a critical factor in soil fertility. For example, the ability of soils to provide Cu to plants depends on the rate at which this element is transformed from a solid phase to a soluble chemical form. Similarly, the weathering of soil solids containing Cd as a trace element will determine the potential hazard of this toxic element to microbes and plants.

1.4 Soil Air and Soil Water

The fluid phases in soil constitute between one and two thirds of the soil volume. The gaseous phase, *soil air,* typically is the same kind of fluid mixture as atmospheric air. Because of biological activity in soil, however, the percentage composition of soil air can differ considerably from that of atmospheric air (781 mL N_2, 209 mL O_2, 9.3 mL Ar, and 0.33 mL CO_2 in 1 L dry air). Well-aerated soil contains 180 to 205 mL O_2 L^{-1} soil air, but this can drop to 100 mL L^{-1}at 1 m below the soil surface, after inundation by rainfall or irrigation, or even to 20 mL L^{-1} in isolated soil microenvironments near plant roots. Similarly, the fractional volume of CO_2 in soil air is typically 3 to 30 mL L^{-1}, but can approach 100 mL L^{-1} at a 1-m depth in the vicinity of plant roots, or after the flooding of soil. This markedly higher CO_2 content of soil air relative to that of the atmosphere has a significant impact on both soil acidity and carbonate chemistry. Soil air also contains variable but important contributions from H_2, NO, N_2O, NH_3, CH_4, and H_2S produced with microbial mediation under conditions of low or trace oxygen content.

Soil water is found principally as a condensed fluid phase, although the content of water vapor in soil air can approach 30 mL L^{-1} in a wet soil. Soil water is a repository for dissolved solids and gases, and for this reason is referred to as the *soil solution.* With respect to dissolved solids, those that dissociate into ions (*electrolytes*) in the soil solution are most important to the chemistry of soils. The nine ion-forming chemical elements with concentrations in uncontaminated soil solutions that typically exceed all others are C (HCO_3^-), N (NO_3^-), Na (Na^+), Mg (Mg^{2+}), Si [$Si(OH)_4^0$], S (SO_4^{2-}), Cl (Cl^-), K (K^+), and Ca (Ca^{2+}), where the principal chemical species of the element appears in parentheses or square brackets. [The neutral species $Si(OH)_4^0$ is silicic acid.] With the exception of Cl, all are macroelements.

The partitioning of gases between soil air and the soil solution is an important process contributing to the cycling of chemical elements in the soil environment. When equilibrium exists between soil air and soil water with respect to the partitioning of a gaseous species between the two phases, and if the concentration of the gas in the soil solution is low, the equilibrium can be described by a form of *Henry's law:*

$$K_H = [A(aq)]/P_A \tag{1.1}$$

where K_H is a parameter with the units moles per cubic meter per atmosphere of pressure, known as the *Henry's law constant,* [A] is the concentration of gas A in the soil solution (measured in moles per cubic meter), and P_A is the partial pressure of A in soil air (measured in atmospheres). Table 1.6 lists values of K_H at 25 °C for 10 gases found in soil air. As an example of the use of this table, consider a flooded soil in which CO_2 and CH_4 are produced under microbial mediation to achieve partial pressures of 14 and 10 kPa respectively, as measured in the headspace of serum bottles used to contain the soil during

Table 1.6
The "Henry's law constant" for 10 soil gases at 25 °C[a].

Gas	K_H (mol m^{-3} atm^{-1})	Gas	K_H (mol m^{-3} atm^{-1})
H_2	0.78	NO	92
CO_2	34.20	N_2	0.65
CH_4	1.41	O_2	1.27
NH_3	5.71×10^4	SO_2	1.36×10^3
N_2O	24.17	H_2S	1.02×10^2

[a] Based on data compiled from Lide, D. R. (ed.) (2004) *CRC handbook of chemistry and physics*, pp. 8-86–8-89. CRC Press, Boca Raton, FL.

incubation. According to Table 1.6, the corresponding concentrations of the two gases in the soil solution are

$$[CO_2\,(aq)] = 34.03\ \frac{\text{mol m}^{-3}}{\text{atm}} \times \left(14\ \text{kPa} \times \frac{1\ \text{atm}}{101.325\ \text{kPa}}\right)$$

$$= \mathbf{4.7\ mol\ m^{-3}}$$

$$[CH_4\,(aq)] = \frac{1.41\ \text{mol m}^{-3}}{\text{atm}} \times 0.0987\ \text{atm} = \mathbf{0.14\ mol\ m^{-3}}$$

(The units appearing here are discussed in the Appendix.) The result for CO_2 is noteworthy, in that the concentration of this gas in a soil solution equilibrated with the ambient atmosphere would be 400 times smaller.

1.5 Soil Mineral Transformations

If soils were not open systems, soil minerals would not weather. It is the continual input and output of percolating water, biomass, and solar energy that makes soils change with the passage of time. These changes are perhaps reflected most dramatically in the development of soil horizons, both in their morphology and in the mineralogy of the soil clay fraction.

Table 1.7 is a broad summary of the changes in clay fraction mineralogy observed during the course of soil profile development. These changes are known collectively as the *Jackson–Sherman weathering stages.*

Early-stage weathering is recognized through the importance of sulfates, carbonates, and primary silicates, other than quartz and muscovite, in the soil clay fraction. These minerals survive only if soils remain very dry, or very cold, or very wet, most of the time—that is, if they lack significant throughputs of water, air, biota, and thermal energy that characterize open systems in nature. Soils in the early stage of weathering include Aridisols, Entisols, and Gelisols at the Order level in the U.S. Soil Taxonomy. Intermediate-stage

Table 1.7
Jackson–Sherman soil weathering stages.

Characteristic minerals in soil clay fraction	Characteristic soil chemical and physical conditions	Characteristic soil properties[a]
Early stage		
Gypsum Carbonates Olivine/pyroxene/amphibole Fe(II)-bearing micas Feldspars	Low water and humus content, very limited leaching Reducing environments, cold environments Limited amount of time for weathering	Minimally weathered soils: arid or very cold regions, waterlogging, recent deposition
Intermediate stage		
Quartz Dioctahedral mica/illite Dioctahedral vermiculite/chlorite Smectite	Retention of Na, K, Ca, Mg, Fe(II), and silica; moderate leaching, alkalinity Parent material rich in Ca, Mg, and Fe(II), but not Fe(II) oxides Silicates easily weathered	Soils in temperate regions: forest or grass cover, well-developed A and B horizons, accumulation of humus and clay minerals
Advanced stage		
Kaolinite Gibbsite Iron oxides Titanium oxides	Removal of Na, K, Ca, Mg, Fe(II), and silica Intensive leaching by fresh water Oxidation of Fe(II) Low pH and humus content	Soils under forest cover with high temperature and precipitation: accumulation of Fe(III) and Al oxides, absence of alkaline earth metals

[a]Soil taxa corresponding to these properties are discussed in Encyclopedia Britannica (2005) *Soil*. Available at www.britannica.com/eb/.

weathering features quartz, muscovite, and layer-type secondary aluminosilicates (*clay minerals*) prominently in the clay fraction. These minerals survive under leaching conditions that do not entirely deplete silica and the major elements, and do not result in the complete oxidation of ferrous iron [Fe(II)], which is incorporated into illite and smectite. Soils at this weathering stage include the Mollisols, Alfisols, and Spodosols. Advanced-stage weathering, on the other hand, is associated with intensive leaching and strongly oxidizing conditions, such that only hydrous oxides of aluminum, ferric iron [Fe(III)], and titanium persist ultimately. Kaolinite will be an important clay mineral if the removal of silica by leaching is not complete, or if there is an invasion of silica-rich waters, as can occur, for example, when siliceous leachate from the upper part of a soil toposequence moves laterally into the profile of a lower part. The Ultisols and Oxisols are representative soil taxa.

The order of increasing persistence of the soil minerals listed in Table 1.7 is downward, both among and within the three stages of weathering. The primary minerals, therefore, tend to occur higher in the list than the secondary minerals, and the former can be linked with the latter by a variety of chemical reactions. Of these reactions, the most important is termed *hydrolysis and protonation,* which may be illustrated by the weathering of the feldspar *albite* or the mica *biotite,* to form the clay mineral *kaolinite* (Table 1.3). For albite the reaction is

$$\underset{\text{(albite)}}{4\ \mathrm{NaAlSi_3O_8}\ (s)} + 4\mathrm{H^+} + 18\ \mathrm{H_2O}\ (\ell) = \underset{\text{(kaolinite)}}{\mathrm{Si_4Al_4O_{10}(OH)_8}\ (s)} + 8\ \mathrm{Si(OH)_4^0} + 4\ \mathrm{Na^+} \quad (1.2)$$

where solid (s) and liquid (ℓ) species are indicated explicitly, all undesignated species being dissolved solutes by default. The corresponding reaction for biotite is

$$\begin{aligned} &2\ \mathrm{K_2[Si_6Al_2]Mg_3Fe(II)_3O_{20}(OH)_4}\ (s) + 16\ \mathrm{H^+} \\ &\quad + 11\ \mathrm{H_2O}\ (\ell) + \frac{3}{2}\mathrm{O_2}\ (g) \\ &= \underset{\text{(kaolinite)}}{\mathrm{Si_4Al_4O_{10}(OH)_8}\ (s)} + \underset{\text{(goethite)}}{6\ \mathrm{FeOOH}\ (s)} + 8\ \mathrm{Si(OH)_4^0} \\ &\quad + 4\ \mathrm{K^+} + 6\ \mathrm{Mg^{2+}} \end{aligned} \quad (1.3)$$

in which the iron oxyhydroxide, goethite, is also formed. In both reactions, which are taken conventionally to proceed from left to right, the dissolution of a primary silicate occurs through chemical reaction with water and protons to form one or more solid-phase products plus dissolved species, which are then subject to leaching out of the soil profile. These two reactions illustrate an *incongruent dissolution,* as opposed to *congruent dissolution,* in which only dissolved species are products. The basic chemical principles underlying the development of eqs. 1.2 and 1.3 are discussed in Special Topic 1, at the end of this chapter.

The incongruent dissolution of biotite, which contains only ferrous iron [Fe(II)], to form goethite, which contains only ferric iron [Fe(III)], illustrates the electron transfer reaction termed *oxidation* (in the case of Eq. 1.3, the oxidation of ferrous iron)—an important process in soil weathering. (Oxidation, the loss of electrons from a chemical species, is discussed in Chapter 6.) Another important weathering reaction is *complexation,* which is the reaction of an anion (or other ligand) with a metal cation to form a species that can be either dissolved or solid phase. In the case of albite weathering by complexation,

$$\underset{\text{(albite)}}{NaAlSi_3O_8\,(s)} + 4\,H^+ + C_2O_4^{2-} + 4\,H_2O\,(\ell) = AlC_2O_4^+ + 3\,Si(OH)_4^0 + Na^+ \qquad (1.4)$$

The organic ligand on the left side of Eq. 1.4, oxalate, is the anion formed by complete dissociation of oxalic acid ($H_2C_2O_4$, ethanedioic acid) at pH $\geq$ 4.2. It complexes Al^{3+} released by the hydrolysis and protonation of albite. The resulting product is shown on the right side of Eq. 1.4 as a soluble complex that prevents the precipitation of kaolinite (i.e., the dissolution process is now congruent). Oxalate is a very common anion in soil solutions associated with the life cycles of microbes, especially fungi, and with the *rhizosphere,* the local soil environment influenced significantly by plant roots. Organic anions produced by microbes thus play a significant role in the weathering of soil minerals, particularly near plant roots, where anion concentrations can be in the moles per cubic meter range.

The three weathering reactions surveyed very briefly in this section provide a chemical basis for the transformation of soil minerals among the Jackson–Sherman weathering stages. With respect to silicates, a *master variable* controlling these transformations is the concentration of silicic acid in the soil solution. As the concentration of $Si(OH)_4^0$ decreases through leaching, the mineralogy of the soil clay fraction passes from the primary minerals of the early stage to the secondary minerals of the intermediate and advanced stages. Should the $Si(OH)_4^0$ concentration increase through an influx of silica, on the other hand, the clay mineralogy can shift upward in Table 1.7. This possible behavior is in fact implied by the equal-to sign in the chemical reactions in eqs. 1.2 through 1.4.

For Further Reading

Churchman, G. J. (2000) The alteration and formation of soil minerals by weathering, pp. F-3–F-76. In: M. E. Sumner (ed.), *Handbook of soil science.* CRC Press, Boca Raton, FL. A detailed, field-oriented discussion of the chemical weathering processes that transform primary minerals in soils.

Dixon, J. B., and D. G. Schulze (eds.). (2002) *Soil mineralogy with environmental applications.* Soil Science Society of America, Madison, WI. A standard reference work on soil mineral structures and chemistry.

Fraústo da Silva, J. J. R., and R. J. P. Williams. (2001) *The biological chemistry of the elements.* Oxford University Press, Oxford. An engaging discussion of the bioinorganic chemistry of the elements essential to (or inimical to) life processes at the cellular level.

Gregory, P. J. (2006) Roots, rhizosphere and the soil: The route to a better understanding of soil science? *Eur. J. Soil Sci.* 57:2–12. A useful historical introduction to the concept of the rhizosphere and its importance to understanding the plant–soil interface.

Kabata–Pendias, A., and H. Pendias. (2001) *Trace elements in soils and plants.* CRC Press, Boca Raton, FL. A compendium of analytical data on trace elements in the lithosphere and biosphere, organized according to the Periodic Table.

Stevenson, F. J., and M. A. Cole. (1999) *Cycles of soil.* Wiley, New York. An in-depth discussion of the biogeochemical cycles of C, N, P, S, and some trace elements, addressed to the interests of soil chemists.

Problems

The more difficult problems are indicated by an asterisk.

1. The table presented here lists area-normalized average soil C content and annual soil C input, along with mean annual precipitation (MAP), for biomes grouped by mean annual temperature (MAT). Analyze this

Biome	Soil C content[a] (mt ha^{-1})	C input rate (mt ha^{-1} y^{-1})	MAP (mm)
MAT = 5°*C*			
Boreal desert	102	0.5	125
Boreal forest (moist)	116	1.9	375
Boreal forest (wet)	193	6.8	1250
MAT = 9°*C*			
Cool desert	99	2.1	125
Cool grassland	133	3.0	375
Cool temperate forest	127	9.1	2250
MAT = 24°*C*			
Warm desert	14	0.4	125
Tropical grassland	54	4.8	375
Tropical forest (dry)	99	4.6	1250
Tropical forest (moist)	114	24.9	2500
Tropical forest (wet)	191	37.3	6000

[a] mt = metric ton = 10^3 kg; ha = hectare = 10^4 m^2

information quantitatively (e.g., perform statistical regression analyses) to discuss correlations between soil C content and the climatic variables MAP and MAT.

2. The average *residence time* of an element in a storage component of its biogeochemical cycle is the ratio of the mass of the element in the storage component to the rate of element output from the component, calculated under the assumption that the rates of output and input are equal (steady-state condition). Calculate the residence times of C in soil humus for the biomes listed in the table in Problem 1. These residence times may be attributed primarily to soil C loss by emission (Fig. 1.1) as CO_2 (*soil respiration*), if a steady state obtains. Discuss correlations between the C residence times and the two climate variables MAT and MAP.

*3. The average age of soil humus determined by ${}^{14}_{6}C$ dating typically ranges from centuries to millennia, which is significantly longer than the residence times of soil C calculated in Problem 2 using the data in the table given with Problem 1. This paradox suggests that soil C has at least two components with widely different turnover rates. Detailed studies of humus degradation at field sites in boreal, temperate, and tropical forests indicated soil C residence times of 220, 12, and 3 years respectively for the three field sites, whereas ${}^{14}_{6}C$ dating yielded soil humus ages of 950, 250, and 1050 years respectively. Measurements of the fraction of soil C attributable to humic substances gave 62%, 78%, and 22% respectively. Use these data to estimate the average age of humic substances in the three soils. (*Hint:* The inverse of the soil C residence time equals the weighted average of the inverses of the two component residence times, whereas the average age of soil C is equal to the weighted average of the ages of the two components. In each case, the weighting factors are the fractions of soil C attributable to the two components. To a good first approximation, the "old" component can be neglected in the first expression, whereas the "young" component can be neglected in the second one. Check this approximation assuming that residence time and ${}^{14}_{6}C$ age are the same for each humus component.)

4. Consult a suitable reference to prepare a list of metals that are essential to the growth of higher plants. Use Table 1.2 to classify these metals into groups of major and trace elements in soils. Use Table 2.1 to calculate the IPs of the metals in their most common valence states, then examine your results for relationships between IP and (a) mean soil content, (b) AMF, and (c) toxicity classification (Fig. 1.3).

5. The table presented here lists the average content of organic C and five metals in agricultural soils of the United States, grouped according to the Order level in U.S. Soil Taxonomy. Compare these data with those in tables 1.1 and 1.2, and examine them for any relationships

between metal content and Jackson–Sherman weathering stage. Classify the metals (as bivalent cations) according to their possible toxicity (Fig. 1.3).

Order	Cd (mg kg^{-1})	Cu (mg kg^{-1})	Ni (mg kg^{-1})	Pb (mg kg^{-1})	Zn (mg kg^{-1})	C (g kg^{-1})
Alfisol	0.112	10.9	12.6	9.6	31.3	8.6
Aridisol	0.304	25.0	24.3	10.6	70.1	6.3
Entisol	0.246	21.1	21.0	10.0	65.5	6.8
Mollisol	0.227	19.1	22.8	10.7	54.4	13.9
Spodosol	0.200	48.3	22.0	10.0	44.1	17.3
Ultisol	0.049	6.2	7.4	8.0	13.8	7.8

6. Careful study of the rhizosphere in a Spodosol under balsam fir and black spruce forest cover showed that rhizosphere pH was somewhat lower (pH 4.8 vs. 5.0), whereas organic C was higher (21 g C kg^{-1} vs. 5 g C kg^{-1}) than in the bulk soil. The table presented here shows the content of four metals in the rhizosphere (R) and bulk (B) soil that could be extracted by $BaCl_2$ ("soluble and weakly adsorbed"), $Na_4P_2O_7$ (sodium pyrophosphate, "organic complexes"), and $NH_5C_2O_4$ (ammonium oxalate, "coprecipitated in a poorly crystalline solid phase").

   a. Explain the differences between major elements and trace elements with respect to trends across the three extractable fractions, irrespective of R and B.
   b. What is the principal chemical factor determining the differences between rhizosphere and bulk soil with respect to "weakly adsorbed" metal? (*Hint:* Compare the average ratio of weakly adsorbed metal with C content between R and B.)

Metal	Soil	$BaCl_2$	Pyrophosphate	Oxalate
Al (g kg^{-1})	R	0.16	5.5	16.1
	B	0.06	4.7	15.4
Fe (g kg^{-1})	R	0.02	2.0	10.2
	B	0.01	1.5	9.3
Cu (mg kg^{-1})	R	4.0	4.7	9.0
	B	1.7	0.9	4.8
Zn (mg kg^{-1})	R	2.0	4.3	5.5
	B	0.4	1.4	3.9

B, bulk; R, rhizosphere.

7. Calculate the average chemical formula and its range of variability for soil humic and fulvic acids using the composition data in the table presented here. Does the H-to-C molar ratio differ significantly between the two humic substances?

Humic substance	C ($g\ kg^{-1}$)	H ($g\ kg^{-1}$)	N ($g\ kg^{-1}$)	S ($g\ kg^{-1}$)	O ($g\ kg^{-1}$)
Humic acid	554 ± 38	48 ± 10	36 ± 13	8 ± 6	360 ± 37
Fulvic acid	453 ± 54	50 ± 10	26 ± 13	13 ± 11	462 ± 52

8. Calculate the corresponding concentrations of CO_2 dissolved in soil water as the CO_2 partial pressure in soil air increases in the order 3.02×10^{-4} (atmospheric CO_2), 0.003, 0.01, 0.05, 0.10 atm (flooded soil).

9. The ideal gas law, $PV = nRT$, can be applied to the constituents of soil air to a good approximation, where P is pressure, V is volume, n is the number of moles, R is the molar gas constant, and T is absolute temperature, as described in the Appendix. Use the ideal gas law to show that Eq. 1.1 can be rewritten in the useful form

$$H = [\mathrm{A(g)}]/[\mathrm{A(aq)}]$$

where [ ] is a concentration in moles per cubic meter, and $H = 10^3/K_H RT$ is a *dimensionless* constant based on $R = 0.08206$ atm L mol^{-1}K^{-1}, and T in Kelvin (K). Prepare a table of H values based on Table 1.6.

*10. The table presented here shows partial pressures of O_2 and N_2O in the pore space of an Alfisol under deciduous forest cover as a function of depth in the soil profile. Calculate the N_2O-to-O_2 molar ratio in soil air and in the soil solution as a function of depth.

Depth (m)	P_{O2}(atm)	P_{N_2O} (10^{-6} atm)
0.1	0.14	2.3
0.2	0.13	3.5
0.4	0.10	7.7
0.6	0.09	20.8

11. Write a balanced chemical reaction for the congruent dissolution of the olivine forsterite (Mg_2SiO_4) by hydrolysis and protonation.

12. Feldspar can weather to form gibbsite instead of kaolinite. Write a balanced chemical reaction for the incongruent dissolution of K-feldspar ($KAlSi_3O_8$) to produce gibbsite by hydrolysis and protonation.

*13. Write a balanced chemical reaction for the incongruent dissolution of the mica muscovite, $[K_2[Si_6Al_2]Al_4O_{20}(OH)_4]$, to form the smectite, $K_{1.08}[Si_{6.92}Al_{1.08}]Al_4O_{20}(OH)_4$, by hydrolysis and protonation. This smectite is an example of the common soil clay mineral *beidellite*.

14. Each formula unit of soil fulvic acid can dissociate about 40 protons at circumneutral pH to become a negatively charged polyanion. With respect to its soil solution chemistry as a complexing ligand, fulvic acid thus can be represented simply by the formula $H_{40}L$, where L^{40-} denotes the chemical formula of the highly charged polyanion (*fulvate*) that remains after 40 protons are deleted from the chemical formula given in Section 1.3. Use this convention to write a balanced chemical reaction for the congruent dissolution of albite by complexation, hydrolysis, and protonation, with a neutral complex between hydrolyzed Al^{3+} (i.e., $AlOH^{2+}$) and fulvate as one of the products.

15. When CO_2 dissolves in the soil solution, it solvates to form the chemical species $CO_2 \cdot H_2O$ and the neutral complex $H_2CO_3^0$ (a very minor species), which together are denoted $H_2CO_3^*$ (*carbonic acid*). Carbonic acid, in turn, dissociates a proton to leave the species HCO_3^- (*bicarbonate ion*). Write a series of balanced chemical reactions that show the formation of $H_2CO_3^*$ from $CO_2(g)$, the formation of bicarbonate from carbonic acid, and the reaction of bicarbonate with Ca-feldspar ($CaSi_2Al_2O_8$) to form calcite and kaolinite. Sum the reactions to develop an overall reaction for the weathering of Ca-feldspar by reaction with $CO_2(g)$.

Special Topic 1: Balancing Chemical Reactions

Chemical reactions like those in eqs. 1.2 through 1.4 must fulfill two general conditions: *mass balance* and *charge balance*. Mass balance requires that the number of moles of each chemical element be the same on both sides of the reaction when written as a chemical equation. Charge balance requires that the net total ionic charge be the same on both sides of the reaction. These constraints can be applied to develop the correct form of a chemical reaction when only the principal product and reactant are given.

As a first example, consider the incongruent dissolution of albite, $NaAlSi_3O_8$, to produce kaolinite, $Si_4Al_4O_{10}(OH)_8$, as in Eq. 1.2. Because 1 mol of reactant albite contains 1 mol Na, 1 mol Al, and 3 mol Si, whereas 1 mol of product kaolinite contains 4 mol Si and Al, these amounts, by mass balance, must appear equally on both sides of the reaction:

$$4\,NaAlSi_3O_8(s) \longrightarrow Si_4Al_4O_{10}(OH)_8(s) + 8\,Si(OH)_4^0 + 4\,Na^+ \quad (S.1.1)$$

where the excess Si has been put into silicic acid, the dominant aqueous species of Si(IV) at pH less than 9. A *mechanism* for the reaction, *hydrolysis and protonation,* is then invoked:

$$4\,NaAlSi_3O_8(s) + H^+ + H_2O(\ell) \longrightarrow Si_4Al_4O_{10}(OH)_8(s) + 8\,Si(OH)_4^0 + 4\,Na^+ \quad (S.1.2)$$

Charge balance requires adding 3 mol protons to the left side to match the four cationic charges on the right side:

$$4\,NaAlSi_3O_8(s) + 4\,H^+ + H_2O(\ell) \longrightarrow Si_4Al_4O_{10}(OH)_8(s) + 8\,Si(OH)_4^0 + 4\,Na^+ \quad (S.1.3)$$

Mass balance for protons now is considered. There are 40 mol H on the right side (8 mol from kaolinite and $4 \times 8 = 32$ mol from silicic acid), but only 6 mol on the left side, so 34 mol H are needed. This need is met by changing the stoichiometric coefficient of water to 18:

$$4\,NaAlSi_3O_8(s) + 4\,H^+ + 18\,H_2O(\ell) = Si_4Al_4O_{10}(OH)_8(s) + 8\,Si(OH)_4^0 + 4\,Na^+ \quad (S.1.4)$$

Note that 50 mol O now appears on both sides of the reaction to give O mass balance.

The same procedure is used to develop the more complex reaction in Eq. 1.3:

$$2\,K_2[Si_6Al_2]Mg_3Fe(II)_3O_{20}(OH)_4(s) \longrightarrow Si_4Al_4O_{10}(OH)_8(s) + 6\,FeOOH(s) + 8\,Si(OH)_4^0 + 4\,K^+ + 6\,Mg^{2+} \quad (S.1.5)$$

bearing in mind that the products have been *selected* to be kaolinite and goethite. In this example, there are *two* mechanisms invoked—*hydrolysis/protonation* and *oxidation:*

$$2\,K_2[Si_6Al_2]Mg_3Fe(II)_3O_{20}(OH)_4(s) + H^+ + H_2O(\ell) + O_2(g) \longrightarrow Si_4Al_4O_{10}(OH)_8(s) + 6\,FeOOH(s) + 8\,Si(OH)_4^0 + 4\,K^+ + 6\,Mg^{2+} \quad (S.1.6)$$

The next step is charge balance. The net ionic charge on the right side of Eq. S.1.6 is 16 cationic charges (6 from Mg, 4 from K, and 6 from Fe), thus requiring 16 as the stoichiometric coefficient of H^+ on the left side:

$$2\,K_2[Si_6Al_2]Mg_3Fe(II)_3O_{20}(OH)_4(s) \;+\; 16\,H^+ \;+\; H_2O(\ell) \;+\; O_2(g)$$
$$\longrightarrow Si_4Al_4O_{10}(OH)_8(s) \;+\; 6\,FeOOH(s) \;+\; 8\,Si(OH)_4^0$$
$$+\; 4\,K^+ \;+\; 6\,Mg^{2+} \qquad \text{(S.1.7)}$$

Mass balance on H in Eq. S.1.7 requires the addition of 20 mol H on the left side ($2 \times 4 + 16 + 2 = 26$ vs. $8 + 6 + 8 \times 4 = 46$), which is accomplished by changing the stoichiometric coefficient of water to 11:

$$2\,K_2[Si_6Al_2]Mg_3Fe(II)_3O_{20}(OH)_4(s) + 16\,H^+ + 11\,H_2O(\ell) + O_2(g)$$
$$\longrightarrow Si_4Al_4O_{10}(OH)_8(s) \;+\; 6\,FeOOH(s) \;+\; 8\,Si(OH)_4^0$$
$$+\,4\,K^+ \;+\; 6\,Mg^{2+} \qquad \text{(S.1.8)}$$

Finally, unlike the case of Eq. S.1.4, mass balance on O is not satisfied unless 1 mol O is added to the left side of Eq. S.1.8 ($2 \times 24 + 11 + 2 = 61$ vs. $18 + 6 \times 2 + 8 \times 4 = 62$):

$$2\,K_2[Si_6Al_2]Mg_3Fe(II)_3O_{20}(OH)_4(s) \;+\; 16\,H^+$$
$$+\; 11\,H_2O(\ell) \;+\; \frac{3}{2}O_2(g)$$
$$= Si_4Al_4O_{10}(OH)_8(s) \;+\; 6\,FeOOH(s) \;+\; 8\,Si(OH)_4^0$$
$$+\; 4\,K^+ \;+\; 6\,Mg^{2+} \qquad \text{(S.1.9)}$$

Note that in both of these examples, reactants and products, as well as the mechanisms of reaction, are *free choices* to be made before imposing the mass and charge balance constraints.

2

Soil Minerals

2.1 Ionic Solids

The chemical elements making up soil minerals occur typically as ionic species with an electron configuration that is unique and stable regardless of whatever other ions may occur in a mineral structure. The attractive interaction between one ion and another of opposite charge nonetheless is strong enough to form a chemical bond, termed an *ionic bond.* Ionic bonds differ from covalent bonds, which involve a significant distortion of the electron configurations (*orbitals*) of the bonding atoms that results in the sharing of electrons. Electron sharing mixes the electronic orbitals of the atoms, so it is not possible to assign to each atom a unique configuration that is the same regardless of the partner with which the covalent bond has formed. This loss of electronic identity leads to a more coherent fusion of the orbitals that makes covalent bonds stronger than ionic bonds.

Ionic and covalent bonds are conceptual idealizations that real chemical bonds only approximate. In general, a chemical bond shows some degree of ionic character and some degree of electron sharing. The Si–O bond, for example, is said to be an even partition between ionic and covalent character, and the Al–O bond is thought to be about 40% covalent, 60% ionic. Aluminum, however, is exceptional in this respect, for almost all the metal–oxygen bonds that occur in soil minerals are strongly ionic. For example, Mg–O and Ca–O bonds are considered 75% to 80% ionic, whereas Na–O and K–O bonds are 80% to 85% ionic. Covalence thus plays a relatively minor role in determining the atomic structure of most soil minerals, aside from the important feature that

Si–O bonds, being 50% covalent, impart particular stability against mineral weathering, as discussed in Section 1.3.

Given this perspective on the chemical bonds in minerals, the two most useful atomic properties of the ions constituting soil minerals should be their valence and radius. Ionic valence is simply the ratio of the electric charge on an ionic species to the charge on the proton. Ionic radius, however, is a less direct concept, because the radius of a single ion in a solid cannot be measured. Ionic radius thus is a *defined quantity* based on the following three assumptions: (1) the radius of the bivalent oxygen ion (O^{2-}) in *all* minerals is 0.140 nm, (2) the sum of cation and anion radii equals the measured interatomic distance between the two ions, and (3) the ionic radius may depend on the coordination number, but otherwise is independent of the type of mineral structure containing the ion. The coordination number is the number of ions that are nearest neighbors of a given ion in a mineral structure. Table 2.1 lists standard cation radii calculated from crystallographic data under these three assumptions. Note that the radii depend on the valence (Z) as well as the coordination number (CN) of the metal cation. The radius decreases as the valence increases and electrons are drawn toward the nucleus, but it increases with increasing coordination number for a given valence. The coordination numbers found for cations in soil minerals are typically 4 or 6, and occasionally 8 or 12. The geometric arrangements of anions that coordination numbers represent are illustrated in Figure 2.1. Each of these arrangements corresponds to a regular geometric solid (a *polyhedron*, as shown in the middle row of Fig. 2.1). It is evident that the strength of the anionic electrostatic field acting on a cation will increase as its coordination number increases. This stronger anionic field draws the "electron cloud" of the cation more into the void space between the anions, thereby causing the cation radius to increase with its coordination number.

Two important physical parameters can be defined using the atomic properties listed in Table 2.1. The first parameter is *ionic potential* (or *IP*),

$$\mathrm{IP} = \frac{\mathrm{Z}}{\mathrm{IR}} \qquad (2.1)$$

which is proportional to the coulomb potential energy at the periphery of a cation, as discussed in Section 1.2. The second parameter is *bond strength* (s), a more subtle concept from Linus Pauling,

$$\mathrm{s} = \frac{|\mathrm{Z}|}{\mathrm{CN}} \qquad (2.2)$$

which is proportional to the electrostatic flux emanating from (or converging toward) an ion along one of the bonds it forms with its nearest neighbors. Given that the number of these latter bonds equals CN, it follows that *the sum of all bond strengths assigned to an ion in a mineral structure is equal to the absolute value of its valence* (i.e., |Z|). This characteristic property of bond strength is a special case of Gauss' law in electrostatics.

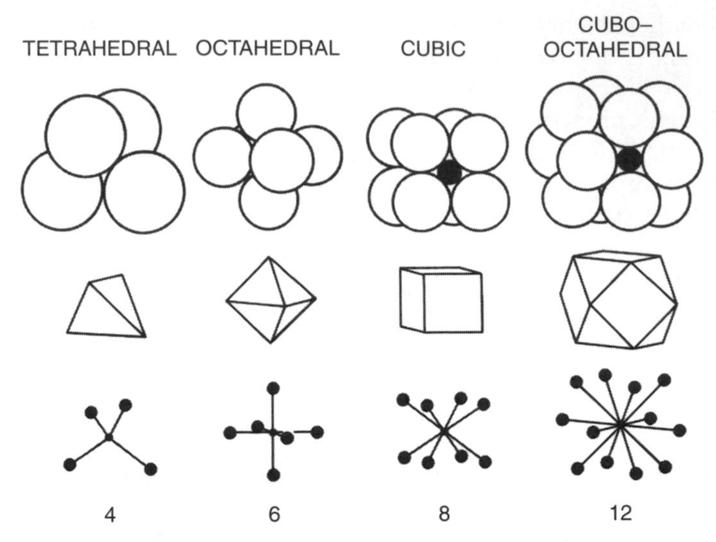

Figure 2.1. The principal coordination numbers for metal cations in soil minerals, illustrated by closely packed anion spheres (top), polyhedra enclosing a metal cation (middle), and "ball-and-stick" drawings (bottom).

The chemical significance of bond strength can be illustrated by an application to the four oxyanions discussed in Section 1.1: NO_3^-, SO_4^{2-}, CO_3^{2-}, and PO_4^{3-}. The strength of the bond between the central cation and one of the peripheral O^{2-} in each of the four oxyanions can be calculated using Eq. 2.2:

$$s = \frac{5}{3} = 1.67 \text{ vu } (NO_3^-) \qquad s = \frac{6}{4} = 1.50 \text{ vu } (SO_4^{2-})$$

$$s = \frac{4}{3} = 1.33 \text{ vu } (CO_3^{2-}) \qquad s = \frac{5}{4} = 1.25 \text{ vu } (PO_4^{3-})$$

where vu means valence unit, a conventional (dimensionless) unit for bond strength. Note that the sum of the bond strengths assigned to each central cation is equal to its valence (e.g., $3 \times 5/3 = 5$, the valence of N in the nitrate anion). But, because none of these bond strengths equals 2.0 (the absolute value of the valence of O^{2-}), any peripheral oxygen ion still has the ability to attract and bind an additional cationic charge external to the oxyanion. This conclusion follows specifically from Gauss' law, mentioned earlier, although it is also evident from the overall negative charge on each oxyanion. It is apparent that the strength of an additional bond formed between a peripheral oxygen ion and any external cationic charge will be smallest for nitrate (i.e., 2.00 – 1.67 = 0.33 vu) and largest for phosphate (i.e., 0.75 vu), with the resultant ordering: $NO_3^- < SO_4^{2-} < CO_3^{2-} < PO_4^{3-}$. This ordering is also the same as observed experimentally for the reactivity of these anions with positively

Table 2.1
Ionic radius (*IR*), coordination number (*CN*), and valence (*Z*) of metal and metalloid cations.[a]

Metal	*Z*	*CN*	*IR* (nm)	Metal	*Z*	*CN*	*IR* (nm)
Li	1	4	0.059	Co	2	6	0.075
	1	6	0.076		3	6	0.061
Na	1	6	0.102	Ni	2	6	0.069
Mg	2	6	0.072	Cu	2	4	0.057
Al	3	4	0.039		2	6	0.073
	3	6	0.054	Zn	2	6	0.074
Si	4	4	0.026	As	3	6	0.058
K	1	6	0.138		5	4	0.034
	1	8	0.151	Sr	2	6	0.118
	1	12	0.164	Zr	4	8	0.084
Ca	2	6	0.100	Mo	6	4	0.041
	2	8	0.112	Ag	1	6	0.115
Ti	4	6	0.061	Cd	2	6	0.095
Cr	3	6	0.062	Cs	1	6	0.167
	6	4	0.026		1	12	0.188
Mn	2	6	0.083	Ba	2	6	0.135
	3	6	0.065	Hg	1	6	0.119
	4	6	0.053		2	6	0.102
Fe	2	6	0.078	Pb	2	6	0.119
	3	6	0.065		4	6	0.078

[a] Shannon, R. D. (1976) Revised effective ionic radii and systematic studies of interatomic distances in halides and chalcogenides. *Acta. Cryst.* **A32**:751–767.

charged sites on particle surfaces (noted in Section 1.1), and it is the order of increasing affinity of the anions for protons in aqueous solution, as indicated by the pH value at which they will bind a single proton. The example given here shows that bond strength can be pictured as the absolute value of an *effective valence* of an ion, assigned to one of its bonds under the constraint that the sum of all such effective valences must equal the absolute value of the actual valence of the ion.

Bond strength usually has only one or two values for cations of the Class A metals discussed in Section 1.2, because these metals typically exhibit only one or two preferred coordination numbers in mineral structures, but bond strength can be quite variable for cations of Class B metals. This happens because of their large polarizability (i.e., large deformability of their electron clouds), which allows them access to a broader range of coordination numbers. A prototypical example is the Class B metal cation Pb^{2+}, for which coordination numbers with O^{2-} ranging from 3 to 12 are observed, with the corresponding bond strengths then varying from 0.67 to 0.17 vu, according to Eq. 2.2. This kind of broad variability and the tendency of cation radii to increase with coordination number, as noted earlier, suggest that an inverse relationship should exist between the ionic radius of Pb^{2+} and its bond strength (an idea also from Linus Pauling). Systematic analyses of thousands of mineral structures have shown that the exponential formula

$$s = \exp\left[27.03\left(R_0 - R\right)\right] \tag{2.3}$$

provides an accurate mathematical representation of how bond strength s decreases with increasing length of a bond (R, in nanometers) between a metal cation and an oxygen ion. Values of the parameter, R_0, the metal cation–oxygen ion bond length that, for a given cation valence, would yield a bond strength equal to 1.0 vu, are listed in Table 2.2 for the metal and metalloid cations in Table 2.1. If the bond strength of Pb^{2+} ranges from 0.67 to 0.17 vu, one finds with Eq. 2.3 and $R_0 = 0.2112$ nm, introduced from Table 2.1, that the corresponding range of the Pb–O bond length in minerals is from 0.226 to 0.277 nm.

Seen the other way around, as a means for calculating bond strength from a measured value of R, Eq. 2.3 provides an alternative to Eq. 2.2. As an example,

Table 2.2
Bond valence parameter R_0 (Eq. 2.3) for metals and metalloids coordinated to oxygen.[a]

Metal	Z	R_0 (nm)	Metal	Z	R_0 (nm)
Li	1	0.1466	Ni	2	0.1654
Na	1	0.1803	Cu	2	0.1679
Mg	2	0.1693	Zn	2	0.1704
Al	3	0.1651	As	3	0.1789
Si	4	0.1624		5	0.1767
K	1	0.2132	Sr	2	0.2118
Ca	2	0.1967	Zr	4	0.1928
Ti	4	0.1815	Mo	6	0.1907
Cr	3	0.1724	Ag	1	0.1842
	6	0.1794	Cd	2	0.1904
Mn	2	0.1790	Cs	1	0.2417
	3	0.1760	Ba	2	0.2285
	4	0.1753	Hg	2	0.1972
Fe	2	0.1734	Pb	2	0.2112
	3	0.1759	Pb	4	0.2042
Co	2	0.1692			
	3	0.1634			

[a]Brown, I. D., and D. Alternatt. (1985) Bond-valence parameters obtained from a systematic analysis of the inorganic crystal structure database. *Acta. Cryst.* **B41**:244.
The average standard deviation of R_0 in this table is 0.0042 nm.

consider Al^{3+}, for which Eq. 2.3 takes on the form

$$s = \exp\left[27.03(0.1651 - R)\right] \tag{2.4}$$

where R is now the length of an Al–O bond in nanometers. In the aluminum oxide mineral corundum (Al_2O_3), Al^{3+} is in octahedral coordination with O^{2-}. Two different Al–O bond lengths are actually observed in this mineral (0.185 nm and 0.197 nm), corresponding to s values of 0.584 and 0.422 vu respectively. These two bond strengths bracket the ideal value of 0.50 calculated with Eq. 2.2 using $Z = 3$, $CN = 6$. When bond strength is calculated with Eq. 2.3 instead of Eq. 2.2, it is termed *bond valence*, not only to avoid confusion with the original Pauling definition, but also to emphasize its chemical interpretation as an effective valence of the ion to which it is assigned.

The electrostatic picture of ionic solids also has significant implications for what kinds of atomic structures these solids can have. The structures of most of the minerals in soils can be rationalized on the physical grounds that the atomic configuration observed is that which tends to minimize the total electrostatic energy. This concept has been formulated in a most useful fashion through a set of descriptive statements known as the *Pauling Rules:*

Rule 1: A polyhedron of anions is formed about each cation. The cation–anion distance is determined by the sum of the respective radii, and the coordination number is determined by the radius ratio of cation to anion.

Minimum radius ratio	Coordination number
1.00	12
0.732	8
0.414	6
0.225	4

Rule 2: In a stable crystal structure, the sum of the strengths of the bonds that reach an anion from adjacent cations is equal to the absolute value of the anion valence.

Rule 3: The cations maintain as large a separation as possible from one another and have anions interspersed between them to screen their charges. In geometric terms, this means that polyhedra tend *not* to share edges or especially faces. If edges are shared, they are shortened relative to the unshared edges.

Rule 4: In a structure comprising different kinds of cation, those of high valence and small coordination number tend not to share polyhedron elements with one another.

Rule 5: The number of essentially different kinds of ion in a crystal structure tends to be as small as possible. Thus, the number of different types of coordination polyhedra in a closely packed array of anions tends to be a minimum.

Pauling Rule 1 is a statement that has the same physical meaning as Figure 2.1. The anion polyhedra mentioned in the rule are shown in the middle of the figure, and the bottom row of "ball-and-stick" cartoons shows the cation–anion bonds with lengths that are determined by the ionic radii. The radius of the smallest sphere that can reside in the central void created by closely packing anions in the four ways shown at the top of the figure can be calculated with the methods of Euclidean geometry. It turns out that this radius is always proportional to the radius of the coordinating anion. For example, in the case of tetrahedral coordination, the smallest cation sphere that can fit inside the four coordinating anions has a radius that is 22.5% of the anion radius, and for six coordinating anions, it is 41.4% of the anion radius. These minimum cation radii are listed as decimal fractions in the table that accompanies Pauling Rule 1. Specific examples of the cation-to-oxygen radius ratio can be calculated with the *IR* data in Table 2.1 and the defined O^{2-} radius of 0.140 nm. Any cation with a coordination number of 6, for example, should have an ionic radius ≥ 0.058 nm ($= 0.414 \times 1.40$). This is the case for all but two of the *IR* values in the table for CN $= 6$, illustrating the important further point that that Pauling Rules are good *approximations* based on a strictly electrostatic viewpoint.

Pauling Rule 2 will be recognized as a restatement of Gauss' law in terms of bond strength defined in Eq. 2.2. For most soil minerals, the anion to which the rule is applied is O^{2-}, although OH^-, CO_3^{2-}, and SO_4^{2-} also figure importantly (Table 1.3). As an example, consider the oxygen ions in quartz (SiO_2), which are coordinated to Si^{4+} ions. The radius of Si^{4+} is 0.026 nm, and its usual coordination number is 4. It follows from Eq. 2.2 that $s = 1.0$ for Si^{4+}. Because the absolute value of the valence of O^{2-} is 2, Pauling Rule 2 then permits only two Si^{4+} to bond to an O^{2-} in SiO_2. This means that each O^{2-} in quartz must serve as the corner of no more than two silica tetrahedra. Hypothetical atomic structures for quartz that would involve, say, O^{2-} at the corners of three tetrahedra linked together are thus ruled out.

A more subtle example of Pauling Rule 2 occurs in the structure of the iron oxyhydroxide mineral goethite (FeOOH). The radius of Fe^{3+} is 0.065 nm and, by Pauling Rule 1, its coordination number with O^{2-} must be 6 ($0.065 \div 0.140 = 0.464 > 0.414 \Rightarrow$ octahedral coordination). Therefore, $s = 0.5$ for Fe^{3+}, and four Fe^{3+} should bond to each O^{2-} in the goethite structure, according to Pauling Rule 2. However, inspection of the goethite structure reveals that each O^{2-} is bonded to *three* Fe^{3+}, not four (Fig. 2.2). The proton in the goethite structure can be used to provide a cation for a fourth bond to O^{2-}, but because there are twice as many O as H in goethite, each proton must be *shared* between two O^{2-} to satisfy Pauling Rule 2. If this is the case, then each proton will be doubly coordinated with O^{2-}, and its corresponding bond strength will be $s = ½ = 0.5$,

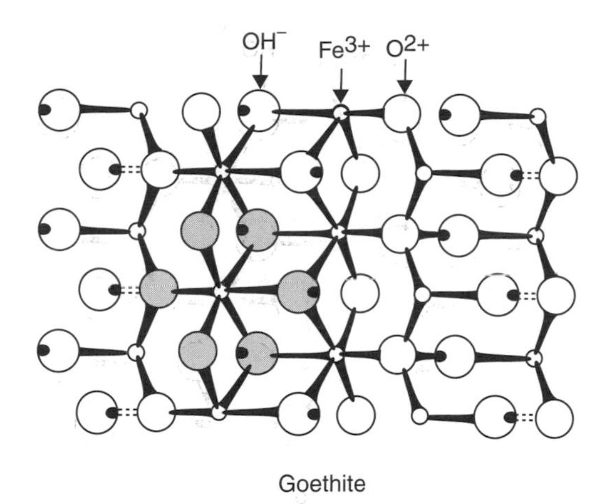

Figure 2.2. "Ball-and-stick" drawing showing the atomic structure of goethite. Note that the coordination number for O^{2-} and OH^- is equal to three.

as required. This sharing of a proton between two oxygens is termed *hydrogen bonding* (Fig. 2.2), by analogy with electron sharing in covalent bonding.

Hydrogen bonds seldom involve the proton placed symmetrically between two oxygens, but instead have one H–O bond significantly shorter (and stronger) than the other. The stronger H–O bond is about 0.095 nm in length and has a bond valence described mathematically by Eq. 2.3, with $R_0 = 0.0882$ nm, thus yielding $s = 0.83$ vu. By Gauss' law, the strength of the weaker bond must be 0.17 vu, because $Z = 1$ for the proton. Corresponding to these deviations from the ideal value, $s = 0.5$ vu, expected for a proton situated at the midpoint between two oxygens that share it, are those of the Fe^{3+} bond valences in goethite, which actually range from 0.377 to 0.600 vu because the Fe–O bond lengths vary from 0.212 to 0.195 nm [Eq. 2.3 with $R_0 = 0.1759$ nm (Table 2.2)]. Pauling Rule 2 can be satisfied either by three long Fe–O bonds combined with the stronger H–O bond or by three short Fe–O bonds combined with the weaker H–O bond.

Pauling Rule 3 reflects coulomb repulsion between cations. The repulsive electrostatic interaction between the cations in a crystal is weakened effectively, or *screened,* by the negatively charged anions in the coordination polyhedra of the cations. If the cations have a large valence, as does, for example, Si^{4+}, then the polyhedra can do no more than share corners if the cations are to be kept as far apart as possible in a structural arrangement that achieves the lowest possible total electrostatic energy. An example of a *sheet* of silica tetrahedra sharing corners is shown in Figure 2.3. If the cation valence is somewhat smaller, as it is for Al^{3+}, the sharing of polyhedron edges becomes possible. Figure 2.3 also shows this kind of sharing for a sheet of octahedra comprising six anions (e.g., O^{2-}) bound to a metal cation (e.g., Al^{3+}). Edge sharing brings

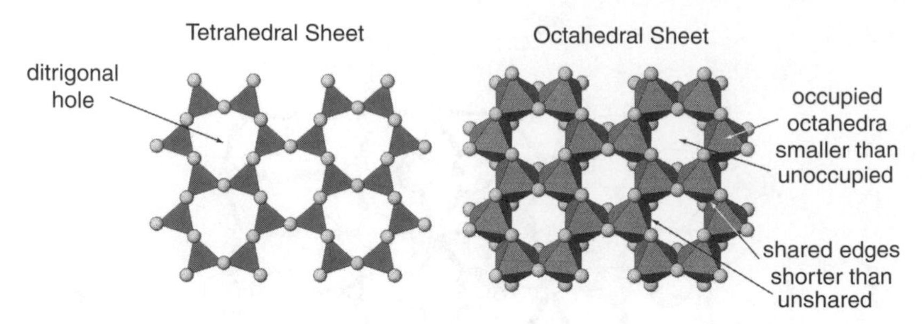

Figure 2.3. (A, B) Sheet structures in soil minerals formed by linking tetrahedra at corners (A) or octahedra along edges (B). Reprinted with permission from Schulze, D. G. (2002) An introduction to soil mineralogy, pp. 1–35. In: J. B. Dixon and D. G. Schulze (eds.), *Soil mineralogy with environmental applications.* Soil Science Society of America, Madison, WI.

the cations closer together than does corner sharing, however, so the task of charge screening by the anions is made more difficult. They respond to this by approaching one another slightly along the shared edge to enhance screening. Doing so, they necessarily shorten the edge relative to unshared edges of the polyhedra (Fig. 2.3), which is why there are short and long Al–O and Fe–O bonds in oxide minerals such as corundum (Al_2O_3) and goethite (FeOOH).

Pauling Rules 4 and 5 continue in the spirit of Rule 3. They reflect the fact that stable ionic crystals containing different kinds of cation cannot tolerate much sharing of the coordination polyhedra or much variability in the type of coordination environment. These and the other three Pauling Rules serve as useful guides to a molecular interpretation of the chemical formulas for soil minerals.

2.2 Primary Silicates

Primary silicates appear in soils as a result of deposition processes and from the physical disintegration of parent rock material. They are to be found mainly in the sand and silt fractions, except for soils at the early to intermediate stages of the Jackson–Sherman weathering sequence (Table 1.7), wherein they can survive in the clay fraction as well. The weathering of primary silicates contributes to the native fertility and electrolyte content of soils. Among the major decomposition products of these minerals are the soluble metal cation species Na^+, Mg^{2+}, K^+, Ca^{2+}, Mn^{2+}, and Fe^{2+} in the soil solution. The metal cations Co^{2+}, Cu^{2+}, and Zn^{2+} occur as trace elements in primary silicates (Table 1.4) and thus are also released to the soil solution by weathering. These *free-cation* species are readily bioavailable and, except for Na^+, are essential to the nutrition of green plants. The major element cations—Na^+, Mg^{2+}, and Ca^{2+}—provide a principal input to the electrolyte content in soil solutions.

The names and chemical formulas of primary silicate minerals important to soils are listed in Table 2.3. The fundamental building block in the atomic

Table 2.3
Names and chemical formulas of primary silicates found in soils.

Name	Chemical formula	Mineral group
Forsterite	Mg_2SiO_4	Olivine
Fayalite	Fe_2SiO_4	Olivine
Chrysolite	$Mg_{1.8}Fe_{0.2}SiO_4$	Olivine
Enstatite	$MgSiO_3$	Pyroxene
Orthoferrosilite	$FeSiO_3$	Pyroxene
Diopside	$CaMgSi_2O_6$	Pyroxene
Tremolite	$Ca_2Mg_5Si_8O_{22}(OH)_2$	Amphibole
Actinolite	$Ca_2Mg_4FeSi_8O_{22}(OH)_2$	Amphibole
Hornblende	$NaCa_2Mg_5Fe_2AlSi_7O_{22}(OH)$	Amphibole
Muscovite	$K_2[Si_6Al_2]Al_4O_{20}(OH)_4$	Mica
Biotite	$K_2[Si_6Al_2]Mg_4Fe_2O_{20}(OH)_4$	Mica
Phlogopite	$K_2[Si_6Al_2]Mg_6O_{20}(OH)_4$	Mica
Orthoclase	$KAlSi_3O_8$	Feldspar
Albite	$NaAlSi_3O_8$	Feldspar
Anorthite	$CaAl_2Si_2O_8$	Feldspar
Quartz	SiO_2	Silica

structures of these minerals is the silica tetrahedron: SiO_4^{4-}. Silica tetrahedra can occur as isolated units, in single or double chains linked together by shared corners (Pauling Rules 2 and 3), in sheets (Fig. 2.3), or in full three-dimensional frameworks. Each mode of occurrence defines a class of primary silicate, as summarized in Figure 2.4.

The *olivines* comprise individual silica tetrahedra in a structure held together with bivalent metal cations like Mg^{2+}, Fe^{2+}, Ca^{2+}, and Mn^{2+} in octahedral coordination. Solid solution (see Section 1.3) of the minerals forsterite and fayalite (Table 2.3) produces a series of mixtures with specific names, such as chrysolite, which contains 10 to 30 mol% fayalite. As discussed in Section 1.3, olivines have the smallest Si-to-O molar ratio among the primary silicates and, therefore, they feature the least amount of covalence in their chemical bonds. Their weathering in soil is relatively rapid (timescale of years), beginning along cracks and defects at the crystal surface to form altered rinds containing oxidized-iron solid phases and smectite (Table 1.3). More extensive leaching can result in congruent dissolution (see Problem 11 in Chapter 1) or can produce kaolinite instead of smectite, the formation of either of these clay minerals requiring a proximate source of Al, because none exists in olivine (except possibly as a trace element).

The *pyroxenes* and *amphiboles* contain either single or double chains of silica tetrahedra that form the repeating unit $Si_2O_6^{4-}$ or $Si_4O_{11}^{6-}$, respectively, with Si-to-O ratios near 0.33 to 0.36 (Fig. 2.4). The amphiboles feature isomorphic substitution of Al^{3+} for Si^{4+} (Table 2.3), and both mineral groups harbor a variety of bivalent metal cations, as well as Na^+, in octahedral

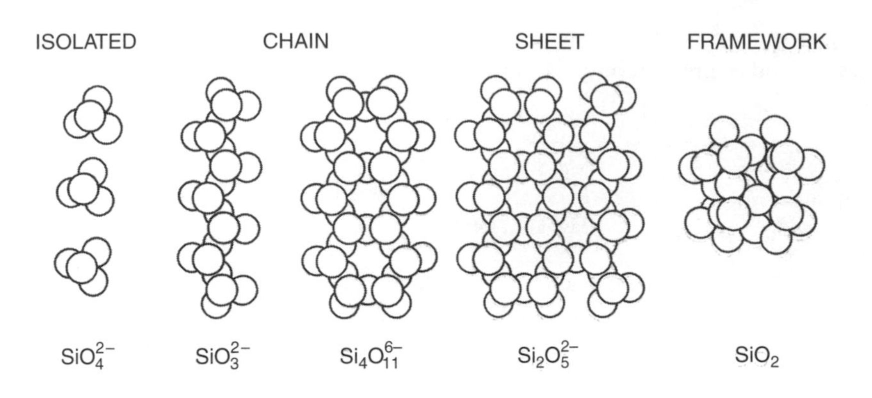

Figure 2.4. Primary silicates classified by the geometric arrangement of their silica tetrahedra.

coordination with O^{2-} to link the silica chains together. The weathering of these silicates is complex, with smectite and kaolinite, along with Al and Fe(III) oxides, being the principal secondary minerals emerging near structural defect sites where mineral dissolution begins. Hydrolysis and protonation, along with oxidation of Fe(II), are the main weathering mechanisms of olivines, pyroxenes, and amphiboles, although complexation (e.g., by oxalate) plays a dominant role when weathering is governed by microorganisms, such as bacteria or fungi.

The *micas* are built up from two sheets of silica tetrahedra ($Si_2O_5^{2-}$ repeating unit) fused to each planar side of a sheet of metal cation octahedra (Fig. 2.3). The octahedral sheet typically contains Al, Mg, and Fe ions coordinated to O^{2-} and OH^-. If the metal cation is trivalent, only two of the three possible cationic sites in the octahedral sheet can be filled to achieve charge balance and the sheet is termed *dioctahedral.* If the metal cation is bivalent, all three possible sites are filled and the sheet is *trioctahedral.* Isomorphic substitution of Al for Si, Fe(III) for Al, and Fe or Al for Mg occurs typically in the micas, along with the many trace element substitutions mentioned in Table 1.4.

Muscovite and biotite are the common soil micas, the former being dioctahedral and the latter trioctahedral (Table 2.3). In both minerals, Al^{3+} substitutes for Si^{4+}. The resulting charge deficit is balanced by K^+, which coordinates to 12 oxygen ions in the cavities of two opposing tetrahedral sheets belonging to a pair of mica layers stacked on top of one another. Thus the K^+ links adjacent mica layers together. It is these linking cations that are removed first as crystallite edges become frayed and, therefore, vulnerable to penetration by water molecules and competing soil solution cations during the initial stage of weathering (Fig. 2.5), which is accelerated by rhizosphere microorganisms that consume K^+ from the vicinal soil solution and release

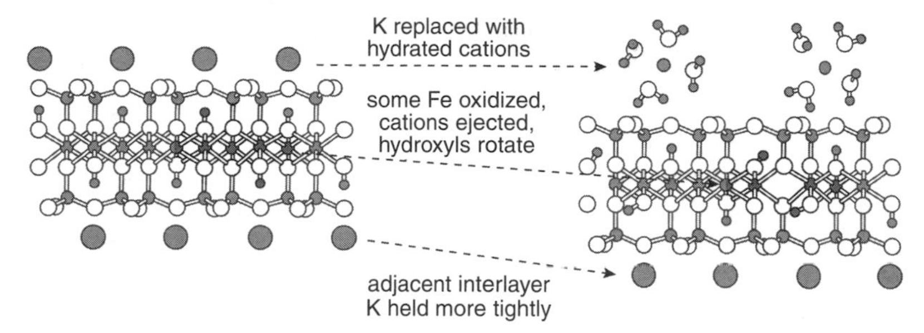

Figure 2.5. Some pathways of the initial stage of weathering of the trioctahedral mica, biotite. There is a loss of interlayer K^+ and oxidation of Fe^{2+} in the octahedral sheet, with consequent rotation of structural OH. Reprinted with permission from Thompson, M. L., and L. Ukrainczyk. (2002) Micas, pp. 431–466. In: J. B. Dixon and D. G. Schulze (eds.), *Soil mineralogy with environmental applications.* Soil Science Society of America, Madison, WI.

organic acids that complex and dislodge Al exposed at crystallite edges. Ferrous iron in biotite is gradually oxidized to ferric iron and ejected to hydrolyze and form an Fe(III) oxyhydroxide precipitate. This, in turn, allows some structural OH groups in the octahedral sheet to rotate toward the now-vacant former Fe(II) sites, the OH protons thereby being rendered less effective at repelling the surviving K^+ between the biotite layers (Fig. 2.5). Under moderate leaching conditions, muscovite transforms to dioctahedral smectite (see Problem 13 in Chapter 1), whereas biotite transforms to trioctahedral vermiculite and goethite (or ferrihydrite). A possible reaction for this latter transformation is

$$\underset{\text{(biotite)}}{K_2[Si_6Al_2]Mg_3\,Fe(II)_3O_{20}(OH)_4(s)} + 2.7\,Mg^{2+} + 3.9\;H_2O(\ell)$$

$$+\,0.75\,O_2(g) = \underset{\text{(vermiculite)}}{K_{1.7}[Si_6Al_2]\,Mg_{5.7}Fe(III)_{0.3}\,O_{20}(OH)_4(s)}$$

$$+\,2.7\,\underset{\text{(goethite)}}{FeOOH(s)} + 0.3\;K^+ + 5.1\;H^+ \qquad (2.5)$$

Note that the layer charge (see Section 1.3), as evidenced by the stoichiometric coefficient of K^+, decreases from 2.0 in biotite to 1.7 in vermiculite because of the oxidation of ferrous iron. Although this layer charge is balanced by K^+ in Eq. 2.5, Mg^{2+} is also a common interlayer cation in trioctahedral vermiculite. Under intensive leaching conditions, biotite will transform to kaolinite and goethite, as illustrated in Eq. 1.3. In this case, silica and Mg^{2+} are lost to the soil solution along with K^+. A comparison of Eqs. 1.3 and 2.5 shows that kaolinite formation is favored by acidity (H^+ is a reactant) and inhibited by soluble Mg^{2+} (a product), whereas vermiculite formation is inhibited by acidity (H^+ is a product) and favored by soluble Mg^{2+} (a reactant).

The atomic structure of the *feldspars* is a continuous, three-dimensional framework of tetrahedra sharing corners, as in quartz, except that some of the tetrahedra contain Al instead of Si, with electroneutrality thus requiring either monovalent or bivalent metal cations to occupy cavities in the framework. These primary minerals, the most abundant in soils, have repeating units of either $AlSi_3\ O_8^-$, with Na^+ or K^+ used for charge balance, or $Al_2\ Si_2\ O_8^{2-}$, with Ca^{2+} used for charge balance (Table 2.3). Solid solution among the three minerals thus formed is extensive, with that between albite and anorthite being known as *plagioclase*, whereas that between albite and orthoclase termed simply *alkali feldspar*. The weathering of these abundant minerals in soils occurs on timescales of millennia.

Figure 2.6 illustrates this last point with measurements of the amounts of hornblende, plagioclase, and K-feldspar remaining (relative to quartz) in the surface (A in Fig. 2.6) and subsurface (B and C in Fig. 2.6) horizons of a soil chronosequence comprising Entisols, Mollisols, Alfisols, and Ultisols, the members of which ranged in age from two centuries to 3000 millennia, as determined by radioactive isotope dating methods. The graph in Figure 2.6 indicates that all three primary silicates were depleted during the first few hundred millennia of weathering and that the overall rate of depletion was in the order of hornblende > plagioclase > K-feldspar, with the surface horizon

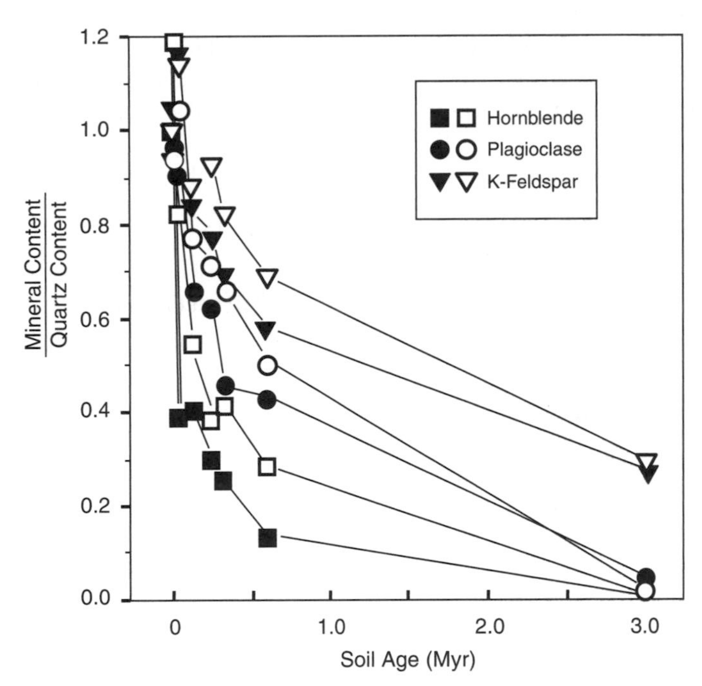

Figure 2.6. Depletion of amphiboles (hornblende) and feldspars (plagioclase and K-feldspar) with time during soil weathering. The ordinate is the content of primary silicate in the soil relative to that of quartz, which is assumed to be conserved. Data from White, A. F., et al. (1995) Chemical weathering rates of a soil chronosequence on granitic alluvium: I. *Geochim. Cosmochim. Acta* **60**:2533–2550.

showing more depletion than subsurface horizons. These trends are in keeping with the smaller Si-to-O ratio in amphiboles than in feldspars (see Section 1.3) and with the more intense weathering expected near the top of a soil profile.

Feldspar dissolution provides metal cations to the soil solution that figure importantly in the neutralization of acidic deposition on soils, the nutrition of plants, and the regulation of CO_2 concentrations. Bacteria and fungi enhance this dissolution process through the production of organic ligands (Eq. 1.4) and protons, particularly in the case of K-feldspar, which then serves as a source of K. Feldspars weather eventually to kaolinite (Eq. 1.2) or gibbsite (see Problem 12 in Chapter 1), but smectite also is a common secondary mineral product (see Problem 13 in Chapter 1):

$$\underset{\text{(orthoclase)}}{5\,KAlSi_3O_8(s)} + 4\,H^+ + 16\,H_2O(\ell) = \underset{\text{(beidellite)}}{K[Si_7Al]Al_4O_{20}(OH)_4(s)} + 8\,Si(OH)_4^0 + 4\,K^+ \tag{2.6}$$

Note the consumption of protons and the production of silicic acid and soluble cations, as also observed in Eq. 1.2.

The general characteristics of primary silicate weathering illustrated by eqs. 1.2, 1.3, 1.4, 2.5, and 2.6 can be summarized as follows:

- Conversion of tetrahedrally coordinated Al to octahedrally coordinated Al
- Oxidation of Fe(II) to Fe(III)
- Consumption of protons and water
- Release of silica and metal cations

In the case of the micas, there is also an important reduction of layer charge accompanying the first two characteristics (Eq. 2.5 and Problem 13 in Chapter 1). From the weathering sequence in Table 1.7, one can conclude that soil development renders tetrahedral Al and ferrous iron unstable in response to continual throughputs of oxygenated fresh water (i.e., rainwater), which provides protons and, in return, receives soluble species of major elements. If these latter elements are not leached, the secondary silicates that characterize the intermediate stage of weathering will form, as in Eq. 2.6. If leaching is extensive, desilicated minerals characteristic of the advanced stage of weathering will begin to predominate in the clay fraction, as in eqs. 1.2 and 1.3.

2.3 Clay Minerals

Clay minerals are layer-type aluminosilicates that predominate in the clay fractions of soils at the intermediate to advanced stages of weathering. These minerals, like the micas, are sandwiches of tetrahedral and octahedral sheet structures like those in Figure 2.3. This bonding together of the tetrahedral and octahedral sheets occurs through the apical oxygen ions in the tetrahedral sheet and produces a distortion of the anion arrangement in the final layer structure formed. The distortion occurs primarily because the apical oxygen ions in the tetrahedral sheet cannot be fit to the corners of the octahedra to form a layer while preserving the ideal hexagonal pattern of the tetrahedra. To fuse the two sheets, pairs of adjacent tetrahedra must rotate and thereby perturb the symmetry of the cavities in the basal plane of the tetrahedral sheet, altering them from hexagonal to ditrigonal (Fig. 2.3). Besides this distortion, the sharing of edges in the octahedral sheet shortens them (Pauling Rule 3, Fig. 2.3). These effects occur in the micas and in the clay minerals, both of whose atomic structures were first worked out by Linus Pauling (see Special Topic 2, at the end of this chapter). The clay minerals are classified into three *layer types*, distinguished by the number of tetrahedral and octahedral sheets combined to form a layer, and further into five *groups*, differentiated by the

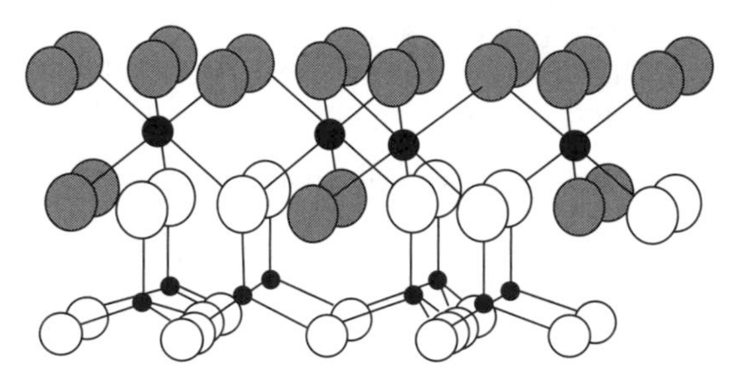

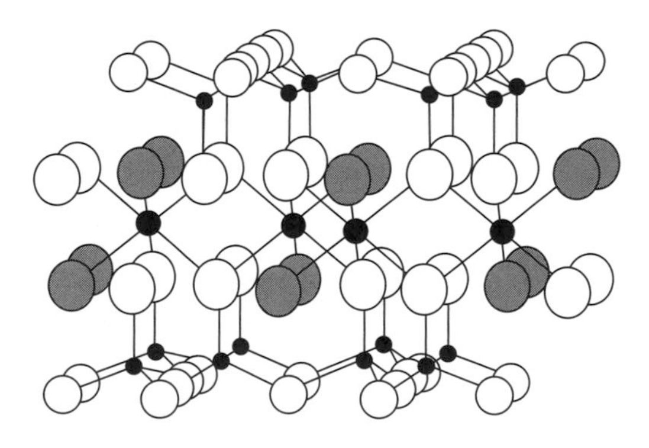

Figure 2.7. "Ball-and-stick" drawings of the atomic structures of 1:1 and 2:1 layer-type clay minerals.

Table 2.4
Clay mineral groups.[a]

Group	Layer type	Layer charge (x)	Typical chemical formula[b]
Kaolinite	1:1	< 0.01	$[Si_4]Al_4O_{10}(OH)_8 \cdot nH_2O$ ($n = 0$ or 4)
Illite	2:1	1.2–1.7	$M_x[Si_{6.8}Al_{1.2}]Al_3Fe_{0.25}Mg_{0.75}O_{20}(OH)_4$
Vermiculite	2:1	1.2–1.8	$M_x[Si_7Al]Al_3Fe_{0.5}Mg_{0.5}O_{20}(OH)_4$
Smectite[c]	2:1	0.4–1.2	$M_x[Si_8]Al_{3.2}Fe_{0.2}Mg_{0.6}O_{20}(OH)_4$
Chlorite	2:1 with hydroxide interlayer	Variable	$(Al(OH)_{2.55})_4\ [Si_{6.8}Al_{1.2}]Al_{3.4}Mg_{0.6}O_{20}(OH)_4$

[a]Guggenheim, S., et al. (2006) Summary of recommendations of nomenclature committees relevant to clay mineralogy. *Clays Clay Miner.* **54**:761–772.
[b][ ] indicates tetrahedral coordination; $n = 0$ is kaolinite and $n = 4$ is halloysite; H_2O = interlayer water; M = monovalent interlayer cation.
[c]Principally montmorillonite and beidellite in soils.

extent and location of isomorphic cation substitutions in the layer. The layer types are shown in Fig. 2.7, whereas the groups are described in Table 2.4.

The 1:1 layer type consists of one tetrahedral and one octahedral sheet. In soil clays, it is represented by the *kaolinite group*, with the generic chemical formula $[Si_4]Al_4O_{10}(OH)_8 \cdot nH_2O$, where the element enclosed in square brackets is in tetrahedral coordination and n is the number of moles of hydration water between layers. As is common for soil clay minerals, the octahedral sheet has two thirds of its cation sites occupied (dioctahedral sheet). Normally there is no isomorphic substitution for Si or Al in kaolinite group minerals, although low substitution of Fe for Al is sometimes observed in Oxisols, and poorly crystalline varieties of kaolinite are thought to have some substitution of Al for Si. Kaolinite group minerals are the most abundant clay minerals in soils worldwide, although, as implied in Table 1.7, they are particularly characteristic of highly weathered soils (Ultisols, Oxisols). Their typical particle size is less than 10 μm in diameter (fine silt and clay fractions), and their specific surface area ranges from 0.5 to $4.0 \times 10^4\ m^2\ kg^{-1}$, with the larger values being measured for poorly crystalline varieties. Aggregates of these clay minerals are observed as stacks of hexagonal plates if the layers are well crystallized, whereas elongated tubes with inside diameters on the order of 15 to 20 μm or more (or sometimes spheroidal particles) are found if the layers are poorly crystallized. (If the repeating structure based on the chemical formula of a solid phase persists throughout a molecular-scale region with a diameter that is at least as large as 3 nm, the solid phase is said to be *crystalline.* If structural regularity does not exist over molecular-scale distances this large, the solid phase is termed *poorly crystalline.*) The subgroup associated with the tubular morphology contains interlayer water ($n = 4$ in the chemical formula) and is named *halloysite* (Table 2.4). Halloysite tends to be found under conditions of active weathering abetted by ample water, but it can dehydrate eventually to form more well-crystallized kaolinite ($n = 0$ in the chemical formula). The tubular morphology is thought to be an alternative structural response to tetrahedral–octahedral sheet misfit, wherein the tetrahedral sheet rolls around the octahedral sheet because interlayer water has prevented the tetrahedra from rotating as they do in kaolinite.

The oxygen ions in the basal plane of the tetrahedral sheet in kaolinite are bonded to a pair of Si^{4+}, whereas the apical oxygen ions are bonded to one Si^{4+} and two Al^{3+} in consonance with Pauling Rule 2 (Fig. 2.7). Similarly, the OH ions in the basal plane of the octahedral sheet are bonded to two Al^{3+}, as are the OH in the interior of the layer. Therefore, if the layer were infinite in lateral extent, it would be completely stable according to the Pauling rules. However, the oxygen and OH ions at the edge surfaces of a finite layer structure will always be missing some of their cation bonding partners, leading to the ability to bind additional cationic charge, as discussed for free oxyanions in Section 2.1. An exposed oxygen ion bound to a single Si^{4+} at an edge, for example, bears an excess charge of –1.0 vu and, therefore, requires a cation partner with a bond valence of 1.0 vu to be stable. This requirement can be met easily if a proton from aqueous solution becomes bound to the oxygen ion (Fig. 2.8). The situation is not as simple for an exposed OH ion, which bears an excess charge of –0.5 vu and thus requires a cation partner with a bond valence of 0.5 vu to be stable. Attraction of a proton from aqueous solution leads to the formation of $OH_2^{1/2+}$ at the edge surface, which is still not stable. The excess positive charge created can be neutralized in principle by a neighboring $OH_2^{1/2-}$, but this possibility clearly will be affected by soil solution pH. It turns out that the

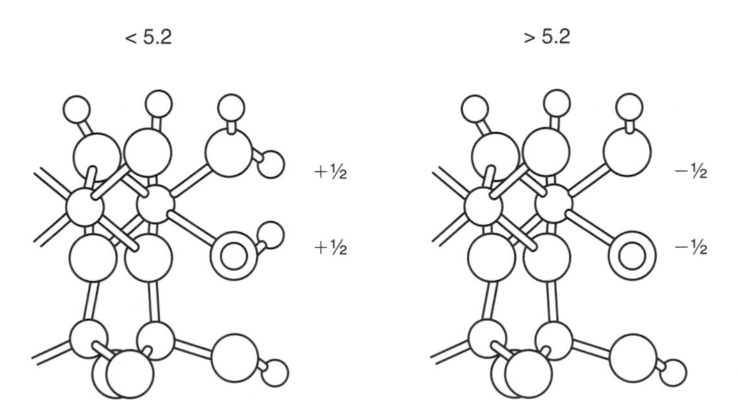

Figure 2.8. Atomic structure at the edge of a kaolinite layer exposed to water. As pH drops below 5.2, exposed Si–O^- and Al–$OH^{-1/2}$ each protonate to form Si–OH^0 and Al–$OH_2^{+1/2}$ respectively.

affinity of the kaolinite edge surface for protons leads to an electrically neutral condition at pH 5.2, with the surface being increasingly positively charged below this pH value and increasingly negatively charged above it. As was the case for the oxyanions considered in Section 2.1, failure to satisfy the Pauling Rules leads to reactivity with protons—in the present case, those in the soil solution contacting the edge surface of kaolinite.

The 2:1 layer type has two tetrahedral sheets that sandwich an octahedral sheet (Fig. 2.7). The three soil clay mineral groups with this structure are *illite, vermiculite,* and *smectite.* If a, b, and c are the stoichiometric coefficients of Si, octahedral Al, and Fe(III) respectively, in the chemical formulas of these groups, then

$$x \equiv \text{moles Al substituting for Si} + \text{moles Mg and Fe(II) substituting for Al} \quad (2.7)$$

$$= (8 - a) + (4 - b - c) = 12 - a - b - c \quad (2.8)$$

is the *layer charge,* the number of moles of excess electron charge per chemical formula that is produced by isomorphic substitutions. As indicated in Table 2.4, the layer charge decreases in the order illite > vermiculite > smectite. The vermiculite group is further distinguished from the smectite group by a greater extent of isomorphic substitution in the tetrahedral sheet. Among the smectites, two subgroups also are distinguished in this way, those for which the substitution of Al for Si exceeds that of Fe(II) or Mg for Al (called *beidellite;* Eq. 2.6), and those for which the reverse is true (called *montmorillonite*). The smectite chemical formula in Table 2.4 represents montmorillonite. In any of these 2:1 minerals, the layer charge is balanced by cations that reside near or in the ditrigonal cavities of the basal plane of the oxygen ions in the tetrahedral sheet (Figs. 2.3 and 2.9). These interlayer cations are represented by M in the chemical formula of smectite (Table 2.4).

The layer charge in Eq. 2.7 is closely related to the *structural charge,* σ_0, defined by the equation

$$\sigma_0 = -(x/\mathrm{M_r}) \times 10^3 \quad (2.9)$$

where x is the layer charge and M_r is the relative molecular mass (see the Appendix). The units of σ_0 are moles of charge per kilogram ($\mathrm{mol_c\,kg^{-1}}$, see the Appendix). The value of $\mathrm{M_r}$ is computed with the chemical formula and known relative molecular masses of each element that appears in the formula. For example, in the case of the smectite with a chemical formula that is given in Table 2.4,

$$M_\mathrm{r} = \underset{\mathrm{Si}}{8\,(28.09)} + \underset{\mathrm{Al}}{3.2\,(27)} + \underset{\mathrm{Fe}}{0.2\,(55.85)} + \underset{\mathrm{Mg}}{0.6\,(24.3)} + \underset{\mathrm{O}}{24\,(16)} + \underset{\mathrm{H}}{4\,(1)} = \mathbf{725\ Da}$$

Therefore, according to Eq. 2.7 and the range of x in Table 2.4, σ_0 varies between –0.7 and –1.7 $mol_c\ kg^{-1}$ for smectites. In a similar way, σ_0 is found to vary from –1.9 to –2.8 $mol_c\ kg^{-1}$ for illites, and from –1.6 to –2.5 $mol_c\ kg^{-1}$ for vermiculites. These minerals are significant sources of negative structural charge in soils.

Particle sizes of the 2:1 clay minerals place them in the clay fraction, with illite and vermiculite typically occurring in larger aggregates of stacked layers than smectite, for which lateral particle dimensions around 100 to 200 nm are characteristic. Specific surface areas of illite average about $10^5\ m^2\ kg^{-1}$, whereas those of vermiculite and smectite can approach $8 \times 10^5\ m^2\ kg^{-1}$, depending on the number of stacked layers in an aggregate. The origin of this latter value, which is very large (equivalent to 80 ha kg^{-1} clay mineral), can be seen by calculating the specific surface area (a_s) of an Avogadro number of unit cells (unit cells are the basic repeating entities in a crystalline solid) forming a layer of the smectite featured in Table 2.4:

$$\begin{aligned} a_s &= \text{surface area per unit cell } \times (N_A/M_r) \times 10^3 \\ &= 0.925\ \text{nm}^2 \text{ per cell } \times 10^{-18} \text{m}^2\ \text{nm}^{-2} \\ &\quad \times \frac{6.022 \times 10^{23}}{725\ \text{g}} \text{ cells } \times 10^3 \text{g kg}^{-1} \\ &= \mathbf{7.6 \times 10^5 m^2\ kg^{-1}} \end{aligned}$$

where $N_A = 6.022 \times 10^{23}$ is the Avogadro constant (also denoted L; see the Appendix) and the surface area of the smectite unit cell is calculated as twice the nominal surface area of one face in the crystallographic ab plane (i.e., twice 0.4627 nm^2), which is valid for a crystal layer with lateral dimensions (100–200 nm) that greatly exceed its thickness (only 1 nm for 2:1 clay minerals). This very large specific surface area pertains to particles that comprise a single crystal layer 100 to 200 nm in diameter. If, instead, n such layers are stacked to build an aggregate, the specific surface area is equal to the value found previously divided by n, because stacking a pair of layers together necessarily consumes the area of one basal surface of each. In aqueous suspensions, $n = 1$ to 3 for smectites with monovalent interlayer cations (e.g., Li^+, Na^+, K^+), whereas dehydrated smectites are found in aggregates with about 10 times as many stacked layers. Thus a_s for smectite aggregates can vary from nearly 80 ha kg^{-1} to around 4 ha kg^{-1}.

The 2:1 layer type with a hydroxide interlayer is represented in soils by vermiculite or smectite with an Al–hydroxy polymer cation in the interlayer regions (Table 2.4), with the collective name for these subgroups being *pedogenic chlorite.* Formation of these clay minerals is mediated by acidic conditions, under which Al^{3+} is released by mineral dissolution, hydrolyzes, and replaces the interlayer cations in vermiculite or smectite, with incomplete hydrolysis resulting in a cationic Al–hydroxy polymer with a fractional stoichiometric coefficient for OH < 3 instead of $Al(OH)_3$. Pedogenic chlorite is characteristic of highly weathered soils, such as Ultisols and Oxisols, but also is found in Alfisols and Spodosols. Whenever a complete, isolated gibbsite sheet [$Al(OH)_3$] forms in the interlayer region, the resulting mineral is termed simply *chlorite.*

Structural disorder in the 2:1 clay minerals listed in Table 2.4 is induced through isomorphic substitutions in their octahedral sheets (tables 1.5 and 2.4). More pronounced structural disorder exists in silica and in aluminosilicates that are freshly precipitated in soils undergoing active weathering, because these solid phases typically are excessively hydrated and poorly crystalline. Even among the more crystalline soil clay minerals, there is also wide variability in nanoscale order, with disorder created by dislocations (microcrevices between offset rows of atoms) and irregular stacking of crystalline unit layers. This kind of disorder exists, for example, in kaolinite and illite group minerals.

Poorly crystalline hydrated aluminosilicates, known collectively as *imogolite* and *allophane* (Table 1.3), are common in the clay fractions of soils formed on volcanic ash deposits (Andisols), but they can also be derived from many other kinds of parent material (e.g., granite or sandstone) under acidic conditions, regardless of temperature regime, if soluble Al and Si concentrations are sufficiently high and Al is not complexed with organic ligands, which interferes with precipitation (Eq. 1.4). Imogolite, having the chemical formula $Si_2Al_4O_{10} \cdot 5\,H_2O$, contains only octahedrally coordinated Al and exhibits a slender tubular particle morphology. The tubes are several micrometers long, with a diameter of about 2 nm, exposing a defective, gibbsitelike outer surface. The specific surface area of imogolite is comparable with—or even greater than—that of smectite. A surface charge develops from unsatisfied oxygen ion bond valences, similar to what occurs in kaolinite group minerals, but the pH value at which imogolite is electrically neutral is much higher (pH $\approx$ 8.4).

Allophane has the general chemical formula $Si_yAl_4O_{6+2y} \cdot n\,H_2O$, where $1.6 \leq y \leq 4$, $n \geq 5$ (Table 1.3). Thus it exhibits Al-to-Si molar ratios both larger and smaller than imogolite ($y = 2$) and it contains more bound water. Its specific surface area is also comparable with that of smectite and, like this latter clay mineral, a structural charge in allophane is possible because of isomorphic substitution of Al for Si in tetrahedral coordination, and charge development from unsatisfied oxygen ion bond valences occurs just as it does in kaolinite and imogolite. The pH value at which the protonation mechanism results in an electrically neutral surface varies inversely with the value of y in the chemical formula, decreasing from about pH 8.0 for $y = 2$ (termed the *proto-imogolite* allophane species) to about pH 5.4 for $y = 4$ (termed the *defect kaolinite* allophane species). Evidently, the increasing presence of Al results in stronger protonation, thus requiring higher pH for electrical neutrality, whereas that of Si has the opposite effect. The atomic structure of allophane is not well understood, but is thought in most cases to consist of fragments of imogolite combined with a 1:1 layer-type aluminosilicate that is riddled with vacant ion sites and doped with Al in tetrahedral coordination. This defective structure promotes a curling of the layer into the form of hollow spheroids 3 to 5 nm in diameter with an outer surface that can contain many microapertures through which molecules or ions in the soil solution might invade. As this structural concept suggests, allophane often is found in association with kaolinite group minerals, especially halloysite.

Poorly crystalline kaolinite group minerals have been observed to precipitate in bacterial *biofilms*, which are layered organic matrices comprising extracellular polymers that enmesh bacterial cells along with nutrients and other chemical compounds. When minerals form in biofilms, the biofilms are termed *geosymbiotic microbial ecosystems* to emphasize the close spatial relationship that exists between the minerals and the microbes. Under highly anaerobic conditions at circumneutral pH in freshwater biofilms that contain a variety of different bacteria and filamentous algae, clay-size, hollow, spheroidal particles identified as poorly crystalline kaolinite group minerals appear to nucleate and grow on bacterial surfaces as a product of feldspar weathering (see Eq. 1.2). Similar observations have been reported for 2:1 layer-type clay minerals under active weathering conditions.

The 2:1 clay minerals, as well as pedogenic chlorite, imogolite, and allophone, all are expected to weather by hydrolysis and protonation to form kaolinite group minerals according to the Jackson–Sherman weathering sequence (Table 1.7):

$$\underset{\text{(beidellite)}}{\mathrm{K[Si_7Al]Al_4O_{20}(OH)_4}}(s) + \mathrm{H^+} + \frac{13}{2}\mathrm{H_2O}(\ell)$$

$$= \frac{5}{4}\underset{\text{(kaolinite)}}{\mathrm{[Si_4]Al_4O_{10}(OH)_8}} + 2\ \mathrm{Si(OH)_4^0} + \mathrm{K^+} \tag{2.10a}$$

$$\underset{\text{(pedogenic chlorite)}}{\mathrm{(Al(OH)_{2.5})_2[Si_7Al]Al_4O_{20}(OH)_4}}(\mathrm{s}) + 10\ \mathrm{H_2O}(\ell) \tag{2.10b}$$

$$= \underset{\text{(kaolinite)}}{\mathrm{[Si_4]Al_4O_{10}(OH)_8}}(\mathrm{s}) + 3\ \underset{\text{(gibbsite)}}{\mathrm{Al(OH)_3}}(s) + 3\ \mathrm{Si(OH)_4^0}$$

$$\underset{\text{(allophane)}}{\mathrm{Si_3Al_4O_{12}} \cdot n\,\mathrm{H_2O}}(\mathrm{s}) + \frac{9}{2}\mathrm{H_2O}(\ell)$$

$$= \frac{3}{4}\ \underset{\text{(kaolinite)}}{\mathrm{[Si_4]Al_4O_{10}(OH)_8}}(\mathrm{s}) + \underset{\text{(gibbsite)}}{\mathrm{Al(OH)_3}}(\mathrm{s}) + n\,\mathrm{H_2O}(\ell) \tag{2.10c}$$

Each of these reactions requires acidic conditions that are favored by freshwater and good drainage. The pedogenic chlorite reacting in Eq. 2.10b is an example of *hydroxy-interlayer beidellite* ($x = 1.0$), whereas *hydroxy-interlayer vermiculite* ($x = 1.8$) is shown in Table 2.4. The allophane reactant in Eq. 2.10c is a *defect kaolinite* species.

2.4 Metal Oxides, Oxyhydroxides, and Hydroxides

Because of their great abundance in the lithosphere (Table 1.2), aluminum, iron, manganese, and titanium form the important oxide, oxyhydroxide, and hydroxide minerals in soils. They represent the *climax mineralogy* of soils, as indicated in Table 1.7. The most significant of these minerals, all of which are characterized by small particle size and low solubility in the normal range of soil pH values, can be found in Table 2.5, with representative atomic structures of some of them depicted in figures 2.9 and 2.10. For each type of metal cation, the Pauling Rules would indicate primarily octahedral coordination with oxygen or hydroxide anions.

Gibbsite [γ–$Al(OH)_3$], the only Al mineral listed in Table 2.5, is found commonly in Oxisols, Ultisols, Inceptisols, and Andisols, forming parallelepipeds 50 to 100 nm in length under conditions of warm climate and intense leaching that lead to Si removal from clay minerals and primary silicates, especially feldspars (see Table 1.7; Eqs. 2.10b, c; and Problem 12 in Chapter 1). Isomorphic substitutions do not appear to occur in this mineral. Inorganic anions, such as carbonate and silica, and organic ligands, including humic substances, disrupt the formation of gibbsite by complexing Al^{3+} (e.g., Eq. 1.7), and promote instead the precipitation of poorly crystalline Al oxyhydroxides with large specific surface areas (10–60 ha kg^{-1}). These highly reactive, disordered polymeric materials contribute significantly to the formation of stable aggregates in soils, often coating particle surfaces or entering between the layers of 2:1 layer-type clay minerals to form hydroxy-interlayer species (Table 2.4 and Eq. 2.9b).

Table 2.5
Metal oxides, oxyhydroxides, and hydroxides found commonly in soils.

Name	Chemical formula[a]	Name	Chemical formula[a]
Rutile	TiO_2	Hematite	α-Fe_2O_3
Birnessite	$M_xMn(IV)_aMn(III)_b\blacktriangle_cO_2$[b]	Lepidocrocite	γ-FeOOH
Ferrihydrite	$Fe_{10}O_{15}\cdot 9\,H_2O$	Lithiophorite	$LiAl_2(OH)_6Mn(IV)_2Mn(III)O_6$
Gibbsite	γ-$Al(OH)_3$	Maghemite[c]	γ-Fe_2O_3
Goethite	α-FeOOH	Magnetite[c]	$FeFe_2O_4$

[a] γ denotes cubic close-packing of anions, whereas α denotes hexagonal close-packing.
[b] M = monovalent interlayer cation, $x = b + 4c$, $a + b + c = 1$, $\blacktriangle$ = cation vacancy.
[c] Some of the Fe(III) is in tetrahedral coordination.

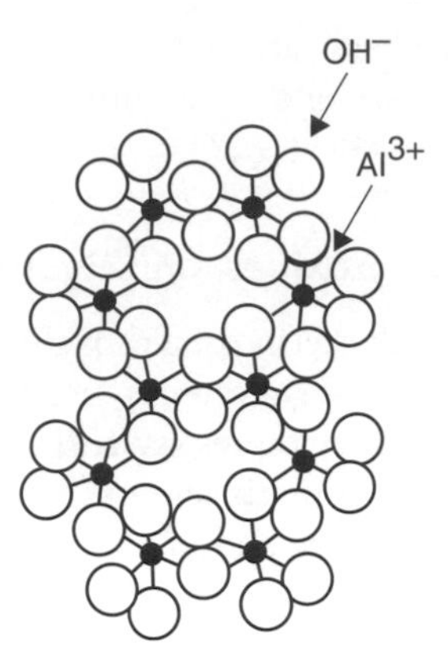

Figure 2.9. "Ball-and-stick" drawing of the atomic structure of gibbsite [γ-Al(OH)$_3$].

Gibbsite is a dioctahedral mineral comprising edge-sharing Al(OH)$_6$ in stacked sheets that are held together as an aggregate by hydrogen bonds that form between opposing OH groups oriented perpendicularly to the basal planes of the sheets. Hydrogen bonds also link the OH groups along the edges of the cavities lying within a single sheet (Fig. 2.9), adding to the distortion of the octahedra that is produced by the sharing of edges (Pauling Rule 3 and Fig. 2.3). According to Pauling Rule 2, the bond strength of Al^{3+} octahedrally coordinated to hydroxide ions is 0.5 and, therefore, each OH^- in gibbsite should be bonded to a pair of Al^{3+}, as indicated in Fig. 2.9 for the bulk structure. At the edges of a sheet, however, pairs of hydroxyls are exposed that

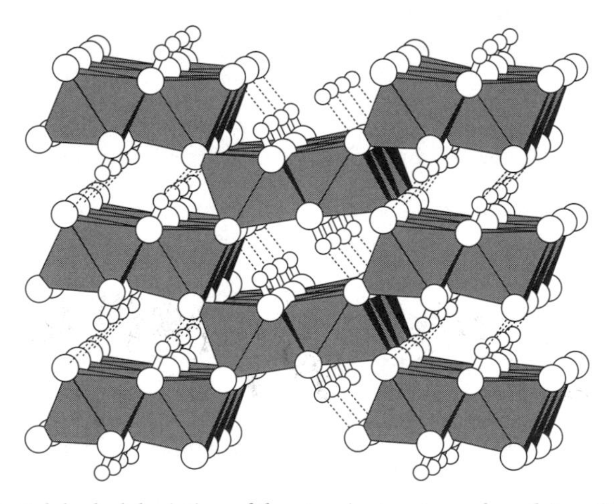

Figure 2.10. Polyhedral depiction of the atomic structure of goethite, with the double chains of Fe(III) octahedra lying perpendicular to the plane of the figure.

have unsatisfied bond valences. These hydroxyls lie along unshared octahedral edges and are located approximately 0.197 nm from the Al^{3+} to which they are coordinated, yielding an associated bond valence of 0.422 vu according to Eq. 2.4. This leaves an unsatisfied bond valence equal to −0.578 vu on each exposed hydroxyl. Adsorption of a proton by one OH, following the paradigm outlined for kaolinite edge surfaces (Fig. 2.8), then yields a more stable configuration of the hydroxyl pairs, which can be stabilized even further by hydrogen bonds with nearby water molecules in the soil solution. The pH value at which an electrically neutral gibbsite edge surface occurs turns out to be about 9.0, implying that this mineral bears a net positive charge over the entire normal range of soil pH values. By contrast, the edge surfaces of clay minerals typically bear a net negative charge above pH 5 to 7, depending on the type of clay mineral, again illustrating the stronger protonation of Al–OH groups relative to Si–OH groups that are exposed on edge surfaces.

Among the iron minerals listed in Table 2.5, *goethite* (α-FeOOH, named in honor of the German polymath, Johann Wolfgang von Goethe, who described iron oxides in the red soils of Sicily during the late 18th century) is the most abundant in soils worldwide, especially in those of temperate climatic zones. Its atomic structure (Fig. 2.10) comprises double chains of edge-sharing, distorted octahedra having equal numbers of O^{2-} and OH^- coordinated to Fe^{3+}, with each double chain then sharing octahedral corners with neighboring double chains. As discussed in Section 2.1, the Pauling Rules, supplemented by the more general concept of bond valence, are satisfied in this structure only because of hydrogen bonding between OH and O (Fig. 2.2). In soils, goethite crystallizes with relatively small particle size, exhibiting specific surface areas that range from 2 to 20 ha kg^{-1}.

Soils in warm, dry climatic zones tend to contain *hematite* (α-Fe_2O_3, named for its red-brown hue) in preference to goethite (which has a yellow-brown hue). Hematite has the same atomic structure as the Al oxide mineral corundum, mentioned in Section 2.1, with both having hexagonal rings of edge-sharing octahedra arranged in stacked sheets that are themselves linked through face-sharing octahedra. All this polyhedral sharing pushes the Fe^{3+} cations closer together and produces considerable structural distortion, as would be predicted from Pauling Rule 3. Hematite particles tend to have rather low specific surface areas (<10 ha kg^{-1}). Substantial isomorphic substitution of Al for Fe can occur in both goethite and hematite, the upper limit for the Al-to-(Al + Fe) molar ratio being 0.33 in goethite and half of this value in hematite. Aluminum substitution in these minerals is favored in soils under acidic conditions that produce abundant soluble Al without the interference of complexation by organic ligands or silica.

If organic ligands—especially humic substances—or soluble silica are at significant concentrations, then the crystallization of goethite or hematite is inhibited and poorly crystalline Fe(III) oxyhydroxides precipitate instead. This situation is especially characteristic of the rhizosphere, resulting in the formation of root-associated Fe(III) mineral mixtures known as *iron plaque.*

Ferrihydrite ($Fe_{10}O_{15} \cdot 9H_2O$, an approximate chemical formula, because up to half of the H may be in hydroxide ions, not water) is the most common of these materials, found typically in soils where biogeochemical weathering is intense, soluble Fe(II) oxidation is rapid, and water content is seasonally high (e.g., Andisols, Inceptisols, and Spodosols). This mineral, often detected along with goethite in soils, exhibits varying degrees of ordering of its Fe(III) octahedra, with many structural defects, and spheroidal particle diameters of a few nanometers, leading to specific surface areas of 20 to 40 ha kg^{-1}. Ferrihydrite can precipitate abiotically from oxic soil solutions at circumneutral pH, but its formation tends to be mediated by bacteria at acidic pH or under anaerobic conditions that slow Fe(II) oxidation significantly. Even at circumneutral pH under oxic conditions, bacterial cell walls can nucleate ferrihydrite (and goethite) precipitation after complexing dissolved Fe^{3+} cations and, in some cases, producing organic polymers that constrain precipitation to occur near the cell surface. Bacteria that thrive within biofilms either in highly acidic oxic environments, or in anaerobic environments at circumneutral pH, can oxidize Fe(II) enzymatically and rapidly enough to produce ferrihydrite at rates well above those for abiotic pathways. The resulting poorly crystalline mineral is encapsulated within extracellular organic polymers that keep it from fouling the bacterial surface while it also serves as fortification against predation of the organism. When polymeric matrices become fully encrusted with ferrihydrite within this geosymbiotic microbial ecosystem, they are eventually abandoned by the bacteria, which then begin to manufacture a new biofilm.

Magnetite [$Fe(II)Fe(III)_2O_4$], a mixed-valence iron oxide, contains Fe^{2+} and half of its Fe^{3+} in octahedral coordination with O^{2-}, with the remaining Fe^{3+} being in tetrahedral coordination. This mineral, named for its magnetic properties, is widespread in soils and can form both abiotically (e.g., inherited from soil parent material, or precipitated during the incongruent dissolution of ferrihydrite promoted by a reaction with dissolved Fe^{2+} cations) and biogenically (e.g., within *magnetotactic* bacteria that utilize this mineral for orientation and migration in the earth's magnetic field, or as a secondary precipitate under anaerobic conditions following the weathering of ferrihydrite by bacteria that oxidize organic matter). *Maghemite* (γ-Fe_2O_3), another magnetic mineral, is also widespread in soils of warm climatic zones, forming through the oxidation of magnetite or from the intense heating of goethite and ferrihydrite, as produced by fire. Like goethite, hematite, and ferrihydrite, Al substitution, with an upper limit as high as found for goethite, occurs in both magnetite and maghemite, the latter commonly arising from a heat-promoted transformation of Al-substituted goethite.

Another mixed-valence mineral that can be formed by either abiotic or bacterially mediated incongruent dissolution of ferrihydrite under anaerobic conditions is *green rust* [$(A^{-\ell} \cdot nH_2O)\ Fe(III)_xFe(II)_y(OH)_{3x+2y-\ell}$], which comprises a ferric–ferrous hydroxide sheet bearing a positive structural charge (because of ferric iron) that is balanced by hydrated anions ($A^{-\ell} \cdot nH_2O$), such as chloride ($x = 1, y = 3, \ell = 1$), sulfate, or carbonate ($x = 2, y = 4, \ell = 2$), which reside in the interlayer region, analogous to the interlayer cations that balance the negative structural charge in 2:1 layer-type clay minerals (see Section 2.3 and Table 2.4). Also in parallel to the 2:1 clay minerals, individual sheets stack to form aggregates, with the stacking arrangement of the sheets being determined by the nature of the interlayer anion. Green rust occurs under alkaline conditions in poorly drained, biologically active soils that are subject to anaerobic conditions because of high water content (hydromorphic soils).

Birnessite [$M_xMn(IV)_aMn(III)_b\blacktriangle_cO_2$, where M is a monovalent interlayer cation, $a + b + c = 1$, and $\blacktriangle$ is an empty cation site in the octahedral sheet] is the most common manganese oxide mineral in soils, where it is often observed in fine-grained coatings on particle surfaces. Like gibbsite, it is a layer-type mineral with sheets that comprise mainly $Mn^{4+}O_6$ octahedra, but with significant isomorphic substitution by Mn^{3+} ($0 \leq b \leq 0.3$) and a generous population of cation vacancies ($0 \leq \blacktriangle \leq 0.2$), both of which induce a negative structural charge. In practice, isomorphic substitutions tend to offset cation vacancies, such that a range of birnessites exists, varying from those with only Mn^{3+} substitution (*triclinic birnessite*) to those with only cation vacancies (*δ-MnO₂* or *vernadite*). The resulting layer charge, $x = b + 4c$, is compensated by protons and hydrated metal cations, including both Mn^{2+} and Mn^{3+}, that reside in the interlayer region, particularly near cation vacancies, each of which bears four electronic charges in the absence of protonation

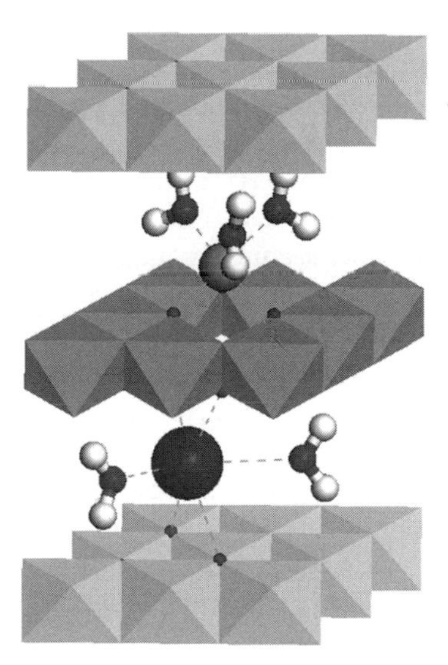

Figure 2.11. Polyhedral depiction of the local atomic structure in birnessite, showing a cation vacancy with charge-balancing, solvated interlayer cations (Mn^{3+} on top and K^+ on the bottom). Visualization courtesy of Dr. Kideok Kwon.

(Fig. 2.11). The layer charge is quite variable, but values near 0.25 are commonly observed, implying $\sigma_o \approx -3$ mol$_c$ kg^{-1} (Eq. 2.8), which is comparable with the negative structural charge observed for 2:1 clay minerals. Birnessite typically forms poorly crystalline particles comprising a small number of stacked, defective sheets less than 10 nm in diameter. Specific surface areas of these particles range from 3 to 25 ha kg^{-1}, a range that is similar to soil goethites.

Birnessites precipitate in soils as a result of the oxidation of dissolved Mn^{2+}, which, if it occurs abiotically, is orders of magnitude slower than that of dissolved Fe^{2+} at circumneutral pH in the presence of oxygen. Bacteria and fungi that catalyze the oxidation of Mn(II) under these conditions enzymatically and rapidly (timescales of hours for bacterial oxidation vs. hundreds of days for abiotic oxidation) are widespread in nature, leading to the conclusion that soil birnessites are primarily of biogenic origin. Similar to bacteriogenic ferrihydrite, birnessites produced by bacteria often are found enmeshed within biofilms, where these highly reactive, poorly crystalline nanoparticles may serve to impede predation and sequester both toxic and nutrient metal cations. Geosymbiotic microbial ecosystems thus play an important role in the biogeochemical cycling of Al, Fe, and Mn in soils and natural waters.

2.5 Carbonates and Sulfates

The important carbonate minerals in soils include calcite ($CaCO_3$), dolomite [$CaMg(CO_3)_2$], nahcolite ($NaHCO_3$), trona [$Na_3H(CO_3)_2 \cdot 2\,H_2O$], and soda ($Na_2CO_3 \cdot 10\,H_2O$). Calcite may be, and dolomite appears often to be, a primary mineral in soils. Secondary calcite that precipitates from soil solutions enriched in soluble Mg coprecipitates with $MgCO_3$ to form *magnesian calcite,* $Ca_{1-y}Mg_yCO_3$, with the stoichiometric coefficient y typically well below 0.10. This mode of formation accounts for much of the secondary Mg carbonate found in arid-zone soils. Like secondary metal oxides and hydroxides, secondary Ca/Mg carbonates can occur as coatings on other minerals, in nodules or hardened layers, and as clay or silt particles. They are important repositories of inorganic C in Aridisols and Mollisols.

Pedogenic calcites are normal weathering products of Ca-bearing primary silicates (pyroxenes, amphiboles, feldspars) as well as primary carbonates. Their formation is favored in the rhizosphere, where bacteria and fungi mediate calcite precipitation, both through nucleation around excreted Ca^{2+} that has been complexed by cell walls and through increases in soil solution pH (>7.2) induced by enzymatically catalyzed reduction of nitrate, Mn, Fe, and sulfate or methane production, the last process being associated with pedogenic dolomite formation. As an example of primary mineral weathering to produce secondary calcite, the feldspar anorthite (Table 2.3) may be considered as follows:

$$\underset{\text{(anorthite)}}{\mathrm{CaAl_2Si_2O_8(s)}} + 0.5\ \mathrm{Mg^{2+}} + 3.5\ \mathrm{Si(OH)_4^0} + \mathrm{CO_2(g)}$$

$$= \underset{\text{(smectite)}}{\mathrm{Ca_{0.5}[Si_{7.5}Al_{0.5}]Al_{3.5}Mg_{0.5}O_{20}(OH)_4(s)}} + \underset{\text{(calcite)}}{\mathrm{CaCO_3(s)}}$$

$$+ 0.5\ \mathrm{Ca^{2+}} + 5\ \mathrm{H_2O}(\ell) \qquad (2.11)$$

This incongruent dissolution reaction takes advantage of soluble Mg and silica available from weathering and of ubiquitous biogenic CO_2 in soils. Note that the reaction products are favored by abundant CO_2, because it is a reactant, and are inhibited by abundant H_2O, because it is one of the products. Thus, calcite formation can be prompted by elevated CO_2 concentration.

The formation of calcite from the dissolution of primary carbonates also is favored by abundant CO_2, but not as a source of dissolved carbonate ions. Instead, carbonic acid that is formed when CO_2 dissolves in the soil solution serves as a source of protons to aid in the dissolution of primary calcite or dolomite:

$$\mathrm{CO_2(g)} + \mathrm{H_2O}(\ell) = \mathrm{H_2CO_3^*} = \mathrm{H^+} + \mathrm{HCO_3^-} \qquad (2.12a)$$

$$\mathrm{CaCO_3(s)} + \mathrm{H^+} = \mathrm{Ca^{2+}} + \mathrm{HCO_3^-} \qquad (2.12b)$$

where $H_2CO_3^*$ conventionally designates the sum of undissociated carbonic acid ($H_2CO_3^0$) and solvated CO_2 ($CO_2 \cdot H_2O$), because these two dissolved species are very difficult to distinguish by chemical analysis (see Problem 15 in Chapter 1). If soil leaching is moderate and followed by drying, the reaction in Eq. 2.11b is reversed and secondary calcite forms. Note that this reversal is favored by high pH (i.e., low proton concentration).

Calcium coprecipitation bivalent with Mn, Fe, Co, Cd, or Pb by sorption onto calcite is not uncommon (see Table 1.5). The trace metals Zn, Cu, and Pb also may coprecipitate with calcite by inclusion as the hydroxycarbonate minerals hydrozincite [$Zn_3(OH)_6(CO_3)_2$], malachite [$Cu_2(OH)_2CO_3$], azurite [$Cu_3(OH)_2(CO_3)_2$], or hydrocerrusite [$Pb_3(OH)_2(CO_3)_2$]. Under anoxic conditions that favor Mn(II), Fe(II), and abundant CO_2, rhodocrosite ($MnCO_3$) and siderite ($FeCO_3$) solid-solution formation is possible—in the absence of inhibiting sorption of humus by the nucleating solid phase, which also retards calcite precipitation. Green rust, the Fe(II)–Fe(III) hydroxy carbonate discussed in Section 2.4, can precipitate under these conditions as well, with CO_3^{2-} then being the interlayer anion.

Like secondary carbonates, Ca, Mg, and Na sulfates tend to accumulate as weathering products in soils that develop under arid to subhumid conditions, where evaporation exceeds rainfall (Table 1.7). The principal minerals in this group are gypsum ($CaSO_4 \cdot 2\,H_2O$), anhydrite ($CaSO_4$), epsomite ($MgSO_4 \cdot 7\,H_2O$), mirabilite ($Na_2SO_4 \cdot 10\,H_2O$), and thenardite (Na_2SO_4). Gypsum, similar to calcite, can dissolve and reprecipitate in a soil profile that is leached by rainwater or irrigation water and can occur as a coating on soil minerals, including calcite. The Na sulfates, like the Na carbonates, form at the top of the soil profile as it dries through evaporation.

In highly acidic soils, sulfate, either produced through sulfide oxidation or introduced by amendments (e.g., gypsum), can react with the abundant Fe and Al in the soil solution to precipitate the minerals schwertmannite [$Fe_8O_8(OH)_6SO_4$], jarosite [$KFe_3(OH)_6(SO_4)_2$], alunite [$KAl_3(OH)_6(SO_4)_2$], basaluminite [$Al_4(OH)_{10}SO_4 \cdot 5\ H_2O$], or jurbanite ($AlOHSO_4 \cdot 5\ H_2O$). These minerals, in turn, may dissolve incongruently to form ferrihydrite and goethite or gibbsite upon further contact with a percolating, less acidic soil solution. Under similar acidic conditions, phosphate minerals such as wavellite [$Al_3(OH)_3(PO_4)_2 \cdot 5\ H_2O$], angellite [$Al_2(OH)_3PO_4$], barandite [$(Al,Fe)PO_4 \cdot 2\ H_2O$], and vivianite [$Fe_3(PO_4)_2 \cdot 8\ H_2O$] have been observed in soils, with the latter requiring anoxic conditions to precipitate, whereas the others require phosphoritic parent materials. As soil pH increases, Ca phosphates such as apatite [$Ca_3(OH,F)(PO_4)_3$] and octacalcium phosphate [$Ca_8H_2(PO_4)_6 \cdot 5\ H_2O$] tend to form, particularly if soluble phosphate has been introduced in abundance by soil amendments or wastewater percolation.

For Further Reading

Banfield, J. F., and K. H. Nealson (eds.). (1997) *Geomicrobiology: Interactions between microbes and minerals.* The Mineralogical Society of America, Washington, DC. The 13 chapters of this edited workshop volume provide a fine introduction to the important roles played by microorganisms in the formation and weathering of minerals in soils and aquatic environments.

Dixon, J. B., and D. G. Schulze (eds.). (2002) *Soil mineralogy with environmental applications.* Soil Science Society of America, Madison, WI. Chapters 6 through 22 of this standard reference work on soil minerals may be read to gain in-depth information about their structures, occurrence, and weathering reactions.

Essington, M. E. (2004) *Soil and water chemistry.* CRC Press, Boca Raton, FL. Chapter 2 of this comprehensive textbook may be consulted to learn more about the atomistic details of soil mineral structures through its many visualizations.

The following three specialized books offer a deeper understanding of the structure and reactivity of minerals in natural soils and aquatic systems, including those affected by pollution:

Cornell, R. M., and U. Schwertmann. (2003) *The iron oxides.* Wiley-VCH Verlag, Weinheim, Germany. This beautifully produced, exhaustive treatise on the iron oxides is an indispensable reference for anyone who wants to know specialized information.

Cotter–Howells, J. D., L. S. Campbell, E. Valsami–Jones, and M. Batchelder. (2000) *Environmental mineralogy.* The Mineralogical Society of Great Britain & Ireland, London. This edited volume provides useful overviews of the microbial mediation of mineral weathering, as well as of mineral structure and reactivity in contaminated soil environments.

Giese, R. F., and C. J. van Oss. (2002) *Colloid and surface properties of clays and related minerals.* Marcel Dekker, New York. A detailed, comprehensive reference on the structure and colloidal properties of the clay minerals.

Problems

The more difficult problems are indicated by an asterisk.

1. Use Pauling Rule 2 to show that, in a stable mineral structure, a corner of an Si–O tetrahedron can be bonded solely to one other Si–O tetrahedron, but not solely to one other Al–O tetrahedron. For the latter case, show that bonding the Si–O tetrahedron to an Al–O tetrahedron and one bivalent cation having CN = 8 will satisfy Pauling Rule 2. [The feldspar mineral anorthite (Table 2.3) is an example.]

*2. Oxygen ions exposed on the edge surfaces of a goethite crystallite can be bonded to one, two, or three Fe^{3+} ions in the bulk structure, depending on how the particle surface has formed. Apply Pauling Rule 2 to estimate the unsatisfied bond valence on each type of exposed O^{2-}, taking the average Fe–O bond length to be 0.204 nm. Then consider whether the formation of singly or doubly protonated species of the three types of surface oxygen ion would stabilize them in the sense of Pauling Rule 2. Which of the protonated species is likely to be a very weak acid (poor proton donor)? Which among them should be the strongest acid? (*Hint:* Review the examples discussed for oxyanions and for the edge surfaces of kaolinite and gibbsite in Sections 2.1, 2.3, and 2.4.)

*3. Oxygen ions on the basal planes of birnessite (Fig. 2.11) are bonded to three Mn^{4+} ions, whereas those exposed on the edge surfaces are bonded either to one or two Mn^{4+}. Use Pauling Rule 2 to examine the stability of these three types of surface oxygen ion, taking the average Mn–O bond length to be 0.192 nm. Consider whether protonation of any of the three will improve its stability.

4. The octahedral cation vacancies in a sheet of birnessite bear two electronic charges on each of two equilateral triangles of oxygen ions, one exposed at the top of the sheet and one at the bottom (see Fig. 2.11). This distribution of negative structural charge suggests that bivalent cations could adsorb on the triangular sites, with one such cation bound to each side of a vacancy to satisfy charge balance. A birnessite produced by a *Pseudomonas* species (soil and freshwater bacterium), with the chemical

formula $Na_{0.15}Mn(III)_{0.17}[Mn(IV)_{0.83}\blacktriangle_{0.17}O_2]$, was observed to adsorb Zn^{2+} to achieve a maximum Zn-to-Mn molar ratio equal to 0.43 ± 0.04. Show that this molar ratio is consistent with Zn^{2+} replacing all interlayer Na^+ and Mn^{3+} in binding to the triangular vacancy sites in the Mn oxide sheets.

5. Calculate the structural charge (σ_0, in moles of charge per kilogram) on the following layer-type minerals, given their chemical formulas. Identify each of the minerals in light of your results.

   (a) $K_{1.5}[Si_7Al]Al_{3.10}Fe(III)_{0.40}Mg_{0.50}O_{20}(OH)_4$

   (b) $Na_{0.78}[Si_8]Al_{2.92}Fe(III)_{0.30}Mg_{0.78}O_{20}(OH)_4$

   (c) $Na_{0.17}Mn(IV)_{0.83}Mn(III)_{0.17}O_2$

6. Alumino-goethite [$Fe_{1-y}Al_yO(OH)$], ferri-kaolinite [$Si_4(Al_{1-y}Fe_y)_4O_{10}(OH)_8$], magnesian calcite [$Ca_{1-y}Mg_yCO_3$], mangano-siderite [$Fe_{1-y}Mn_yCo_3$], and barrandite [$Al_{1-y}Fe_yPO_4 \cdot 2\,H_2O$] are examples of coprecipitated soil minerals, with the metal having the stoichiometric coefficient y being in the minor component. For each of these solids, rewrite the chemical formula to indicate $1-y$ moles of the major component mineral combined with y moles of the minor component mineral. [The minor component AlO(OH) in alumino-goethite is known as diaspore when it occurs as a pure solid phase, and the two components of barrandite are known as variscite (Al) and strengite (Fe).]

7. The table presented here lists mass-normalized steady-state congruent dissolution rates at pH 5 and 25 °C for three silicate minerals of importance in soils. These data can be used to calculate an *intrinsic dissolution timescale,*

   $$\tau_{dis} = (M_r \times \text{dissolution rate})^{-1}$$

   where M_r is the relative molecular mass of the dissolving mineral and the dissolution rate is in units of moles per gram per second. The value of τ_{dis} characterizes the timescale on which 1 mole of a mineral will dissolve in water. Calculate τ_{dis} in years for the three minerals, then compare your results with the trends expressed in Table 1.7.

Mineral	Dissolution rate (mol g^{-1} s^{-1})
Forsterite	5.7×10^{-11}
Hornblende	4.3×10^{-14}
Quartz	2.1×10^{-16}

8. Using the notation in Problem 6, write a balanced chemical reaction for the incongruent dissolution of ferri-kaolinite having 2 mol% Fe(III)

substituted for Al. The principal products are goethite, gibbsite, and silicic acid.

9. Orthoclase can weather to form kaolinite and gibbsite under humid tropical conditions. Select a weathering mechanism, then write a balanced chemical reaction for this transformation.

*10. The weathering of biotite as shown in Eq. 2.5 is typical of temperate humid regions. In tropical humid regions, the clay mineral product is typically kaolinite, not vermiculite. Develop a balanced chemical reaction for the weathering of biotite to form kaolinite and goethite by hydrolysis and protonation.

*11. Develop a single chemical equation that describes a reaction among trona, nahcolite, and $CO_2(g)$. Which of the two Na carbonates would be favored by increasing the CO_2 partial pressure in soil?

12. Gypsum is added to an acidic soil containing the Al-saturated beidellite featured in Eq. 2.10b. Develop a chemical equation that describes the formation of Ca-saturated beidellite and jurbanite from the incongruent dissolution of gypsum in the presence of Al-beidellite. This reaction could improve soil fertility by providing exchangeable and soluble Ca as well as by reducing the bioavailability of Al through precipitation.

13. Develop a balanced chemical reaction for the transformation of schwertmannite to goethite.

*14. Generalize Eq. 2.10c to be a weathering reaction for allophane having the general chemical formula given in Section 2.3. Determine the threshold value of the stoichiometric coefficient y above which more kaolinite than gibbsite will be produced by the weathering of allophane.

*15. Combine Eqs. 2.10b and 2.10c to derive a chemical reaction for the weathering of allophane, $Si_yAl_4O_{6+2y} \cdot nH_2O$, to form pedogenic chlorite. What conditions favor this reaction? (*Hint:* The value of y varies from 1.6 to 4.0.)

Special Topic 2: The Discovery of the Structures of Clay Minerals

Near the end of his long life, Linus Pauling published an informal account of his research—which took place more than 75 years ago—on the atomic structures of clay minerals and oxides [reprinted with permission from the newsletter of The Clay Minerals Society (pp. 25–27, CMS News, September 1990)]. Pauling, the only person to receive two unshared Nobel Prizes, was perhaps the greatest physical chemist of the past century. His life achievements related to crystallography were recorded by Pauling himself in the first and fifth chapters of a testimonial volume, The Chemical Bond, *edited by A. Zewail (Academic Press, New York, 1992), but the newsletter article provides a more focused tale of direct relevance to the present chapter. Note that Pauling was only 28 when he formulated his rule for stable crystal structures. Sterling B. Hendricks, mentioned in the article as Pauling's first graduate student, went on to a distinguished career with the U.S. Department of Agriculture in clay mineralogy and, later, plant physiology. His breakthrough article in 1930 (with William H. Fry) on the crystal structures of soil colloids has been reprinted in a celebratory issue of the journal,* Soil Science (Hendricks, S. B., and W. H. Fry (2006) The results of X-ray and microscopical examinations of soil colloids. *Soil Science, Supplement to Volume 171, June 2006, pp. S51–S73).*

I have been interested in the clay minerals for nearly eighty years, and I was pleased when Patricia Jo Eberl wrote to me, asking me to write an account of the discovery of their structure.

My interest in minerals began in 1913, when I was 12 years old, a year before it shifted to chemistry. At that time I collected a few minerals and read books on mineralogy. Then in the fall of 1922, a couple of months after I had entered the Division of Chemistry and Chemical Engineering at the California Institute of Technology as a graduate student and had been taught X-ray crystallography by Roscoe Gilkey Dickinson, the first person to have obtained a Ph.D. degree from the California Institute of Technology (1920). I determined with Dickinson the crystal structure of a mineral, molybdenite. This mineral was interesting as the first one to be found in which a metal atom, with ligancy 6, is surrounded by atoms at the corners of a trigonal prism, rather than at the corners of an octahedron.

The X-ray-diffraction method of determining the structures of crystals was a marvelous method. It was not then very powerful, however; nevertheless during the period around 1922, many crystal structures, the simpler ones, were discovered and thoroughly investigated. For example, Sterling B. Hendricks and I made a careful redetermination of the structure of hematite and corundum that had been investigated earlier by W.L. Bragg (later Sir Lawrence Bragg), who when he was a student had discovered the "Bragg equation." Sterling Hendricks was my first graduate student. The X-ray laboratory of the California Institute of Technology, which had been set up in 1917, was turned over to me by Dickinson in 1924. By 1927 I had become impatient, as a result of having had to abandon the study of many minerals and other inorganic crystals because of the limited power of X-ray crystallography, at that time, to locate the atoms. Bragg had in 1926, in his effort to determine the structures of some silicate minerals, formulated the hypothesis that, in these crystals, the structure was often to some extent determined by having the large anions of oxygen arranged in cubic close packing or hexagonal close packing, with the metal ions in the interstices. I had the idea that the use of auxiliary information of this sort could make the X-ray technique more powerful. From studying the known structures of two forms of titanium dioxide, rutile and anatase, I recognized that they were similar in a remarkable way. In each structure there are octahedra of six oxygen ions around a titanium ion. (At that time I overemphasized the ionic character of bonds in the oxide minerals.) In rutile each octahedron shares two edges with adjacent octahedra, and in anatase each octahedron shares four edges with adjacent octahedra. I surmised that in brookite, the third form of titanium dioxide, there would also be octahedra, with each octahedron sharing three edges with adjacent octahedra, and I formulated two structures satisfying this hypothesis, and with all of the octahedra in each structure crystallographically equivalent.

My second graduate student, James Holmes Sturdivant (Ph.D. 1928), made X-ray photographs of brookite and found that the dimensions of the orthorhombic unit cell agreed reasonably well with those that I had predicted from the interatomic distances in rutile and anatase, in which the shared edges of the octahedral are shortened to about 2.50 Å from the average value about 2.8 Å, and that the intensities of the diffraction maxima were in reasonable agreement with those predicted for one of the two structures, which is now accepted as the structure of brookite. I also used the idea, based on the ionic radii that I had published in the *Journal of the American Chemical Society* in 1927, that in topaz, $Al_2SiO_4F_2$, there would be AlO_4F_2 octahedra and SiO_4 tetrahedra, and in this way was able to locate atoms in this orthorhombic crystal.

In 1929, after having studied some other minerals and applied this method of predicting their structures and then checking by comparison with the X-ray data, I published two papers on a set of principles determining the structure of complex ionic crystals. One of these rules is the Valence Rule. The valence of a cation is divided equally among the bonds to the surrounding anions, and the sum of the bond strengths of the bonds to each anion should be close to its negative valence, usually within one quarter of a valence unit. In the papers I started the argument by mentioning Bragg's use of the idea that the oxygen (and fluorine) ions are often arranged in a close-packed structure, but it turned out that for many silicates this arrangement does not occur, whereas the principles of the coordination theory are satisfied.

At that time, 1929, I became interested in the structure of mica, and a few months later, of the chlorites and the clay minerals. I had become interested in mica when I was 12 years old, and had studied the large grains of mica in samples of granite that I had collected, and had also observed that sheets of mica were used as windows in the wood-burning stove in the house in which I had lived with my parents and my two sisters. I read a paper that Mauguin had published in 1927, in which he gave the dimensions $a = 5.17$ Å, $b = 8.94$ Å,

$c = 20.01$ Å, with $\beta = 96°$ for the monoclinic (pseudohexagonal) unit cell of structure of muscovite. I also made Laue photographs and rotation photographs of a beautiful blue-green translucent specimen of fuchsite, a variety of muscovite containing some chromium, and verified Mauguin's dimensions.

The crystal of fuchsite had been given to me, along with about a thousand other mineral specimens, in 1928, by my friend J. Robert Oppenheimer, who had obtained them, mainly by purchase from dealers, when he was a boy. Oppenheimer's first published paper, written when he was about 16 years old, was in the field of mineralogy. He later got his bachelor's degree in chemistry from Harvard University and then a Ph.D. in physics from Göttingen. Many of my early X-ray studies of minerals were made with specimens from the Oppenheimer collection, and I still take pleasure in examining some of the more striking specimens.

I recognized at once that the layers clearly indicated to be present in mica by the pronounced basal cleavage contained close-packed layers of oxygen atoms, and that the dimensions were similar to octahedral layers in hydrargillite and brucite and also tetrahedral layers in beta-tridymite and beta-cristobalite, the dimensions for hydrargillite (now called gibbsite) and the two forms of silica being equal to those for the mica sheets to within about two percent. With the rules about the structure of complex ionic crystals as a guide, the structure of mica could at once be formulated as consisting of a layer of aluminum octahedra condensed with two layers of silicon tetrahedra, one on each side, with these triple layers superimposed with potassium ions in between. Calculation of the intensities of the X-ray diffraction maxima out to the 18th order from the basal plane gave results agreeing well with the observed intensities, so that there was little doubt that this structure was correct for mica. I pointed out in my paper, which was communicated to the National Academy of Sciences on January 16, 1930, and published a month later [February issue, (1930) *Proc. Nat. Acad. Sci. USA* **16**:123–129] that clintonite, a brittle mica, has a similar structure, with the triple layers held together by calcium ions instead of potassium ions, and that the correspondingly stronger forces bring the layers closer together, the separation of adjacent layers being 9.5 to 9.6 Å in place of the value of 9.9 to 10.1 Å for the micas. I also pointed out that talc and pyrophyllite have the same structure, but with the layers electrically neutral, and held together only by stray electrical forces. As a result these crystals are very soft, feeling soapy to the touch, whereas to separate the layers in mica, it is necessary to break the bonds of the univalent potassium ions, so that the micas are not so soft, thin plates being sufficiently elastic to straighten out after being bent, and that the separation of layers in the brittle micas involves breaking the stronger bonds of bipositive calcium ions, these minerals then being harder and brittle instead of elastic, but still showing perfect basal cleavage. I also mentioned the significance of the sequence of hardness in relation to the strength of the bonds: talc and pyrophyllite, 1–2 on the Mohs hardness scale, the micas, 2–3, and the brittle micas, 3.5–6.

I then made Laue photographs and oscillation photographs of specimens of penninite and clinochlore, and found a monoclinic unit of structure with $a = 5.2\text{–}5.3$ Å, $b = 9.2\text{–}9.3$ Å, $c = 14.3\text{–}14.4$ Å, and monoclinic angle of 96° 50′. It was clear from the dimensions and the pronounced basal cleavage that the chlorites consisted of layers somewhat similar to those found in mica. At first I tried to formulate a single layer made of two octahedral and two tetrahedral layers, but I soon recognized that there are layers similar to the mica layers, with, however, layers similar to the brucite or hydrargillite layers, but with a positive electrical charge interspersed between them, in place of the potassium ions in mica. I then communicated a paper to the *Proceedings of the National Academy of Sciences* on July 9, 1930, while my wife and I and our eldest son, Linus Jr. (then five years old) were in Europe. This paper was published two months later [Pauling, L. (1930) The structure of chlorites. *Proc. Nat. Acad. Sci. USA* **16**:578–582], with the title "The Structure of Chlorites." There was good agreement between the calculated intensities of X-ray maxima out to the 26th order from the basal plane and the observed intensities.

In this paper I also proposed a structure for kaolinite, consisting of an octahedral layer with a silicon tetrahedral layer on only one side. I also mentioned that with this unsymmetrical layer there would be a tendency for the layer to curve, one face becoming concave and the other convex, and that this tendency would in general not be overcome by the relatively weak forces operated between adjacent layers. I did not predict that jelly roll structures of clay minerals would be found (and perhaps already had been reported at that time; I am not sure about when they were discovered), but I used the argument that unsymmetrical layers probably would be curved, and only in some clay minerals, kaolinite, would the tendency to curve be overcome by the forces between layers. I also discussed briefly the possibility that a clay mineral similar to chlorite, but with a neutral brucite layer, might exist, and I suggested the possibility that more complex minerals might be discovered, with alternation between the mica structure and the chlorite structure.

It now seems to me to be odd that I should have published the mica paper without mentioning talc and pyrophyllite in the title, and the chlorite paper without mentioning kaolinite in the title. Also, each of these papers ends with the statement that a detailed account of the investigation would be published in the journal *Zeitschrift für Kristallographie*, and in fact no such detailed accounts were published.

I made many more X-ray photographs of specimens of micas and chlorites, and had my graduate student Jack Sherman make many such photographs. This work was never completed, however, partially because Jack Sherman soon became tired of the experimental work and began making quantum mechanical calculations with me, and I also became much involved during 1930 and later years in working on the quantum mechanics of the chemical bond and on a new method that we were starting to use in our laboratory, the determination of the structure of gas molecules by the diffraction of electrons. It was, of course, poor judgment on my part to say that detailed discussions would be published later.

My first graduate student, Sterling Hendricks, after he left Pasadena, carried out a number of investigations of the micas and the chlorites, as well as of other minerals. Jack Sherman continued to make calculations, and his X-ray studies of the micas remain his only effort in this field (never published). I, however, together with my students and associates, made many more studies of the crystal structure of minerals, and I have retained my interest in this field up to the present time. In fact, my most recent mineral paper, published together with my son-in-law Barclay Kamb [(1982) *American Mineralogist* **62**:817–821] is on the crystal structure of lithiophorite, which is a clay mineral. The structure that we assigned to lithiophorite, $Al_{14}Li_6Mn_{21}(OH)_{84}$, involves alternative brucite (octahedral) layers of two kinds. One layer has the composition $Al_{14}Li(OH)_{42}$, with one octahedron in 21 vacant, and the other layer has the composition $Mn_3^{2+}Mn_{18}^{4+}O_{42}$. The hexagonal unit has $a = 13.37$ Å and $c = 28.20$ Å, space group $P3_1$. The determination of this structure involved the application of structural principles in a somewhat new way, which might be useful in the consideration of other complex clay minerals. The new way consists in consideration of transfer of charge through hydrogen bonds in relation to the electroneutrality principle.

At the present time my work in X-ray and electron diffraction by crystals relates to intermetallic compounds, especially the so-called quasicrystals, and the structures of metals under high pressure. I may, however, get interested in the clay minerals again, since I remember how much excitement and pleasure I had in 1929 and early 1930 when I was working on the micas, chlorites, and related substances.

Linus Pauling
Palo Alto, California

3

Soil Humus

3.1 Biomolecules

Soils are biological milieux teeming with microorganisms. Ten grams of fertile soil may contain a population of bacteria alone exceeding the world population of human beings, with the number of different bacterial species present exceeding one million. One kilogram of uncontaminated soil serves as habitat for up to 10 trillion bacteria, 10 billion actinomycetes, and one billion fungi. Even the microfauna population (e.g., protozoa) can approach one billion in a kilogram of soil. These microorganisms play essential roles in *humification*, the transformation of plant, microbial, and animal litter into *humus* (Section 1.1). Humus formed in soils and sediments is the largest repository of organic C on the planet (four times that of the biosphere), producing annual CO_2 emissions through microbial respiration that are about an order of magnitude larger than those currently attributable to fossil fuel combustion. Clearly the biogeochemistry of humus is of major importance to the cycling of C and, therefore, to that of N, S, P, and most of the metal elements discussed in Chapter 1.

Biomolecules are the compounds in humus synthesized to sustain directly the life cycles of the soil biomass. They are usually the products of litter degradation and microbial metabolism, ranging in complexity from low-molecular mass organic acids to extracellular enzymes. Organic acids are among the best characterized biomolecules. Table 3.1 lists five aliphatic organic acids that are found commonly associated with microbial activity or rhizosphere chemistry. These acids contain the molecular unit R–COOH, where COOH is the

Table 3.1
Common aliphatic organic acids in soils.

Name	Chemical Formula	pH_{dis}[a]
Formic acid	HCOOH	3.8
Acetic acid	CH_3COOH	4.8
Oxalic acid	HOOCCOOH	1.3
Tartaric acid	H O | HOOC—C—COOH H | O H	3.0
Citric acid	COOH H | H HOOC C—C—C COOH H | H O H	3.1

[a]The pH value at which the most acidic carboxyl group has a 50% probability to be dissociated in aqueous solution.

carboxyl group and R represents H or an organic moiety such as CH_3 or even another carboxylic unit. The carboxyl group can dissociate its proton easily in the normal range of soil pH (see the third column of Table 3.1) and so is an example of a Brønsted acid. The dissociated proton can attack soil minerals to provoke their decomposition (see eqs. 1.2–1.4), whereas the carboxylate anion (COO^-) can form soluble complexes with metal cations released by mineral weathering (see Eq. 1.4). The total concentration of organic acids in the soil solution ranges from 0.01 to 5 mol m^{-3}, which is quite large relative to trace metal concentrations (≤ 1 mmol m^{-3}). These acids have very short lifetimes in soil (perhaps hours), but they are produced continually throughout the life cycles of microorganisms and plants.

Formic acid (methanoic acid), the first entry in Table 3.1, is a monocarboxylic acid produced by bacteria and found in the root exudate of corn. Acetic acid (ethanoic acid) also is produced microbially—especially under anaerobic conditions—and is found in the root exudates of grasses and herbs. Formic and acetic acid concentrations in the soil solution range from 2 to 5 mol m^{-3}. Oxalic acid (ethanedioic acid), ubiquitous soils, and tartaric acid (D-2,3-dihydroxybutanedioic acid) are dicarboxylic acids produced by fungi and excreted by the roots of cereals; their soil solution concentrations range from 0.05 to 1 mol m^{-3}. The tricarboxylic citric acid 2-hydroxypropane-1,2,3-tricarboxylic acid also is produced by fungi and is excreted by plant roots. Its soil solution concentration is less than 0.05 mol m^{-3}. Besides these aliphatic organic acids, soil solutions contain aromatic acids with a fundamental structural unit that is a benzene ring. To this ring, carboxyl (benzene carboxylic acids) or hydroxyl (phenolic acids) groups can be bonded in a variety of arrangements. The soil solution concentration of these acids is in the range 0.05 to 0.3 mol m^{-3}.

Organic acids with the chemical formula

$$\begin{array}{c} \text{H} \\ \text{R—C—COOH} \\ \text{NH}_2 \end{array}$$

are *amino acids.* These acids, with concentration in the soil solution that is typically in the range 0.05 to 0.6 mol m^{-3}, can account for as much as one half the N in soil humus. Several of the most abundant amino acids in soils are listed in Table 3.2. Glycine and alanine are examples of *neutral* amino acids, for which the side-chain unit R contains neither the carboxyl group nor the *amino group,* NH_2. The name *neutral* is apt because the COOH group contributes a negative charge by dissociating a proton, whereas NH_2 contributes a positive charge by accepting a proton to become NH_3^+. Neutral amino acids account for about two thirds of soil amino acids. *Acidic* amino acids, for which R includes a carboxyl group (aspartic and glutamic acids), and *basic* amino acids, for which R includes an amino group (arginine and lysine), account for about equal portions of the remaining one third. Amino acids can combine according to the reaction

$$\begin{array}{ccc} \begin{array}{c} \text{H} \\ \text{R—C—COOH} \\ \text{NH}_2 \end{array} & + & \begin{array}{c} \text{H} \\ \text{R}'\text{—C—COOH} \\ \text{NH}_2 \end{array} \end{array}$$

$$\longrightarrow \begin{array}{c} \quad\ \text{H}\quad \text{O}\quad\ \ \text{R} \\ \text{R—C—C—N—CH—COOH} + \text{H}_2\text{O} \\ \quad \text{NH}_2 \quad\ \ \text{H} \end{array} \tag{3.1}$$

to form a *peptide,*

$$\begin{array}{c} \qquad\qquad\quad \text{R}' \\ \text{H}\quad \text{O}\qquad\ \ | \\ \text{R—C—C—N—CH} \\ \text{NH}_2 \quad\ \text{H} \end{array}$$

the fundamental repeating unit in proteins. Because the peptide group is repeated, proteins are polymers, and because water is a product in peptide formation (Eq. 3.1), proteins are specifically *condensation polymers* of amino acids. Peptides of varying composition and structure are the dominant chemical form of amino acids in soils.

Table 3.2
Common amino acids in soils.

Name	Chemical formula
Glycine	$H-\underset{H}{C}(NH_2)-COOH$
Alanine	$CH_3-\underset{H}{C}(NH_2)-COOH$
Aspartic acid	$HOOC-CH_2-CH(NH_2)-COOH$
Glutamic acid	$HOOC-CH_2-CH_2-\underset{H}{C}(NH_2)-COOH$
Arginine	$NH_2-C(=NH)-NH-CH_2-CH_2-CH_2-CH(NH_2)-COOH$
Lysine	$NH_2-CH_2-CH_2-CH_2-CH_2-CH(NH_2)-COOH$

Another class of important and highly specialized biomolecule is represented by the *siderophores,* which are low-molecular mass compounds synthesized by bacteria, fungi, and grasses to scavenge and compete for Fe(III) in minerals and other sources of nutrient Fe under oxic, Fe-limited conditions. Nearly 500 different siderophore compounds have been identified and characterized. Microbial siderophores complex Fe(III) with hydroxamate, catecholate, and hydroxycarboxylate functional groups. *Hydroxamate* (HO–N–C=O) groups are found mainly in siderophores produced by fungi, actinomycetes, and some bacteria, whereas *catecholate* (aromatic acid with two adjacent OH on the benzene ring) and *hydroxycarboxylate* (HO–C–COOH) groups are found mainly in siderophores produced by certain bacteria (notably, pseudomonads) and by fungi. The concentrations of these siderophores in the soil solution are estimated to be in the nanomolar range. Almost all siderophores contain three complexing functional groups that bind Fe^{3+} in octahedral coordination with O ligands. These functional groups typically are located along a relatively long molecular chain that constitutes the siderophore "backbone" and thus can act more or less independently as they form complexes of remarkably high stability. Siderophores are known to complex both bivalent metal cations and trivalent metal cations besides Fe^{3+}—particularly, Al^{3+}, Co^{3+}, and Mn^{3+}. These additional complexes are believed to play roles in reducing metal toxicity to microorganisms as well as in facilitating their uptake of metals.

Carbohydrates, biopolymers of plant and microbial origin that can account for up to one half of the organic C in soil humus, include the *monosaccharides* listed in Figure 3.1. The monosaccharides have a ring structure with a characteristic substituent group and arrangement of hydroxyls. In glucose, galactose, and mannose, the substituent group is CH_2OH, whereas in xylose it is H, in glucuronic acid it is COOH, and in glucosamine it is NH_2. (Note the close structural relationship among glucose, glucuronic acid, and glucosamine in Fig. 3.1.) Xylose is a monosaccharide of plant origin, whereas galactose, mannose, and glucosamine are of microbial origin. Glucose and the other monosaccharides in Figure 3.1 are rapidly metabolized by microorganisms in soil. However, monosaccharides polymerize to form *polysaccharides*. For example, two glucose units can link together through oxygen at the site of HOH in each to form a repeating unit of *cellulose* after eliminating water. Thus cellulose, the major carbohydrate found in plants, is a *condensation polymer* of glucose. It can account for up to one sixth of the organic C in soil.

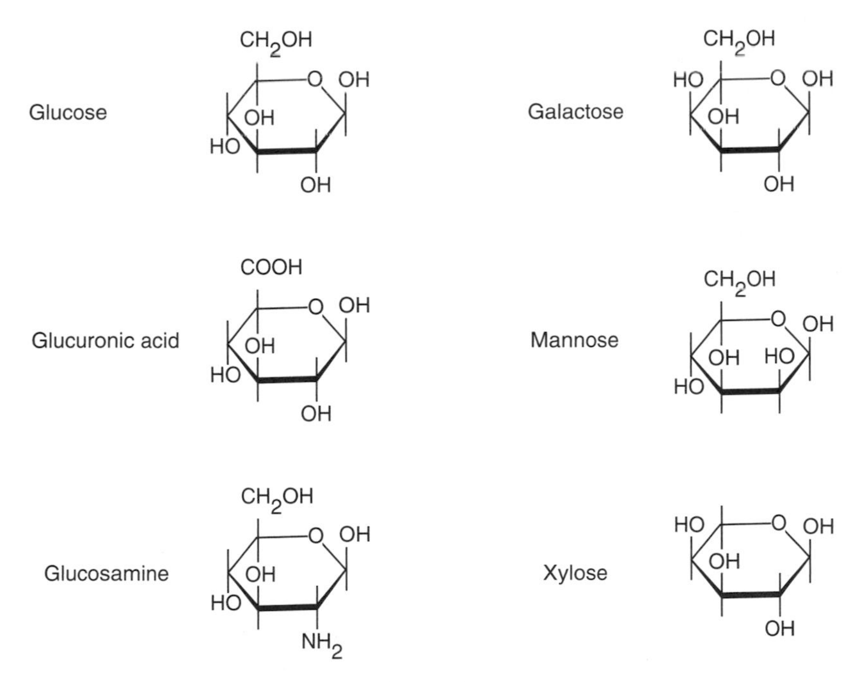

Figure 3.1. Common monosaccharides in soils.

The biomolecules just described are among the most abundant in soils, but by no means do they exhaust the long list of organic compounds produced by living organisms in the soil environment. Organic P compounds, which can account for up to 80% of soil P, occur principally in the form of inositol phosphates (benzene rings with H_2PO_4 bound through O to the ring carbon atoms), and organic S compounds, which can account for nearly all the soil S, occur principally as S-containing amino and phenolic acids and polysaccharides. The chemistry of biomolecules of low relative molecular mass, such as siderophores and those listed in Tables 3.1 and 3.2, has a strong influence on acid–base and metal complexation reactions in soils, whereas the chemistry of biopolymers such as polysaccharides influences the surface and colloid chemistry of soils through adsorption reactions with the solid particles in soil.

3.2 Humic Substances

In simple terms, *humic substances* are organic compounds in humus not synthesized directly to sustain the life cycles of the soil biomass (Section 1.3). More specifically, they are dark-colored, biologically refractory, heterogeneous organic compounds produced as by-products of microbial metabolism. They may account for up to 80% of soil humus (and up to half of aquatic humus), and differ from the biomolecules present in humus because of their long-term persistence (see Problem 3 in Chapter 1) and their molecular architecture. This broad concept of humic substances implies neither a particular pathway of formation and resulting set of organic compounds, nor a characteristic relative molecular mass and associated chemical reactivity. However, it does exclude exogenous materials such as kerogen, a complex hydrocarbon mixture that constitutes nearly all the organic matter in sedimentary rocks, and *black carbon*, an equally complex mixture of organic compounds produced by combustion processes, including fossil fuel burning and fire. These two organic mixtures typically enter soils from parent material and atmospheric deposition respectively.

The chemical properties of humic substances are often investigated after fractionation of soil humus based on solubility characteristics. Organic material that has been solubilized by mixing soil with a 500 mol m^{-3} NaOH solution is separated from the insoluble material (termed *humin*) and brought to pH 1 with concentrated HCl. The precipitate that forms after this acidification is called *humic acid*, whereas the remaining, soluble organic material is called *fulvic acid*. Repeated extractions of this type are often done on the humin and humic acid fractions to enhance separation. The humic and fulvic acids recovered also are subjected to centrifugation and ion exchange resin treatments to remove inorganic constituents and loosely associated biomolecules.

The average chemical composition of soil humic and fulvic acids worldwide is summarized in Table 3.3. Except for the content of S (for which the number of available measurements is about one third the number available for

Table 3.3
Mean content (measured in grams per kilogram) of nonmetal elements in soil humic substances worldwide.[a]

Substance	C	H	N	S	O	H/C	O/C
Humic acid	554 ± 38	48 ± 10	36 ± 13	8 ± 6	360 ± 37	1.04 ± 0.25	0.50 ± 0.09
Fulvic acid	453 ± 54	50 ± 10	26 ± 13	13 ± 11	462 ± 52	1.35 ± 0.34	0.78 ± 0.16

[a]Rice, J. A., and P. MacCarthy. (1991) Statistical evaluation of the elemental composition of humic substances. *Org. Geochem.* 17:635.

the other elements), these data do not differ greatly from the average chemical composition of aquatic humic and fulvic acids or those extracted from peat deposits. Overall the remarkably small standard deviations around the mean values listed in Table 3.3 suggest that humification processes in soil yield characteristic refractory organic products in the two fractions, irrespective of environmental conditions. The average chemical formulas for humic and fulvic acid given in Section 1.3 were developed from the composition data in Table 3.3 (see Problem 7 in Chapter 1). On the basis of a formula unit containing 1 mol H, for which there is no statistically significant difference in content between the two fractions, the average relative molecular mass of humic acid would be larger than that of fulvic acid. Detailed statistical analyses indicate that there is more C and N but less O per unit mass in humic acid compared with fulvic acid. Thus the molar ratios H-to-C and O-to-C both are larger in fulvic acid than they are in humic acid, implying that the latter is the more aromatic (see Section 1.3) and less polar humic substance. Noninvasive spectroscopic methods have proved useful in obtaining a fingerprint of the distribution of C in the two fractions, which supports these inferences. On average, about half the C in soil fulvic acids is associated with polar O-containing moieties, whereas a quarter of the C is associated with aromatic moieties. For humic acids, on the other hand, about one third of the C is aromatic, whereas polar C accounts for about 40% of the total.

Careful spectroscopic examination of humic substances in aqueous solution, after treatment with organic acids and solvents to provoke disaggregation, indicates that humic and fulvic acids are in fact assemblies (*supramolecular associations*) of many diverse components having rather low relative molecular masses (< 2000 Da). These components appear to be held together mainly by hydrogen bonds and hydrophobic interactions (Section 3.4). Thus, the average relative molecular mass of humic substances, particularly humic acid, characterizes a supramolecular association, not a polymer in the sense of the protein and carbohydrate structures discussed in Section 3.1. Fulvic acid, with its more polar nature, is less likely than humic acid to engage in hydrophobic interactions and thus may be pictured in aqueous solution more simply as a dynamic mixture of molecularly small polar components with an association that is largely unaffected by pH, consistent with its defining solubility property.

In keeping with these observations, the carboxyl content of humic acids tends to range from 3 to 5 mol kg^{-1}, whereas that for fulvic acids ranges from 4 to 8 mol kg^{-1}. The phenolic OH content of both humic and fulvic acids ranges from 1 to 4 mol kg^{-1}. These two classes of functional group provide essentially all the Brønsted acidity of humic substances, which, as indicated by their ranges of carboxyl content, is significantly larger for fulvic acids than it is for humic acids. Because most of this acidity is reactive below pH 7 (Table 3.1), and protonation of their amino groups is limited, humic substances bear a net negative charge in all but the most acidic soils. Besides these important O-containing functional groups, a variety of moieties derived from the microbial degradation of biopolymers are found in humic substances. These moieties include fragments of polysaccharides (which are also O containing and account for up to one fourth of the C in humic substances), peptides [the principal chemical form of N in humic substances, also O containing, and characterized by the *amide group* (HN−C=O)], lipids (organic molecules of relatively low water solubility with mixed hydrophilic–hydrophobic character), and lignin (a polymer comprising aromatic alcohols that feature a three-C chain attached to a benzene ring). Alkyl moieties in humic substances, which account for about one fourth of their total C, may be contributed by many of these biopolymeric fragments. They tend to increase in importance with increasing molecular mass and to become associated with hydrophobic domains.

Thus, humic substances emerge from a slow process of biological decomposition, oxidation, and condensation as characteristic organic mixtures having two fundamental properties:

1. *Supramolecular association:* self-organized assemblies of diverse low-molecular mass organic compounds that have either a predominantly hydrophilic (fulvic acid) or hydrophilic–hydrophobic (humic acid) nature, with the latter being mediated in aqueous solution by hydrogen bonds and hydrophobic interactions.
2. *Biomolecular provenance:* identifiable biopolymeric fragments that form an integral part of a labile molecular architecture and that govern both conformational behavior and chemical reactivity.

3.3 Cation Exchange Reactions

Soil humus plays a major role in the buffering of both proton and metal cation concentrations in the soil solution. The mechanistic basis for this buffer capacity is *cation exchange*. A cation exchange reaction involving dissociable protons in soil humus and a cation like Ca^{2+} in the soil solution can be written as

$$SH_2(s) + Ca^{2+} = SCa(s) + 2H^+ \quad (3.2)$$

where SH_2 represents an amount of particulate humus (S) bearing 2 mol dissociable protons, and SCa is the same amount of humus bearing 1 mol exchangeable Ca^{2+}. The symbol S^{2-} then would represent an amount of particulate humus bearing 2 mol negative charge that can be neutralized by cations drawn from the soil solution.

The prospect of interpreting S in Eq. 3.2 at the level of detail typical for minerals or biopolymers is dimmed by the need to consider, in the case of humus, many competing cation exchange reactions involving charged organic fragments. Even if the molecular architecture of each possible cation–humus association were worked out, the use of Eq. 3.2 for them would entail the determination of a large number of chemical parameters—too many for the set of data usually available from a cation exchange experiment to provide. For this reason, and because of the complicated way the structural characteristics described in sections 3.1 and 3.2 influence humus reactivity, the modeling of cation exchange reactions involving soil humus always interprets Eq. 3.2 in some *average* sense. This perspective is emphasized by expressing the $H^+ - Ca^{2+}$ cation exchange reaction in an alternate form:

$$2{\equiv}SOH(s) + Ca^{2+} = ({\equiv}SO)_2Ca(s) + 2H^+ \tag{3.3}$$

In this case, $\equiv$SOH represents an amount of acidic functional groups in humus bearing 1 mol dissociable protons, and $(\equiv SO)_2Ca$ is twice this amount. Equations 3.2 and 3.3 are equivalent ways to represent the same cation exchange process, and neither has any particular structural implication. Equations 3.2 and 3.3 do not imply, for example, that a "humus anion" exists with either the valence -1 or -2. The choice of which equation to use is a matter of personal preference, because both satisfy general requirements of mass and charge balance (see Special Topic 1 in Chapter 1).

The *cation exchange capacity* (CEC) of soil humus is the maximum number of moles of proton charge dissociable from unit mass of solid-phase humus under given conditions of temperature, pressure, and aqueous solution composition, including the humus concentration. A method widely used to measure CEC for humus involves determining the moles of protons exchanged in the reaction

$$2{\equiv}SOH(s) + Ba^{2+} = ({\equiv}SO)_2Ba(s) + 2H^+ \tag{3.4}$$

where the Ba^{2+} ions are supplied in a 100 mol m^{-3} $Ba(OH)_2$ or $BaCl_2$ solution at a selected pH value. Measurements of this kind indicate that the CEC of humic acids ranges typically between 5 and 9 $mol_c\ kg^{-1}$, whereas for peat materials it ranges from 1 to 4 $mol_c\ kg^{-1}$. The CEC range observed for humic acids is consistent with the ranges of carboxyl and phenolic OH content given in Section 3.2.

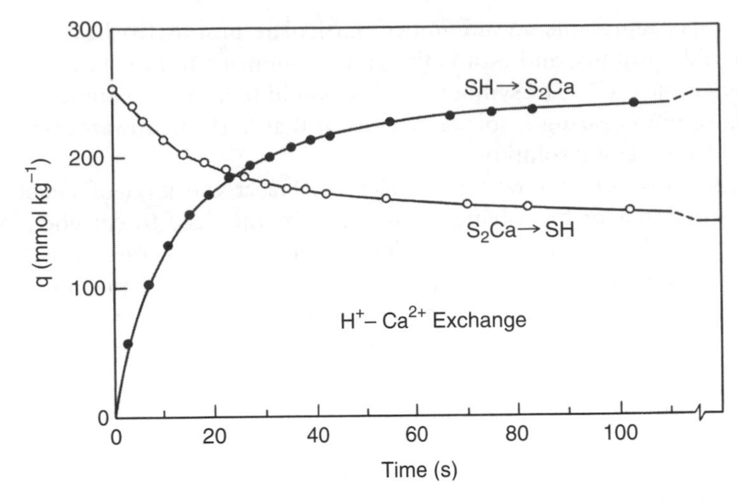

Figure 3.2. Graphs of the moles of adsorbed Ca charge versus time for the cation exchange reaction in Eq. 3.4 with Ca^{2+} as the metal cation replacing H^+ on a sphagnum peat. Filled circles depict the forward (left to right) direction, whereas open circles depict the backward (right to left) direction. Data from Bunzl, K., et al. (1976) Kinetics of ion exchange in soil organic matter. IV. *J. Soil Sci.* **27**:32.

The kinetics of $H^+ - Ca^{2+}$ exchange are illustrated in Figure 3.2 for a suspension of sphagnum peat. The data show the time development of the formation of $(\equiv SO)_2Ca$ (filled circles) after the addition of 50 μmol Ca^{2+} charge and the depletion of $(\equiv SO)_2Ca$ (open circles) after the addition of 50 μmol H^+ charge to a suspension containing 0.1 g peat. It is apparent that the exchange process is relatively rapid. Note that the reaction in Eq. 3.3 proceeds from left to right more readily than from right to left, starting from comparable initial conditions. Additional experiments and data analysis showed that the approximately exponential time dependence of the graphs in Fig. 3.2 can be described by a *film diffusion mechanism.* The basic concept of this mechanism is that the rate of cation exchange is controlled by diffusion of the exchanging ions through a thin (2–50 μm) immobile film of solution surrounding a humus particle in suspension. Film diffusion, discussed in Special Topic 3 at the end of this chapter, is a common process invoked to interpret the observed rates of cation exchange on soil particles.

When the metal cation replacing a proton on soil humus is monovalent and Class A, with low IP (Section 1.2), it is often considered a *background electrolyte ion* in the analysis of proton exchange data. This is done on the hypothesis that all such monovalent metal cations have a much lower affinity for humus than the proton. Attention is then focused on the species $\equiv SOH$. Experimental measurements of the number of moles of strong acid or strong base added to a suspension (or solution) of humus to provoke cation exchange are combined with pH measurements (a combination termed a *titration*) to calculate the *apparent net proton charge:*

$$\delta\sigma_{H,titr} = \frac{(n_A - [H^+]V) - (n_B - [OH^-]V)}{m_s} \tag{3.5}$$

where n_A is the number of moles of strong acid (like HCl) added, and n_B is the number of moles of strong base (like NaOH) added to bring a suspension (or solution) to the volume V with a "free" aqueous proton concentration equal to $[H^+]$ moles per unit volume. The concentration of $[H^+]$ can be determined through a pH measurement, as can that of $[OH^-]$. (Usually, $[OH^-] \approx 10^{-14}/[H^+]$ in dilute solutions, if concentrations are in moles per cubic decimeter.) The numerator in Eq. 3.5 is the difference between H^+ bound and OH^- bound by the humus sample, with *bound* calculated as the difference between moles of added ion and moles of free ion. (Note that bound OH^- is equivalent to dissociated H^+.) After division by m_s, the dry mass of humus, one has computed the *apparent* net proton charge. To convert this quantity to the true net proton charge, σ_H, two steps must be taken. First, corrections must be made for unwanted side reactions involving the added protons or hydroxide ions. These include the formation or dissociation of proton complexes (e.g., HCO_3^- formed from CO_3^{2-} or the reverse reaction, see Problem 15 in Chapter 1) and the dissolution of any minerals present by protonation (Section 1.5) or hydroxide reaction, because none of these reactions involves humus. If only the first type of reaction is occurring, it can be taken into account by a *blank titration* of an aliquot of the aqueous solution contacting the humus sample obtained by separation prior to the addition of strong acid or base. An apparent net proton charge is calculated for this solution using Eq. 3.5 and is then subtracted from $\delta\sigma_{H,titr}$ for the humus suspension (or solution). If mineral dissolution reactions do occur during a titration of humus, they must be taken into account through careful monitoring of the soluble dissolution products (e.g., Al^{3+}) and consideration of both the protonation of the mineral leading to dissolution and the reactions of the soluble dissolution products (e.g., the hydrolysis of Al^{3+}, which produces free protons).

The second step required to convert an apparent net proton charge to σ_H is the establishment of a datum for the blank-corrected $\delta\sigma_{H,titr}$ at some pH value. This must be done because the apparent net proton charge is, by definition, *measured relative to its initial unknown value* in the humus suspension (or solution) prior to the addition of strong acid or base. If $\delta\sigma_{H,titr}$ exhibits a well-defined plateau at low pH, corresponding to the complete protonation of all acidic functional groups, this plateau value can be taken as a datum to be subtracted from all measured values of $\delta\sigma_{H,titr}$ to obtain σ_H, which then will approach zero as pH decreases to the value at which the plateau begins. Alternatively, if the content of carboxyl plus phenolic OH groups has been measured directly, and if it is assumed that only these two acidic functional groups contribute to σ_H, then their combined content may be subtracted from the apparent net proton charge to obtain a true value. In this case, the datum occurs at the pH value where $\delta\sigma_{H,titr}$ is equal and opposite to the combined content of carboxyl and phenolic OH groups. These two examples illustrate the point that the conversion of a blank-corrected $\delta\sigma_{H,titr}$ to obtain σ_H can be problematic.

Figure 3.3 shows a graph of a blank-corrected $\delta\sigma_{H,titr}$ versus –log[H^+] based on the base titration of a purified humic acid extracted from peat. Potassium hydroxide was added incrementally to increase pH and produce the exchange reaction

$$\equiv SOK + H^+ = \equiv SOH + K^+ \qquad (3.6)$$

The pH values measured were converted to –log[H^+] in the KNO_3 solutions used as a background electrolyte, where [H^+] is in moles per cubic decimeter (liter). Equation 3.5 was used to compute $\delta\sigma_{H,titr}$ at each value of –log[H^+], after which blank titration corrections were performed. The graph in Figure 3.3 thus depicts the blank-corrected apparent net proton charge of the humic acid sample at several ionic strengths.

Three characteristic features of the net proton charge on humus are evident in Figure 3.3: (1) negative values over a broad range of pH; (2) the absence of well-defined plateaus, inflection points, or other signatures of different classes of acidic functional group as observed typically in the titration curves of well-defined organic acids; and (3) a tendency to become more negative in value with increasing concentration of the background electrolyte. The first-named property indicates a dominant contribution of proton dissociation over the normal range of pH in soils, whereas Property 2 implies that the

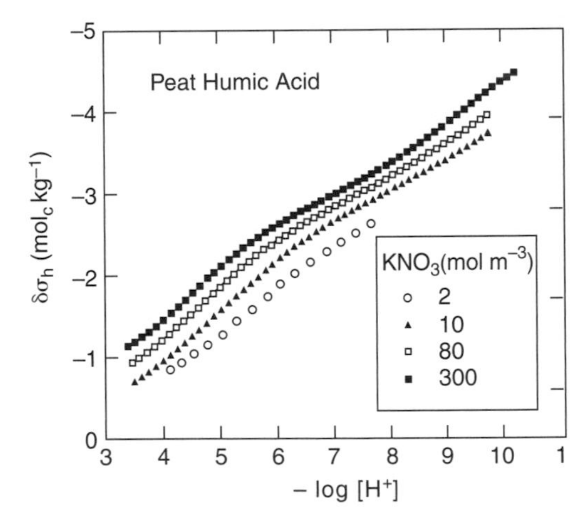

Figure 3.3. Graph of the apparent net proton charge on a peat humic acid versus –log[H^+] at several ionic strengths. Reprinted with permission from Kinniburgh, D. G., et al. (1996) Metal ion binding by humic acid: Application of the NICA–Donnan model. *Environ. Sci. Technol.* **30**:1687–1698.

acidic functional groups present in humus dissociate protons in overlapping ranges of pH, as opposed to exhibiting widely separated characteristic pH values for proton dissociation. Property 3 is consistent with the cation exchange reaction in Eq. 3.6 being driven to the left as the concentration of K^+ increases. Note that the change in net proton charge between pH 3 and 10 is larger than the maximum structural charge observed in 2:1 clay minerals and Mn oxides (Sections 2.3 and 2.4). Changes twice as large as this are observed in similar titration measurements for solutions of fulvic acid.

The *acid-neutralizing capacity* (ANC) of humus in suspension or solution is equal to the concentration of its dissociated acidic functional groups:

$$\mathrm{ANC} = -\sigma_{\mathrm{H}} c_{\mathrm{s}} \quad (\sigma_{\mathrm{H}} \leq 0) \tag{3.7}$$

where c_s is the humus concentration in kilograms per cubic decimeter. Clearly, ANC will increase from zero, at some low value of pH, to the CEC of humus, expressed as a concentration in moles per cubic decimeter (liter), at high pH. The change in ANC with pH (strictly, the derivative dANC/dpH) is called the *buffer intensity,* β_H. If the ANC increases greatly as pH increases, then the solution constituents have a large increase in their capacity to bind and thus neutralize protons; this corresponds to a large buffer intensity. Speaking generally, one can estimate the buffer intensity to be greatest when σ_H changes most rapidly with $-\log[H^+]$. In Figure 3.3, this occurs in the range $4 < -\log[H^+] < 6$, which is typical for soil humus materials. (Note that β_H does not depend on the datum selected for σ_H.) It is for this reason that soil humus is so important in the buffering of acidic soils.

3.4 Reactions with Organic Molecules

The organic compounds that react with soil humus are derived from pesticides, pharmaceuticals, industrial wastes, fertilizers, green manures, and their degradation products. Humus in solid form, either as a colloid or as a coating on mineral surfaces, can immobilize these compounds by adsorption and, in some instances, detoxify or deactivate them. Soluble humus, typically the fulvic acid fraction, can form complexes with organic compounds that then may travel freely with percolating water into the soil profile. Toxic organic materials that otherwise might be localized near the land surface can be transported by this mechanism. Similar transport may occur for organic molecules adsorbed by mobilized humus colloids.

Soil humus reacts by cation exchange with organic molecules that contain N atoms bearing a positive charge. These kinds of structures occur in both aliphatic and aromatic compounds, the latter being common in pesticide and pharmaceutical preparations. The general reaction scheme is analogous to Eq. 3.6:

$$\equiv\mathrm{SOH(s)} + \mathrm{R} - \overset{+}{\mathrm{N}} = \equiv\mathrm{SON} - \mathrm{R(s)} + \mathrm{H}^+ \tag{3.8}$$

where R represents an organic unit bonded to the N atom. Spectroscopic studies of this reaction indicate that some *electron transfer* from humus to the N compound takes place, thereby enhancing the stability of the humus–organic complex. Humus also contains electron-deficient aromatic moieties, such as quinones or other benzene rings with highly polar substituents, that can attract and bind electron-rich molecules, such as polycyclic aromatic hydrocarbons (PAH; two or more fused benzene rings), to form stable *charge-transfer complexes.*

Organic molecules that become positively charged when protonated can react with COOH groups in soil humus by *proton transfer* from the latter to the former. Basic amino acids, like arginine (Table 3.2), with two "protonatable" NH_2 groups, are good examples of these compounds, as are the s-triazine herbicides, which contain protonatable N substituents on an aromatic ring. Protonated functional groups like COOH and NH also can form *hydrogen bonds* with electronegative atoms such as O, N, and F. As an example, the $C = O$ group in the phenylcarbamate and substituted urea pesticides can form a hydrogen bond (denoted . . .) with NH in soil organic matter, $C = O \ldots HN$, and NH groups in the imidazolinone herbicides can form hydrogen bonds with $C = O$ groups in humus. (Hydrogen bonds of this type also form in peptides.) Humus contains carboxyl, hydroxyl, carbonyl, and amino groups in a broad variety of molecular environments that lead to a spectrum of possibilities for hydrogen bonding within its own supramolecular structure and with exogenous organic compounds. The additive effect of these interactions makes hydrogen bonding an important reaction mechanism, despite its relatively low bonding energy.

Much of the supramolecular architecture of soil humus is not electrically charged. This nonionic structure can nevertheless react strongly with the uncharged part of an organic molecule through *van der Waals interactions.* The van der Waals interaction involves weak bonding between polar units, which may be either permanent (like OH and $C=O$) or induced momentarily by the presence of a neighboring molecule. The induced van der Waals interaction is the result of correlations between fluctuating polarization created in the "electron clouds" of two nonpolar molecules that approach one another closely. Although the average polarization induced in each molecule by the other is zero (otherwise they would not be nonpolar molecules), the negative correlations between the two induced polarizations do not average to zero. These correlations produce a net attractive interaction between the two molecules at very small distances (around 0.1 nm). The van der Waals interaction between two molecules is very weak, but when many molecules in a supramolecular structure like humus interact simultaneously, the van der Waals component is additive and, therefore, strong.

The interaction between uncharged molecules (or uncharged portions of molecules) and soil humus is often stronger than the interaction between these kinds of molecules and soil water, resulting in their exit from the soil solution to become adsorbed by humus. This occurs for two distinct reasons. First, water molecules interacting with a nonpolar molecule in the soil solution are confronted by a lack of electronegative atoms with which to form a hydrogen bond, so they cannot orient their very polar OH toward the molecule in ways that are compatible with the tetrahedral coordination they engage in the bulk liquid structure. Instead, the water molecules must form a network of hydrogen bonds that point roughly parallel to the surface of the nonpolar molecule, thereby enclosing it in a kind of cage structure (*hydrophobic effect*). The resultant disruption of the tetrahedral ordering in liquid water and the cost in energy to produce the anomalous cage result in a low water solubility of the nonpolar molecule. The second reason for a stronger interaction with humus is the presence of nonpolar moieties in the latter. From the perspective of minimizing disruption of the normal liquid water structure, it is optimal to have a nonpolar molecule adsorb on a nonpolar domain of humus, so that fewer water molecules are needed to accommodate to the two than when they are far apart. Although van der Waals interactions between nonpolar molecules are approximately of the same strength as those between water molecules, or those between nonpolar molecules and water molecules, the gain to the latter in not having to form as extensive a cage structure produces a strong tendency for nonpolar units to bind together in the presence of liquid water (*hydrophobic interaction*).

The relationship between the hydrophobic effect and water solubility can be quantified by two important properties of uncharged organic molecules: the number of *chlorine substituents* (N_{Cl}) and the *solvent-excluding area* (SES). Chlorine is a highly electronegative atom that, upon replacing H on a carbon atom, can then withdraw significant electron charge density carbon-carbon bonds in chain or ring structures, thus rendering them less polar and more hydrophobic. Solvent-excluding area (the same as the total surface area for a nonpolar molecule) provides a measure of the size of the interface across which no hydrogen bonds cross, which is created when the hydrophobic effect occurs. This interface disrupts the structure of liquid water, leading to cage formation that is inimical to high water solubility. These ideas are summarized qualitatively in Figure 3.4, which gives ranges of water solubility observed for several classes of toxicologically important organic compounds. Solubility is seen to decrease as either the number of Cl or molecular size increases across a given class.

Statistical correlations have been worked out that express these trends in quantitative form and serve as useful predictors. For example, the common logarithm of water solubility [S, expressed in moles per cubic decimeter (liter)] for chlorinated benzenes has been shown to decrease linearly with N_{Cl}:

$$\log S = -0.6608\, N_{Cl} - 1.7203 \quad (R^2 = 0.98) \tag{3.9a}$$

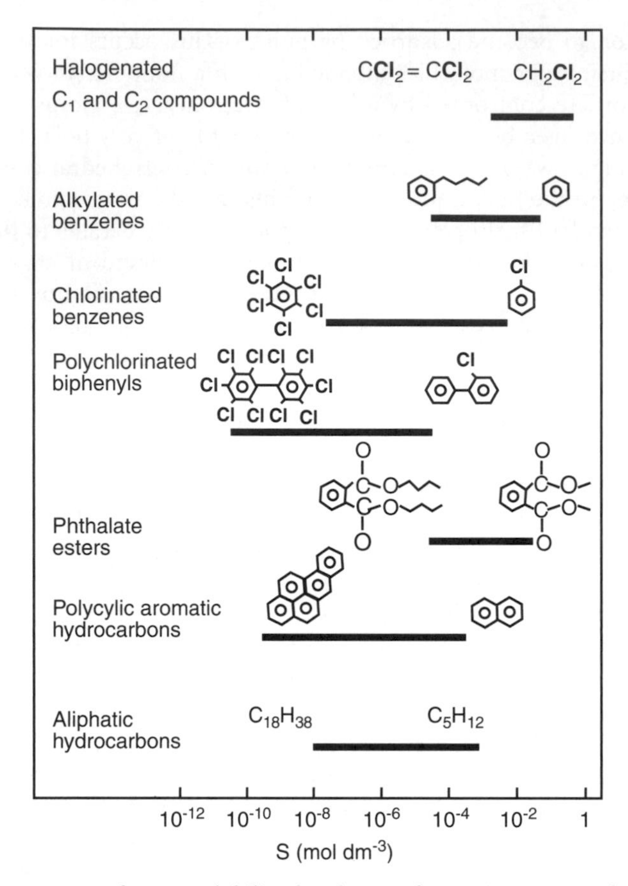

Figure 3.4. Ranges of water solubility for classes of organic compounds of varying hydrophobicity produced by increasing chlorine substitution or solvent-excluding area. Data and format from Schwarzenbach, R. P., P. M. Gschwend, and D. M. Imboden. (2003) *Environmental organic chemistry.* John Wiley, Hoboken, NJ.

and that for polycyclic aromatic hydrocarbons has been shown to decrease linearly with SES:

$$\log S = -4.27\text{SES} + 3.07 \quad (R^2 = 0.998) \tag{3.9b}$$

where SES is expressed in square nanometers. Taking as a simple—but telling—case, the benzene molecule, with a measured log water solubility (in the units of S presented earlier) at 25 °C is −1.64. Equation 3.9a yields −1.72 (with $N_{Cl} = 0$), whereas Eq. 3.9b yields −1.63 using SES equal to the total surface area of the benzene molecule, 1.1 nm^2. Both of these solubility estimates are in agreement with the measured value.

The relationship between the hydrophobic interaction and water solubility is often described quantitatively by a linear partition equation analogous to

Henry's law (Section 1.4), with soil humus instead of air playing the role of the nonaqueous phase:

$$K_{oc} = \frac{(n/f_{oc})}{[A(aq)]} \qquad (3.10)$$

where n is the number of moles of an organic compound A that is adsorbed by 1 kg *soil* with an organic carbon content that is equal to f_{oc} (measured in kilograms organic C per kilogram), thus making the quantity n/f_{oc} the number of moles of A adsorbed per kilogram of soil organic C. The constant parameter K_{oc} may be termed the *Chiou distribution coefficient,* with units of liters per kilogram of organic C. By hypothesis, this parameter is not dependent (or very weakly dependent) on the chemical properties of soil humus (i.e., division by f_{oc} on the right side of Eq. 3.10 is hypothesized to remove all such dependence by normalizing n to the content of organic C in a soil). Perhaps remarkably, this hypothesis has been verified rather well (i.e., K_{oc} calculated with Eq. 3.10 varying within a factor of about two) in careful studies involving a variety of soils interacting with a single hydrophobic organic compound, such as dichlorobenzene or carbon tetrachloride, making the Chiou distribution coefficient a very useful model parameter.

Equation 3.10 describes the partitioning of compound A between two phases: soil humus and the soil solution. This partitioning, in the case of organic compounds like those shown in Figure 3.4, is expected to favor soil humus because of the hydrophobic effect. Because the latter is inversely related to water solubility, it is reasonable to expect that the Chiou distribution coefficient also will be inversely related to water solubility. Such a statistical correlation often has been observed and is of the general mathematical form

$$\log K_{oc} = a - b \log S \qquad (3.11)$$

where a and b are empirical parameters that in principle depend on the class of organic compounds under consideration. One useful correlation that holds for a broad variety of organic compounds and predicts the value of $\log K_{oc}$ within ± 0.45 (i.e., predicts K_{oc} values within a factor of about three) has a = 3.95 and b = 0.62, with K_{oc} in units of cubic decimeter (liter) per kilogram and S in units of grams per cubic meter. For example, the industrial pollutant 1,4-dichlorobenzene has a water solubility of 83 g m^{-3} and, therefore, Eq. 3.11 predicts

$$\log K_{oc} = 3.95 - 0.62 \log 83 = \mathbf{2.76} \qquad (3.12)$$

compared with an observed value of 2.74 (i.e., $K_{oc} = 550$ L kg_{oc}^{-1}). By comparison, benzene has the much larger water solubility of 1780 g m^{-3} (note the dramatic effect of the chlorine substituents!), corresponding to $\log K_{oc} = 1.93$, using Eq. 3.11 with the values of a and b given earlier. This is also the observed value, equivalent to $K_{oc} = 85\ L\ kg_{oc}^{-1}$. *Increasing water solubility corresponds to decreasing partitioning of nonpolar compounds into soil humus.* The Chiou distribution coefficient is a quantitative parameter that captures this trend accurately.

3.5 Reactions with Soil Minerals

Soil humus in itself is not biologically refractory. Laboratory experiments with fungi, bacteria, enzymes, and chemical oxidants indicate clearly that humus in aqueous extracts—even its aromatic components—can be degraded readily under aerobic conditions over periods of days to weeks. Evidently anaerobic conditions and, more significantly, interactions with soil particles are essential in protecting humus from microbial attack and conversion to CO_2. Numerous circumstantial studies of the biodegradability of humus in temperate-zone soils support this idea, with reports of organic C content and mean age of humus increasing with decreasing particle size. Soil humus found in silt-size particles tends to have C-to-N ratios well above the average soil value of 8 (Section 1.1), whereas that in clay-size particles does conform to the average value, indicating that protection mechanisms must be operating in the latter that are either not available or not effective in the former. Encapsulation and, therefore, physical isolation along with the attendant anaerobic conditions likely is the principal mechanism by which humus survives in soil silt fractions, whereas this mechanism plus strong adsorption reactions with minerals likely contribute to the long life of humus observed for soil clay fractions. Modulating these trends is the spectrum of inherent differing susceptibility to microbial degradation of the components of humus themselves, with biopolymer fragments placed at the high end of the spectrum, alkyl O-containing moieties placed in the middle of the spectrum, and hydrophobic moieties placed at the low end.

The low C-to-N ratio of soil clay fractions suggests that peptidic moieties are involved importantly in reactions of humus with soil minerals having very small particle size. These moieties may engage in cation exchange with acidic surface OH groups (Eq. 3.8, with the reactant $\equiv$SOH now interpreted as a mineral surface OH) or they may bind through proton transfer, hydrogen bonding, and van der Waals interaction mechanisms, as described in Section 3.4, but with humus moieties now being the "organic molecule" and a soil mineral being the adsorbing solid phase. The latter two modes of interaction also apply to the other components of humus, and in particular to humic substances.

Another important reaction mechanism that extends to any humus component of suitable composition is *bridging complexation,* in which anionic or polar functional groups (e.g., carboxylates or carbonyls) become bound to a metal cation adsorbed by a negatively charged mineral surface (e.g., negative structural charge on clay minerals and Mn oxides or ionized surface OH). If one or more water molecules is superposed between the adsorbed cation and the polar organic functional group, the mechanism is termed *outer-sphere bridging complexation* (Fig. 3.5, also termed *water bridging*), whereas if the adsorbed cation is bound directly to the polar organic functional group, it is termed *inner-sphere bridging complexation* (Fig. 3.6, also termed *cation bridging*). As a rule, monovalent adsorbed cations form outer-sphere bridging complexes with polar functional groups, whereas bivalent adsorbed cations tend to form both types of complex, although Class B metal cations (Section 1.2) are likely to form inner-sphere complexes exclusively. Of the six modes of interaction described, the weakest are cation exchange, proton transfer, and outer-sphere bridging complexation; the strongest are hydrogen bonding, van der Waals interactions, and inner-sphere bridging complexation. For humic substances, van der Waals interactions and, in particular, hydrophobic interactions with the atoms in a mineral surface can be quite strong and relatively long range, resulting in the formation of very stable complexes. These latter interaction mechanisms are especially apparent when the binding humus moieties are large molecular fragments or when chemical conditions are such that they suppress the ionization of acidic functional groups either in humus or on the mineral surface—for example, when the pH value results in no net surface charge on the latter (sections 2.3 and 2.4).

Studies of soil humus retention and recalcitrance (the latter being indicated by resistance to chemical oxidants) consistently show that these two properties are positively correlated with the content of poorly crystalline Al

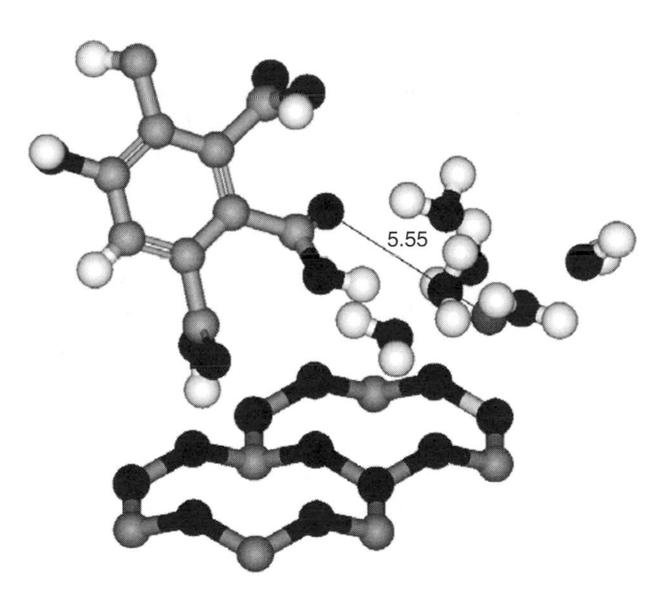

Figure 3.5. Outer-sphere bridging complexation of a cation adsorbed on a clay mineral surface by a carbonyl group in humus, with the cation–carbonyl O distance shown in Ångstroms. Visualization courtesy of Dr. Rebecca Sutton.

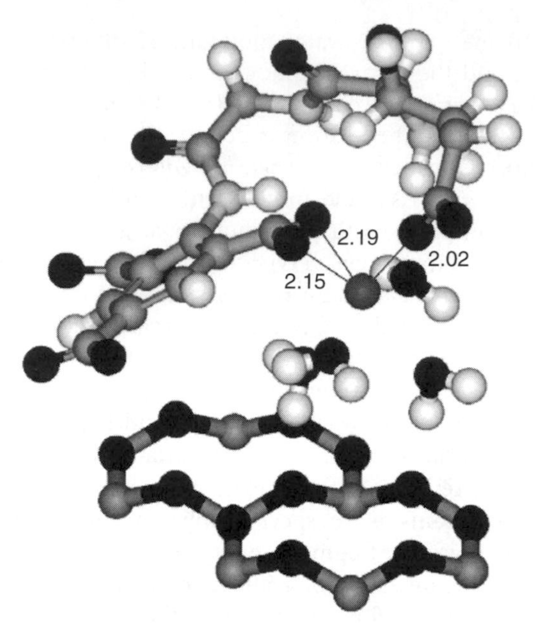

Figure 3.6. Inner-sphere bridging complexation of a cation adsorbed on a clay mineral surface by carboxyl groups in humus, with cation–O distances shown in Ångstroms. Visualization courtesy of Dr. Rebecca Sutton.

and Fe oxyhydroxides (conventionally estimated by an extraction with ammonium oxalate). A similar correlation is found for allophanic minerals, which are inherently poorly crystalline. This relationship is an expected result of the relatively large specific surface area of poorly crystalline metal oxyhydroxides and aluminosilicates, which allows greater adsorption of humus per unit mass of solid phase, and the abundance of acidic surface OH groups on these minerals (sections 2.3 and 2.4), which promotes greater reactivity per unit mass of solid phase.

A mechanistic basis for the relationship is provided by *ligand exchange,* a chemical reaction in which direct bond formation takes place between an O-containing functional group in humus, typically carboxylate, and either Al(III) or Fe(III) at the surface of a poorly crystalline Al or Fe aluminosilicate or oxyhydroxide mineral. This reaction involves stronger chemical bonds than those that occur even in the inner-sphere bridging complexation reaction because the metal ion involved is part of the mineral structure, not an adsorbed species. The general reaction scheme for ligand exchange can be expressed by two chemical equations:

$$\equiv\text{MOH(s)} + \text{H}^+ = \equiv\text{MOH}_2^+\text{(s)} \tag{3.13a}$$

$$\equiv\text{MOH}_2^+\text{(s)} + \text{S} - \text{COO}^- = \equiv\text{MOOC} - \text{S(s)} + \text{H}_2\text{O}(\ell) \tag{3.13b}$$

where, by analogy with Eq. 3.3, ≡MOH(s) represents 1 mol reactive surface OH bound to a metal M (M = Al or Fe) in an aluminosilicate or oxyhydroxide mineral structure and, similarly to the first reactant in Eq. 3.2, $S-COO^-$ represents an amount of dissolved humus bearing 1 mol carboxylate groups. The protonation step is analogous to the reversible protonation step illustrated in Fig. 2.8 for an $Al-OH_2^{1/2+}$ on the edge surface of kaolinite (see the discussion of acidic surface OH groups in sections 2.3 and 2.4). It creates a positively charged water molecule at the mineral surface, an unstable surface species with an instability that makes the ligand exchange (H_2O for COO^-) in Eq. 3.12b more likely. Thus, ligand exchange is favored at pH values below which the mineral surface bears a net positive charge (e.g., less than pH 5.4–up to 8.0 for allophanic minerals, as discussed in Section 2.3). The species ≡MOOC–S on the right side of Eq. 3.12b is similar in structure to the inner-sphere bridging complex depicted in Figure 3.6, with the important difference that the metal M involved is bound into the mineral structure. This fact and the trivalent charge on the metal ion leads to a very strong complex between the humus carboxylate moiety $S-COO^-$ and the mineral surface. If the humus moiety complexed has hydrophobic domains now exposed to the soil solution, it is likely that they will serve to bind to similar organic moieties in dissolved humus through hydrophobic interactions (and, if polar units are included, through hydrogen bonding), thus nucleating the construction of a supramolecular association of humus components that is strongly anchored to the mineral surface. This association evidently would grow in a disorganized layered fashion as dissolved humus moieties continued to adsorb onto anchored humus moieties. Humus nearest the mineral surface would thus be effectively protected from microbial attack, whereas that exposed to the soil solution at the top of a multilayer patch would be susceptible to desorption and to microbial attack.

For Further Reading

Chiou, C. T. (2002) *Partition and adsorption of organic contaminants in environmental systems.* Wiley-Interscience, Hoboken, NJ. Chapter 7 of this useful monograph describes the theory and application of the Chiou distribution coefficient, including its estimation from statistical correlation equations.

Clapp, C. E., M. H. B. Hayes, N. Senesi, P. R. Bloom, and P. M. Jardine (eds.). (2001) *Humic substances and chemical contaminants.* Soil Science Society of America, Madison WI. The 23 chapters of this edited workshop/symposium offer a comprehensive survey of humic substance structure, reactivity, and transport in soils.

Essington, M. E. (2004) *Soil and water chemistry.* CRC Press, Boca Raton, FL. Chapter 4 of this textbook gives a comprehensive survey of humus structure and reactivity, including an introduction to the spectroscopic techniques used commonly to examine them. Chapter 7 provides an introduction to the concept and application of the Chiou distribution coefficient.

The following five technical journal articles offer probing, advanced reviews of the evolving concepts of humic substance structure and preservation in natural soils.

Allison, S. D. (2006) Brown ground: A soil carbon analogue for the green world hypothesis? *Amer. Naturalist* **167**:619–627.

Baldock, J. A., and J. O. Skjemstad. (2000) Role of the soil matrix and minerals in protecting natural organic materials against biological attack. *Org. Geochem.* **31**:697–710.

Burdon, J. (2001). Are the traditional concepts of the structures of humic substances realistic? *Soil Sci.* **166**:752–769.

Piccolo, A. (2001) The supramolecular structure of humic substances. *Soil Sci.* **166**:810–832.

Sutton, R., and G. Sposito. (2005) Molecular structure in soil humic substances: The new view. *Environ. Sci. Technol.* **39**:9009–9015.

Problems

The more difficult problems are indicated by an asterisk.

1. Develop a reaction analogous to that for peptide formation in Eq. 3.1 to demonstrate that cellulose is a condensation polymer of glucose.

*2. The *Soil Science* article by Piccolo (see "For Further Reading") offers many lines of experimental evidence for humic acids to be pictured as supramolecular associations of diverse components having relatively small molecular size. On pages 820 to 821 of his article, Piccolo describes the results of a study in which fractions of a humic acid treated with acetic acid (Table 3.1) at pH 3.5 were compared regarding the types of C they contained (e.g., aromatic C) with fractions of the humic acid not so treated.

   a. Give a chemical definition of *supramolecular association*. Be sure to cite your source.
   b. Piccolo states five findings concerning the composition of the fractions of the humic acid treated with acetic acid that he concludes are evidence for a supramolecular association. What are these five pieces of evidence?

*3. The table presented here shows the mean content of nonmetal elements in freshwater humic substances worldwide, taken from the same source as the data in Table 3.3. Given the standard deviations for the mean values, it is possible to compare them pairwise to determine whether statistically significant differences exist between the compositions of soil and freshwater humic substances. This can be done with a two-sided t test applied at a chosen level of significance—say, $P < 0.01$ (less than one chance in 100 that the two compound mean values are equal, if t is large enough).

a. Examine the two sets of composition data for significant differences in C, H, and O content ($P < 0.01$). Take n = 215 for soil humic acids; n = 56 for freshwater humic acids.
b. Compare the molar ratios O-to-C and H-to-C regarding whether they are significantly different between soil and freshwater humic acids ($P < 0.01$).
c. Is it accurate to state that soil humic acids are less polar and more aromatic than freshwater humic acids?

(*Hint:* Use available software or online programs to perform the t tests.)

C ($g\ kg^{-1}$)	H ($g\ kg^{-1}$)	N ($g\ kg^{-1}$)	S ($g\ kg^{-1}$)	O ($g\ kg^{-1}$)	O/C	H/C
512 ± 30	47 ± 16	26 ± 6	19 ± 14	404 ± 38	0.60 ± 0.08	1.12 ± 0.17

4. Combine Eqs. 3.3 and 3.6 to derive a chemical equation for the cation exchange of Ca^{2+} for Na^+ on humus.

5. Apply the concept of cation exchange to explain why the pH of an acidic suspension of humic acid with a 100 mol m^{-3} background electrolyte solution would be expected to be lower than that of a suspension with no background electrolyte solution.

*6. The net proton charge for humic substances is described mathematically by the two-term model equation

$$\sigma_H = -\frac{b_1}{1 + (K_1[H^+])^{p_1}} - \frac{b_2}{1 + (K_2[H^+])^{p_2}}$$

where b_1 is the carboxyl content and b_2 is the phenolic OH content. The parameter K_i (i = 1, 2) represents the average affinity of either carboxyl groups or phenolic OH groups for protons, whereas the parameter p_i (i = 1, 2) represents the variability within the two acidic functional groups regarding affinity, with $0 \leq p_i \leq 1$, with the upper limit $p_i = 1.0$ signifying no variability. Examination of a large database for humic acids titrated at low background electrolyte concentration produced the parameter estimates

$$b_1 = 3.15\ mol_c\ kg^{-1}, K_1 = 10^{2.93}\ L\ mol^{-1}, p_1 = 0.50$$

$$b_2 = 2.55\ mol_c\ kg^{-1}, K_2 = 10^{8.00}\ L\ mol^{-1}, p_2 = 0.26$$

The values of $K_i (i = 1, 2)$ imply that half the carboxyl groups are dissociated at pH 3, whereas half the phenolic OH groups dissociate at pH 8. Both groups exhibit broad distributions of affinity for protons, because $p_i (i = 1, 2)$ is not close to 1.0.

a. Prepare a graph of σ_H versus $-\log[H^+]$ as in Figure 3.3.
b. Calculate ANC (in micromoles of charge per liter) in the pH range 4 to 7 for a suspension of humic acid having a solids concentration of 30 mg L^{-1}. (Assume that $[H^+] \approx 10^{-pH}$.)

7. The *total acidity* (TA) of solid-phase humus is defined by the equation

$$CEC \equiv TA - \sigma_H \quad (\sigma_H \leq 0)$$

Thus, TA is a quantitative measure of the capacity of humus to donate protons under given conditions of temperature, pressure, and soil solution composition. Use the model equation and parameter values given in Problem 6 to prepare a graph of the ratio TA-to-CEC versus $-\log[H^+]$ over the pH range 3 to 9. (Take pH $\approx -\log[H^+]$.)

*8. Use the model equation and parameter values given in Problem 6 to calculate the buffer intensity of the humic acid suspension in the range 4 to 7. (*Hint:* The first derivative with respect to x of $10^{-\beta x}$ is $-(\ln 10)\beta 10^{-\beta x}$, where ln 10 = 2.303. Use this result and the assumption that $[H^+] \approx 10^{-pH}$ with $\beta_H \equiv dANC/dpH$.)

*9. An *adsorption edge* is a graph of the moles of an adsorbed cation per unit mass of solid phase versus pH (or $-\log[H^+]$). Given that σ_H for the humic acid with the titration behavior illustrated in Figure 3.3 can be described by the model expression in Problem 6, plot the adsorption edge for Na^+ (Eq. 3.6) using the following parameter values:

$$b_1 = 3.1 \text{mol}_c\, \text{kg}^{-1}, K_1 = 10^{2.8}\, \text{L mol}^{-1}, p_1 = 0.48$$

$$b_2 = 2.7 \text{mol}_c\, \text{kg}^{-1}, K_2 = 10^{8.00}\, \text{L mol}^{-1}, p_2 = 0.24$$

Take pH $\approx -\log[H^+]$ to lie in the range 4 to 9. The parameter pH_{50} is defined as the pH value at which the moles of adsorbed cation per unit mass equal one half the maximal CEC. Estimate pH_{50} from your adsorption edge. (*Hint:* Show that the maximal CEC is equal to $b_1 + b_2$.)

10. The table presented here lists water solubilities for two groups of organic pollutant known to contaminate soils. Use these data to estimate the Chiou distribution coefficient (L kg_{oc}^{-1}) for each pollutant. Look up the chemical structures of the pollutants, then use this information to classify them and explain the differences in log K_{oc} among them.

Pollutant	S(g m^{-3})	Pollutant	S (g m^{-3})
Chlorobenzene	484	Naphthalene	112
1,4-Dichlorobenzene	83	Phenanthrene	6.2
1,3,5-Trichlorobenzene	5.3	Anthracene	6.2
1,2,3,5-Tetrachlorobenzene	3.6	Pyrene	0.90
Pentachlorobenzene	0.65	Benzanthracene	0.25
Hexachlorobenzene	5×10^{-3}	Benzopyrene	4.9×10^{-2}

11. The reported water solubility of the organic pollutant 1,2-dichlorobenzene varies from 93 to 148 g m^{-3}. Shown in the following table are measured values of the Chiou distribution coefficient for this pollutant on a variety of soils with varying organic C content. Estimate log K_{OC} for 1,2-dichlorobenzene based on its solubility, then compare your result with the average of the measured log K_{OC}, taking into account the standard deviation for both your estimate and the average.

Soil	K_{OC} (L kg_{OC}^{-1})	Soil	K_{OC} (L kg_{OC}^{-1})
Anoka	261	Pierre	319
Burleigh	263	Piketon	263
Cathedral	407	Renslow	340
Columbus	308	Sanhedrin	344
Elliot	252	Spinks	318
Marlette	223	Wellsboro	383
Manchester	230	West-Central Iowa	248
Oliver	277	Woodburn	296

12. Ciprofloxacin ("Cipro") is a fluoroquinolone antibiotic that is becoming widely distributed in soils through wastewater sludge disposal on land. Concentrations of this antibiotic below 10 mg m^{-3} in the soil solution are deemed acceptable, and some authorities have indicated a similar threshold of 0.01 mg kg^{-1} for its soil content. Given its water solubility of 30 g L^{-1}, what minimum soil organic C content is required to have both the soil solution concentration and the soil content of Cipro at safe levels? Are there soils for which this organic C content is typical?

13. *Ceriodaphnia dubia* is a freshwater invertebrate used for acute toxicity tests involving pesticides and metals. When this organism is exposed to the organophosphate pesticide chlorpyrifos at a concentration of 82 $\mu g m^{-3}$, only 20% survival is found after 24 hours of exposure. If humic acid is added, however, the percentage survival increases in a roughly proportional manner, with 92% survival noted at a humic acid concentration of 100 $g m^{-3}$. The addition of humic acid at a concentration of 30 g m^{-3}

produced a survival rate of 50%, implying that the free concentration of the pesticide was at its LC_{50} value. Given that log K_{oc} = 3.79 for chlorpyrifos, and that the humic acid has a C content of 578 g kg^{-1}, calculate LC_{50} for *C. dubia* exposure to the pesticide added at 82 $\mu g\ m^{-3}$. Reported values of LC_{50} range from 60 to 100 $\mu g\ m^{-3}$.

14. The desorption of PAHs from river sediments has been observed to follow an exponential time dependence

$$n(t) = n_0 \exp(-k_{des}t)$$

where n_0 is an initial value of the amount adsorbed per unit mass of sediments (see Eq. 3.10) and k_{des} is a rate coefficient for the desorption process. Studies with a variety of PAHs, such as those listed in Problem 10, show that the value of k_{des} is correlated negatively with K_{oc} for the PAH compounds

$$\log k_{des} = -0.98 \log K_{oc} - 0.104$$

where k_{des} is in units of day^{-1}. As shown in Eq. S.3.11 (in Special Topic 3 at the end of this chapter), the rate coefficient for an exponential time dependence defines a *half-life* for the decline in n(t). Calculate this half-life for each of the PAHs listed in Problem 10. This parameter defines an intrinsic timescale for their desorption from the sediments.

*15. The table presented here shows measured values of recalcitrant humus content and poorly crystalline Al and Fe oxyhydroxide content for the fine clay fraction (< 0.2 μm) in subsurface horizons in a dozen acidic temperate-zone soils. *Recalcitrant humus* is defined as the humus remaining in a soil sample after oxidation by NaOCl at pH 8. *Poorly crystalline* Al and Fe oxyhydroxides are quantified by the content of Al and Fe extracted by an ammonium oxalate/oxalic acid mixture at pH 3. Apply linear regression analysis to these data to determine whether recalcitrant humus is positively correlated with poorly crystalline Al and Fe oxyhydroxides, as discussed in Section 3.5. Be sure to include 95% confidence intervals

Soil	Recal. humus ($g\ kg^{-1}$)	Poor crys. oxides ($g\ kg^{-1}$)	Soil	Recal. humus ($g\ kg^{-1}$)	Poor crys. oxides ($g\ kg^{-1}$)
1	17.9	25.1	7	97.0	91.9
2	9.1	10.0	8	42.5	68.9
3	6.0	8.0	9	39.9	62.7
4	12.0	6.8	10	35.9	63.3
5	6.7	6.0	11	76.0	72.4
6	11.1	18.6	12	40.0	63.6

Abbreviations: Poor crys. oxides, poorly crystalline oxides; Recal. humus, recalcitrant humus.

on the y-intercept and slope of your regression line. Is there a threshold content of poorly crystalline Al and Fe oxyhydroxides required for the presence of recalcitrant humus? (*Hint:* Consider the 95% confidence intervals in responding to this last question.)

Special Topic 3: Film Diffusion Kinetics in Cation Exchange

An adsorption reaction that involves chemical species in aqueous solution must also involve a step in which these species move toward a reactive site on a particle surface. For example, the Ca^{2+} and H^+ species that appear in the cation exchange reaction in Eq. 3.3 cannot react with the exposed surface site, $\equiv SO^-$, until they exit the bulk aqueous solution phase to come into contact with $\equiv SO^-$. Thus, the kinetics of surface reactions such as cation exchange cannot be described solely in terms of surface site interactions unless the transport step is very rapid when compared with the site interaction step. If, on the contrary, the timescale for the transport step is either comparable with or much longer than that for chemical reaction, the kinetics of adsorption will reflect *transport control,* not *reaction control. Rate laws must then be formulated with parameters that represent physical, not chemical, processes.*

This point can be appreciated more quantitatively after consideration of an important (but simple) model of transport-controlled kinetics: the *film diffusion process.* This process involves the movement of chemical species from a bulk aqueous solution phase through a quiescent boundary layer (*Nernst film*) to a particle surface. The thickness of this boundary layer, δ, will be larger for particles that bind water strongly and smaller for aqueous solution phases that are well stirred. The film diffusion mechanism in cation exchange is based on two assumptions: (1) that a thin film of inhomogeneous solution separates an exchanger particle surface from a homogeneous bulk solution and (2) that the exchanging cations diffuse across the film much more rapidly than the concentrations of these ions change in the bulk solution.

If we call j the rate at which a chemical species like Ca^{2+} arrives at a particle surface, expressed per unit area of the latter (termed the *flux* to the particle surface, in units of moles per square meter per second), and if diffusion is the mechanism by which the species makes its way through the boundary layer, the *Fick rate law* can be invoked to describe the process:

$$j = \frac{D}{\delta}([i]_{bulk} - [i]_{surf}) \tag{S.3.1}$$

where $[i]_{bulk}$ is the concentration of species i in the bulk aqueous solution phase, $[i]_{surf}$ is its concentration at the boundary layer–particle interface, and D is its *diffusion coefficient* (units of square meters per second) in the boundary layer. The Fick rate law is based on the premise that a difference in adsorptive concentration across the boundary layer "drives" the adsorptive to move through the layer. The transport parameter D is a quantitative measure of the effectiveness of the processes that respond to the concentration difference to bring the species to the particle surface. The rate of adsorption (units of moles per cubic meter per second) based on Eq. S.3.1 is equal to the product of the flux, the particle specific surface area, and the particle concentration in the aqueous solution phase:

$$\text{rate of adsorption } (\textit{physical}) = (\mathrm{D}a_s c_s/\delta)([\mathrm{i}]_{\mathrm{bulk}} - [\mathrm{i}]_{\mathrm{surf}}) \qquad (\text{S.3.2a})$$

where a_s is the specific surface area of the particle and c_s is its concentration in the aqueous phase.

A simple rate law for the binding of the species i to a single reactive site on a particle surface can be developed as the difference between a rate of adsorption, proportional to the concentration of i at the boundary layer–particle interface and to that of the reactive site, and a rate of desorption proportional to the concentration of the bound species:

$$\text{Rate of adsorption } (\textit{chemical}) = k_{ads}[\mathrm{i}]_{\mathrm{surf}}[\equiv\mathrm{SO}^-] - k_{des}[\equiv\mathrm{SOi}] \qquad (\text{S.3.2b})$$

where k_{ads} and k_{des} are *rate coefficients* for the two opposing chemical processes. If mass is to be conserved during the overall adsorption process, the right sides of Eqs. S.3.2a and S.3.2b must be equal:

$$(\mathrm{D}a_s c_s/\delta)([\mathrm{i}]_{\mathrm{bulk}} - [\mathrm{i}]_{\mathrm{surf}}) = k_{ads}[\mathrm{i}]_{\mathrm{surf}}[\equiv\mathrm{SO}^-] - k_{des}[\equiv\mathrm{SOi}] \qquad (\text{S.3.3})$$

The film diffusion process thus supplies species i at a rate that is matched by the subsequent chemical reaction through adjustment of the value of $[\mathrm{i}]_{\mathrm{surf}}$ to a steady-state value determined by the mass balance condition in Eq. S.3.3:

$$[\mathrm{i}]_{\mathrm{surf}} = \frac{k_{diff}[\mathrm{i}]_{\mathrm{bulk}} + k_{des}[\equiv\mathrm{SOi}]}{k_{diff} + k_{ads}[\equiv\mathrm{SO}^-]} \qquad (\text{S.3.4})$$

where

$$k_{diff} \equiv \mathrm{D}a_s c_s/\delta \qquad (\text{S.3.5})$$

is a *film diffusion rate coefficient.* Equation S.3.4 can be substituted into either of Eqs. S.3.2a or S.3.2b to calculate the overall rate of adsorption. If Eq. S.3.2a is selected, the final result is

$$\text{rate of adsorption} = k_{diff}\left[\frac{k_{ads}[\mathrm{i}]_{\mathrm{bulk}}[\equiv\mathrm{SO}^-] - k_{des}[\equiv\mathrm{SOi}]}{k_{diff} + k_{ads}[\equiv\mathrm{SO}^-]}\right] \qquad (\text{S.3.6})$$

A comparison between the kinetics of film diffusion and chemical reaction can be made by examining the denominator in Eq. S.3.6. Under the condition $k_{diff} \gg k_{ads}[\equiv\mathrm{SO}^-]$, transport through the boundary layer is much more rapid than the adsorption reaction, and Eq. S.3.6 takes the approximate form:

$$\text{rate of adsorption} \underset{k_{diff}\uparrow\infty}{\sim} k_{ads}\,[\mathrm{i}]_{\mathrm{bulk}}\,[\equiv\mathrm{SO}^-] - k_{des}\,[\equiv\mathrm{SOi}] \qquad (\text{S.3.7})$$

which is like the rate law appearing in Eq. S.3.2b, but is expressed in terms of the *bulk* concentration of the species i. In this limiting case, the kinetics are fully *reaction controlled.* Under the opposite condition, $k_{diff} \ll k_{ads}[SR]$, transport through the boundary layer is very slow compared with the chemical reaction, and Eq. S.3.6 takes the approximate limiting form

$$\text{rate of adsorption} \underset{k_{diff}\downarrow\infty}{\sim} k_{diff}\left([i]_{bulk} - \frac{k_{des}\,[\equiv SOi]}{k_{ads}\left[\equiv SO^{-}\right]}\right) \qquad (S.3.8)$$

The significance of the second term on the right side of Eq. S.3.8 is seen after setting the left side of Eq. S.3.2b equal to zero and solving for $[i]_{surf}$:

$$[i]_{surf}^{eq} = \frac{k_{des}\,[\equiv SOi]}{k_{ads}\left[\equiv SO^{-}\right]} \qquad (S.3.9)$$

which gives the concentration of i produced at the particle surface when the adsorption–desorption reaction has come to equilibrium (rate = 0). Thus Eq. S.3.8 can be expressed in the more useful form

$$\text{rate of adsorption} \underset{k_{diff}\downarrow\infty}{\sim} k_{diff}\left([i]_{bulk} - [i]_{surf}^{eq}\right) \qquad (S.3.10)$$

In this limiting case, the chemical reaction produces a steady value of $[i]_{surf}$ and the kinetics are wholly *transport controlled.* Measurement of the rate of adsorption accordingly would provide little or no chemical information about the process.

The rate law in Eq. S.3.10 is of the mathematical form that leads to an exponential time dependence of $[i]_{bulk}$, with a time derivative that may be equated to minus the rate at which the bulk concentration of species i decreases as it engages in adsorption. The rate coefficient k_{diff} is related to the *half-life* for the exponential decline in $[i]_{bulk}$ through the conventional expression

$$t_{1/2} = \frac{0.693}{k_{diff}} = 0.693\left(\frac{\delta}{D a_s c_s}\right) \qquad (S.3.11)$$

Typical ranges of value for the parameters on the right side of Eq. S.3.11 are $a_s = 10^2$ to 10^3 m^2 kg^{-1}, $D = 10^{-9}$ m^2 s^{-1}, $\delta = 10^{-7}$ to 10^{-5} m, and $c_s = 10^{-2}$ to 10^{-1} kg m^{-3}. These values lead to $t_{1/2}$ on the order of seconds to hours.

4

The Soil Solution

4.1 Sampling the Soil Solution

The soil solution was introduced in Section 1.4 as a liquid water repository for dissolved solids and gases. Speaking more precisely, one can define the soil solution as the aqueous liquid phase in soil with a composition that is influenced by exchanges of matter and energy with soil air, soil solid phases, the biota, and the gravitational field of the earth (Fig. 1.2). This more precise chemical concept identifies the soil solution as an *open system* (Section 1.1), and its designation as a *phase* means two things: (1) that it has uniform macroscopic properties (e.g., temperature and composition) and (2) that it can be isolated from the soil profile and investigated experimentally in the laboratory.

Uniformity of macroscopic properties obviously cannot be attributed to the entire aqueous phase in a soil profile, but instead to a sufficiently small element of volume in the profile (e.g., a soil ped or clod). Spatial variability in the chemical properties of the soil solution at the pedon or landscape scale is axiomatic, and temporal variability, even in a volume element the size of a ped, is commonplace because of diurnal fluctuations and seasonal changes punctuated by direct influence of the biota. On both larger and smaller timescales than those typified by the variability of solar inputs, temporal variation in the properties of the soil solution also occurs because of the kinetics of its chemical reactions.

The problem of isolating a sample of the soil solution without artifacts (a much more difficult task than isolation of a sample from the water column in a river or lake) has not yet been solved, but several techniques for removing the aqueous phase from soil to the laboratory have been established as operational compromises between chemical verisimilitude and analytical convenience. Among these techniques, the most widely applied in situ methods are drainage water collection and vacuum extraction, whereas the common ex situ methods include fluid displacement and extraction by vacuum, applied pressure, or centrifugation. The in situ techniques are influenced by whatever disturbance to a soil profile and, therefore, natural aggregate structure and water flow patterns, has occurred because of their installation. More subtly, they yield a sample of the soil solution that has a largely undefined "support volume" (the multiply connected, three-dimensional soil unit with pore space that provides the aqueous sample), and they differ in whether they provide the *flux composition* or the *resident composition* of a soil solution. A flux composition, which is relevant to long-term chemical weathering (Section 1.5) and, more broadly, to solute transport in soils, is measured in an aqueous sample obtained by natural flow of the soil solution into a collector (e.g., a pan lysimeter). A resident composition, which is relevant to nutrient uptake by the biota in soil (Fig. 1.2), is obtained by removing an aqueous sample instantaneously into a collector, an operation that can be only approximated by vacuum extraction. If for no other reason than the difference in the regions of pore space sampled (e.g., the macropores vs. the macropores plus mesopores), the flux composition will usually deviate significantly from the resident composition of a soil solution. This deviation can become acute if a soil profile receives periodic intense throughputs of water or exhibits a pronounced soil structure with its attendant spectrum of timescales over which water carries solutes around and within aggregates.

The ex situ methods perforce sample a disturbed soil, even if they use soil cores, but they inherently permit more control with regard to the sampling of the water-containing pore space. Fluid displacement utilizes either a miscible solution replacing the indigenous soil solution as it flows down a column, or a dense, unreactive immiscible liquid that replaces soil solution while being forced through a soil sample by centrifugation. High yield and low contamination of the soil solution sample, which need not be water saturated, are possible with this method. In the vacuum extraction method, the aqueous phase of a soil (in situ, as discussed earlier, or a disturbed sample saturated previously with water in the laboratory) is withdrawn through a filter by vacuum. This method suffers from both negative and positive interferences caused by the filter (principally from adsorption–desorption reactions with dissolved constituents) when the extracted solution passes through it. There are also uncertainties associated with the effect of vacuum extraction on the chemical reactions between dissolved constituents and soil solid phases. If the soil sample has been saturated with water prior to extraction, the composition of the extract also may differ considerably from that of a soil solution at ambient water content. Despite these difficulties—which are shared with the applied pressure extraction and centrifugation methods—the vacuum extraction technique, once standardized, is convenient for routine analyses. It usually provides aqueous solutions with a composition that reflects something of the reactions between the soil solution and solid soil constituents that occur in nature.

For any of the common methods of obtaining soil solutions, however, there is still the problem of the inherent *porescale heterogeneity* in soil aqueous phases caused by the electrical charge on soil particles, discussed in sections 2.3, 2.4, and 3.3. This charge creates poorly defined zones of accumulation or depletion of ions in the soil solution near soil particle surfaces, with accumulation occurring for ions with a charge sign that is opposite that of the neighboring surface, and exclusion for those with a charge sign that is the same. Because of this phenomenon, successive increments of, say, a vacuum-extracted soil solution that represent different regions near soil particle surfaces will not have the same composition.

Standard laboratory procedures have been compiled in *Methods of Soil Analysis* (see "For Further Reading" at the end of this chapter) for the determination of the chemical composition of extracted soil solutions. These data, which provide total concentrations of dissolved (i.e., filterable under designated conditions) constituents, pH, conductivity, and so on, make up the primary information needed for the quantitative description of soil solutions, at known temperature and pressure, according to the principles of chemical kinetics and thermodynamics.

4.2 Soluble Complexes

A *complex* is a unit comprising a central ion or molecule that is bound to one or more other ions or molecules such that a stable molecular association is maintained. Examples of soluble complexes formed in the soil solution are given frequently throughout Chapters 1 to 3, mainly as proton complexes in which anions are the central species and protons are the binding species (e.g., bicarbonate, HCO_3^-, mentioned in Section 1.4 as a principal chemical form of C found in soil solutions). Other important soluble complexes mentioned prominently are the mineral weathering products $Si(OH)_4^0$, silicic acid, and $AlC_2O_4^+$, an oxalate complex of Al^{3+}. In these latter complexes, OH^- and $C_2O_4^{2-}$ are the binding species, termed *ligands,* as noted in Section 1.5 in conjunction with the definition of complexation as a key mineral weathering reaction (Eq. 1.4). (*Ligand* is usually applied solely to binding species that are anions or neutral molecules, but it is applicable as well to cationic binding species like the protons in bicarbonate or $H_2PO_4^-$.) If two or more functional groups in a single ligand are bound to a metal cation to form a complex, it is termed a *chelate.* The $AlC_2O_4^+$ species is a chelate in which two COO^- groups in the oxalate ligand are bound to Al^{3+}. The chelates of Fe^{3+} formed by siderophores (Section 3.1) involve three functional groups in the ligand and are especially stable. The propensity of a ligand to coordinate around a metal cation using multiple donor atoms is called its *denticity.* Trihydroxamate siderophores coordinate around a metal cation using both of the O atoms in each of their –O–N–C=O functional groups. Because there are three hydroxamate groups in these ligands, they are hexadentate, which is optimal for the octahedral coordination with O preferred by most metal cations of interest in soils (Section 2.1). *As a general rule, the higher its denticity, the more likely a ligand is to form a very strong complex with a metal cation.* Note that denticity is strictly a property of ligands, not the complexes they form. Siderophores are almost always hexadentate ligands, but their complexes with metal cations are not termed hexadentate. The appropriate term for the complex is based on the coordination number of its metal cation center, which is octahedral in the case of most metal–siderophore complexes. Thus, for example, trihydroxamate siderophores are hexadentate ligands that form octahedral complexes with Fe^{3+}.

If the central ion or molecule and the ligands forming a complex are in direct contact, the complex is termed *inner-sphere,* whereas if one or more water molecules is interposed between the central ion or molecule and the ligands bound to it, the complex is *outer-sphere.* These two concepts were applied in Section 3.5 to bridging complexes between organic ligands and metal cations adsorbed on a mineral surface (figs. 3.5 and 3.6), and to the complex formed through ligand exchange (Eq. 3.12) between an organic ligand and a metal cation bound into a mineral structure. Similarly, the soluble complex $AlC_2O_4^+$, which predominates at low concentrations in acidic oxalate solutions, turns out to be inner-sphere, as is the complex that forms between oxalate and Al(III) bound into the structure of the Al oxide corundum (Section 2.1), the result of a ligand exchange reaction. The octahedral complex between Al^{3+} and water molecules, $Al(H_2O)_6^{3+}$, also is inner-sphere, but conventionally the term *solvation complex* is applied to it instead. The *free-ion* species introduced in Section 1.1 and represented throughout Chapters 1 through 3 by notation such as Al^{3+} or NO_3^-, are actually solvation complexes, reflecting the ubiquitous interactions between charged species and water molecules (dipoles) in aqueous solution (Section 1.2). Inner-sphere complexes usually are much more stable than outer-sphere complexes because the latter cannot easily involve ionic or covalent bonding (Section 2.1) between the central metal cation and ligand, whereas the former can. Thus the "driving force" for inner-sphere complexes is the energy gained through strong bond formation between the central metal cation and ligand. For outer-sphere complexes, the energy gain from bond formation is not so large and the driving force instead involves the disorder induced in the coordination shell about the central metal cation by the binding of the ligand, such as the disruption of the hydration shell that occurs when an anion coordinates to a metal cation through its solvation complex to form an electrostatic bond.

Table 4.1 lists the principal metal complexes found in well-aerated soil solutions. The ordering of free-ion and complex species in each row from left to right is roughly according to decreasing concentration typical for either acidic or alkaline soils. A normal soil solution will easily contain 100 to 200 different soluble complexes, many of them involving metal cations. The main effect of pH on these complexes, evident in Table 4.1, is to favor free metal cations and protonated anions at low pH, and carbonate or hydroxyl complexes at high pH.

Metal complex formation is typically a very fast reaction (microsecond to millisecond timescales) if humus ligands are not involved. Other complexation reactions of importance in soil solutions, however, exhibit slower reaction kinetics. A useful example is provided by the reaction of dissolved CO_2 with water to form the neutral proton complex $H_2CO_3^0$ (Problem 15 in Chapter 1):

$$CO_2 + H_2O(\ell) = H_2CO_3^0 \tag{4.1}$$

where both CO_2 and $H_2CO_3^0$ are *dissolved* species. (The species denoted CO_2 is a *free-molecule* species, a solvation complex of C.) The net rate of formation of $H_2CO_3^0$ can be expressed mathematically by the time derivative of its concentration, $d[H_2CO_3^0]/dt$, where the square brackets represent a concentration in moles per cubic decimeter (liter). It is common practice to *assume* that the observed rate of a reaction like complex formation can be modeled

Table 4.1
Principal metal species in soil solutions.

	Principal species	
Cation	Acidic soils	Alkaline soils
Na^+	Na^+	Na^+
Mg^{2+}	Mg^{2+}	Mg^{2+}
Al^{3+}	org[a], $Al(OH)_n^{3-n}$	$Al(OH)_4^-$
Si^{4+}	$Si(OH)_4^0$	$Si(OH)_4^0$
K^+	K^+	K^+
Ca^{2+}	Ca^{2+}	Ca^{2+}, $CaHCO_3^+$, org[a]
Cr^{3+}	$CrOH^{2+}$	$Cr(OH)_4^-$
Cr^{6+}	$HCrO_4^-$	CrO_4^{2-}
Mn^{2+}	Mn^{2+}	Mn^{2+}, $MnHCO_3^+$
Fe^{2+}	Fe^{2+}	$FeCO_3^0$, Fe^{2+}, $FeHCO_3^+$
Fe^{3+}	$FeOH^{2+}$, $Fe(OH)_3^0$, org[a]	$Fe(OH)_3^0$, org[a]
Ni^{2+}	Ni^{2+}	$NiCO_3^0$, $NiHCO_3^+$, Ni^{2+}
Cu^{2+}	org[a]	$CuCO_3^0$, org[a]
Zn^{2+}	Zn^{2+}	$ZnHCO_3^+$, org[a], Zn^{2+}
Mo^{6+}	$HMoO_4^-$	$HMoO_4^-$, MoO_4^{2-}
Cd^{2+}	Cd^{2+}, $CdCl^+$	Cd^{2+}, $CdCl^+$, $CdHCO_3^+$
Pb^{2+}	Pb^{2+}, org[a]	$PbCO_3^0$, $PbHCO_3^+$, org

[a]Organic complexes.

mathematically by the difference of two terms:

$$\frac{d\left[H_2CO_3^0\right]}{dt} = R_f - R_b \tag{4.2}$$

where R_f and R_b each are functions of the composition of the solution in which the reaction in Eq. 4.1 takes place, as well as being dependent on temperature and pressure. It is to be emphasized that Eq. 4.2 need not have any direct relationship to the mechanism by which $H_2CO_3^0$ actually forms. For example, there may be intermediate chemical species that do not appear in the reaction in Eq. 4.2, but nonetheless help to determine the observed rate and prevent it from being modeled mechanistically by a simple difference expression. Whenever Eq. 4.2 is appropriate, however, R_f and R_b are interpreted as the respective rates of formation (*forward reaction*) and dissociation (*backward reaction*) of $H_2CO_3^0$. It is common practice also to *assume* that these two rates are proportional to powers of the concentrations of the reactants and products in the reaction (Eq. 4.1):

$$\frac{d\left[H_2CO_3^0\right]}{dt} = k_f\,[CO_2]^{\alpha}\,[H_2O]^{\beta} - k_b\left[H_2CO_3^0\right]^{\delta} \tag{4.3a}$$

where k_f, k_b, α, β, and δ are parameters. The exponents α, β, and δ are each termed the *partial order* of the reaction in Eq. 4.1 with respect to the associated species (e.g., αth order with respect to CO_2). The sum $(\alpha + \beta)$ is the *order of the forward reaction,* whereas δ is the *order of the backward reaction.* The parameters k_f and k_b are the *rate coefficients* of the formation (forward) and dissociation (backward) reactions respectively. Each of the five parameters in Eq. 4.3a may depend on solution composition, temperature, and pressure. Note that the units of the two rate coefficients will depend on the values of the partial orders of the reaction.

Equation 4.3a is termed a *rate law,* a mathematical model of the net rate of a reaction containing parameters that must be determined experimentally. Partial reaction orders can be measured directly by observing the dependence of the rate on the concentration of a reactant or product in a series of experiments designed to maintain that concentration at a predetermined value (e.g., reactant added in large excess relative to other species in the reaction). In the particular case of Eq. 4.3a, the reactant $H_2O(\ell)$ is always at a much higher concentration (55.4 mol dm^{-3}) than is CO_2, and the rate law is conventionally simplified by combining the H_2O concentration with the forward rate coefficient:

$$\frac{d\left[H_2CO_3^0\right]}{dt} = k_f^*\,[CO_2]^{\alpha} - k_b\left[H_2CO_3^0\right]^{\delta} \tag{4.3b}$$

where $k_f^* \equiv k_f[H_2O]^{\beta}$ is termed a *pseudo rate coefficient.* In this model form, k_b is a *δ-order backward rate coefficient* and k_f^* is a *pseudo α-order forward rate coefficient.*

Rate laws are often simplified further by *assuming* that a partial reaction order is the same as the stoichiometric coefficient of the associated chemical species in a reaction. In the current example, this assumption yields $\alpha = \delta = 1$ [i.e., the (pseudo) forward and the backward rate coefficients are both first order]:

$$\frac{d\left[H_2CO_3^0\right]}{dt} = k_f^* \left[CO_2\right] - k_b \left[H_2CO_3^0\right] \tag{4.3c}$$

This formulation is useful because it permits a constraint to be imposed on the two rate coefficients. At equilibrium, the left side of Eq. 4.3c is equal to zero and the equation can be rearranged to yield

$$\frac{k_f^*}{k_b} = \frac{\left[H_2CO_3^0\right]_e}{\left[CO_2\right]_e} \equiv K_c \tag{4.4}$$

where $[\ ]_e$ is a concentration measured *at equilibrium*. The parameter defined by the ratio of equilibrium concentrations is called a *conditional equilibrium constant* for the reaction. It is "conditional" because it depends on solution composition, temperature, and pressure, just as the two rate coefficients do. At a given equilibrium solution composition, temperature, and pressure, K_c can be measured independently of kinetics and, therefore, applied to constrain the values of the rate coefficients as indicated in Eq. 4.4.

Measured values of the two rate coefficients in Eq. 4.3c at 25 °C in pure water range from 0.025 to 0.040 s^{-1} for k_f^* and from 10 to 28 s^{-1} for k_b. As discussed in Special Topic 3 (Chapter 3), each of these two first-order rate coefficients defines a half-life or *intrinsic timescale* for the process it represents (either formation or dissociation of $H_2CO_3^0$). The timescale for dissociation follows from Eq. 4.3c after dropping the first term on the right side: $t_{1/2}$ (dissociation) $= 0.693/k_b$. The timescale for the formation reaction follows similarly after dropping the second term on the right side of Eq. 4.3c and rewriting the left side as $-d\,[CO_2]/dt$, as implied by Eq. 4.1: $t_{1/2}(\text{formation}) = 0.693/k_f^*$. Evidently the intrinsic timescale for the formation of $H_2CO_3^0$ ranges from 17 to 27 s, whereas that for dissociation of the complex is much smaller, ranging from 27 to 70 ms. These data indicate that the complex $H_2CO_3^0$ is labile relative to the reactant species, hydrated CO_2. According to Eq. 4.4, the conditional stability constant for the formation of $H_2CO_3^0$ at 25 °C should range in value from 1 to 4×10^{-3}; directly measured values range from 1.0 to 2.9×10^{-3}. It follows from the value of K_c and Eq. 4.4 that about 99.7% of carbonic acid, $H_2CO_3^*$, which comprises both hydrated CO_2 and the neutral complex $H_2CO_3^0$, is in fact hydrated CO_2.

The concept of a half-life (or intrinsic timescale) can be extended to reactions that are not first order. Table 4.2 summarizes graphical relationships that produce straight lines for concentration measured as a decreasing function of time during a chemical reaction that is far from equilibrium. The model

Table 4.2
Graphical analysis of Eq. 4.5.

Reaction order (b)	Plotting variables	Slope	y-Intercept	Half-life[a]
Zero	[A] vs. time	–K	$[A]_0$	$[A]_0/2K$
One	ln[A] vs. time	–K	$\ln[A]_0$	0.693/K
Two	l/[A] vs. time	+K	$1/[A]_0$	$1/K[A]_0$

[a]Valid only for positive-valued K, with $[A]_0$ equal to the initial concentration of species A.

rate law underlying the graphical relationships has the generic form

$$-\left(\frac{d\,[A]}{dt}\right) = K\,[A]^b\ (K > 0) \tag{4.5}$$

where A is a chemical species and b is the partial reaction order. Note that the parameter K in Eq. 4.5 may be a pseudo b-order rate coefficient, the product of a higher order rate coefficient with a concentration (maintained constant during an experiment) raised to a power. The parameter b, like α, β, or δ in Eq. 4.3a, need not be the same as the stoichiometric coefficient of species A in the chemical reaction investigated, because rate laws are strictly empirical. Table 4.2, then, lists the half-life of a reaction according to its order. This parameter is equal to the time required for the concentration of species A to decrease to one half its initial value.

4.3 Chemical Speciation

The total concentrations of dissolved constituents in a soil solution represent the sum of "free" (i.e., solvation complex) and complexed forms of the constituents that are stable enough to be considered well-defined chemical species. The distribution of a given constituent among its possible chemical forms can be described with conditional stability constants, like that in Eq. 4.4, if complex formation and dissociation reactions are at equilibrium. This requirement of stable states is often met on timescales of interest in natural soils: both ion exchange (Section 3.3) and soluble-complex formation are usually fast reactions. On the other hand, certain oxidation–reduction and precipitation–dissolution reactions are so unfavorable kinetically that the reactants can be assumed to be perfectly stable species on the timescale of a laboratory or a field experiment. But these generalizations can fail in important special cases. The half-lives for metal complex formation and dissociation reactions in aqueous solution at concentrations typical for soils actually range over about 15 orders of magnitude, from 10^{-9} s for the dissociation of outer-sphere complexes to 10^6 s for the formation of certain inner-sphere complexes. The two extremes of this spectrum of timescales present no practical limitations on the applicability of conditional stability constants to soil solutions, whereas the range of

10^2 to 10^4 s (e.g., the formation of the inner-sphere complex AlF^{2+}) requires careful consideration of equilibration timescales.

The way in which conditional stability constants are used to calculate the distribution of chemical species can be illustrated conveniently by consideration of the forms of dissolved Al in an acidic soil solution. Suppose that the pH of the soil solution is 4.0 and that the total concentration of Al is 10 mmol m^{-3}. The concentrations of the complex-forming ligands sulfate and oxalate have the values 50 and 10 mmol m^{-3} respectively. The significant complexes between these ligands and Al are $AlSO_4^+$ and $ALOx^+$, where Ox refers to oxalate (see Eq. 1.4 and Section 3.1). These complexes are not the only ones formed with Al, SO_4, or Ox, nor are the two ligands the only ones that form Al complexes in the soil solution, but they will serve to introduce chemical speciation calculations in a relatively simple manner.

According to the speciation concept, the total concentration of Al (Al_T, as determined, for example, by atomic emission spectrometry or by the 8-hydroxyquinoline method to exclude polymeric species) is the sum of free and complexed forms:

$$Al_T = [Al^{3+}] + [AlOH^{2+}] + [AlSO_4^+] + [AlOx^+] \tag{4.6}$$

where the square brackets denote species concentrations in moles per cubic decimeter (liter). (The hydroxy species $AlOH^{2+}$ is also an important one at pH 4.) Each of the complex species in Eq. 4.6 can be described by a conditional stability constant:

$$K_{1c} = \frac{[AlOH^{2+}]}{[Al^{3+}][OH^-]} \approx 10^{8.5} mol^{-1} \ dm^3 \tag{4.7a}$$

$$K_{2c} = \frac{[AlSO_4^+]}{[Al^{3+}][SO_4^{2-}]} \approx 10^{3.9} mol^{-1} \ dm^3 \tag{4.7b}$$

$$K_{3c} = \frac{[AlOx^+]}{[Al^{3+}][Ox^{2-}]} \approx 10^{6.0} mol^{-1} \ dm^3 \tag{4.7c}$$

Common to each of the stability constant expressions is the concentration of the free-ion species Al^{3+}. Therefore, Eq. 4.6 can be factorized in the form

$$\begin{aligned} Al_T &= [Al^{3+}]\left\{1 + \frac{[AlOH^{2+}]}{[Al^{3+}]} + \frac{[AlSO_4^+]}{[Al^{3+}]} + \frac{[AlOx^+]}{[Al^{3+}]}\right\} \\ &= [Al^{3+}]\{1 + K_{1c}[OH^-] + K_{2c}[SO_4^{2-}] + K_{3c}[Ox^{2-}]\} \end{aligned} \tag{4.8}$$

The ratio of $[Al^{3+}]$ to Al_T, termed the *distribution coefficient* for the species Al^{3+}, can be calculated with Eq. 4.8 if the concentrations of the *free-ion species* of the four complexing ligands are known or can be estimated:

$$\alpha_{Al} \equiv \frac{[Al^{3+}]}{Al_T} = \{1 + K_{1c}[OH^-] + K_{2c}[SO_4^{2-}] + K_{3c}[Ox^{2-}]\}^{-1} \quad (4.9)$$

For OH^-, one can readily estimate the free-ion concentration using the pH value:

$$[OH^-] = \frac{K_{wc}}{[H^+]} \approx \frac{10^{-14}}{10^{-pH}} = 10^{pH-14} \text{ mol dm}^{-3} \quad (4.10)$$

where K_{wc} is the ionization product of liquid water under the conditions that exist in soil solution (hence the subscript c). For dilute solutions at 25 °C and under 1 atm pressure, $K_{wc} \approx 10^{-14}$mol^2 dm^{-6} and $[H^+] \approx 10^{-pH}$ *numerically.* Thus, $[OH^-] \approx 10^{-10}$mol dm^{-3} in the current example (pH 4).

For the other ligands in Eq. 4.9, the free-ion concentrations cannot be calculated so directly. Given the large value of K_{3c} relative to K_{2c}, it is reasonable to expect that $[AlOx^+]$ will be nearly equal to Al_T and Ox_T in the current example. Thus, in a first approximation, Eq. 4.7c can be used to estimate α_{Al}:

$$\frac{[AlOx^+]}{[Al^{3+}][Ox^{2-}]} = \frac{\alpha_{AlOx}}{\alpha_{Al}\alpha_{Ox}Ox_T} \approx \frac{1}{\alpha_{Al}^2 Ox_T} = K_{3c} = 10^{6.0} \text{ dm}^3 \text{ mol}^{-1} \quad (4.11)$$

where

$$\alpha_{AlOx} \equiv \frac{[AlOx^+]}{Al_T} \qquad \alpha_{Ox} \equiv \frac{[Ox^{2-}]}{Ox_T} \quad (4.12a)$$

are the distribution coefficients for $AlOx^+$ and Ox^{2-}, respectively, and Ox_T is the total oxalate concentration. In Eq. 4.11, it has been assumed that $\alpha_{AlOx} \approx 1$ and $\alpha_{Al} \approx \alpha_{Ox}$, with the result that

$$\alpha_{Al}^2 \approx (K_{3c}Ox_T)^{-1} = 10^{-1}$$

and $\alpha_{Al} \approx 0.32$. Thus, about 30% of Al_T is in the form of Al^{3+}. This approximate result can be used to estimate the distribution coefficients for each inorganic complex:

$$\alpha_{AlOH} \equiv \frac{[AlOH^{2+}]}{Al_T} = \alpha_{Al}\frac{[AlOH^{2+}]}{[Al^{3+}]} = \alpha_{Al} K_{1c}[OH^-] \approx 10^{-2} \quad (4.12b)$$

$$\alpha_{AlSO4} \equiv \frac{[AlSO_4^+]}{Al_T} = \alpha_{Al}\frac{[AlSO_4^+]}{[Al^{3+}]} = \alpha_{Al} K_{2c}[SO_4^{2-}] \approx 0.13 \quad (4.12c)$$

where the free-ion sulfate concentration has been equated with the total sulfate concentration in Eq. 4.12c.

The assumption that $\alpha_{AlOx} \approx 1$ is not consistent with the large value estimated for α_{Al}. This estimate can be refined by considering the *ligand speciation* in more detail:

$$SO_{4T} = [SO_4^{2-}] + [AlSO_4^+] = [SO_4^{2-}]\{1 + K_{2c}[Al^{3+}]\} \tag{4.13a}$$

$$Ox_T = [Ox^{2-}] + [AlOx^+] = [Ox^-]\{1 + K_{3c}[Al^{3+}]\} \tag{4.13b}$$

where use has been made again of Eqs. 4.7a through 4.7c. Given $[Al^{3+}] = \alpha_{Al}Al_T \approx 3.2 \times 10^{-6}$ mol dm^{-3}, the ligand distribution coefficients are estimated as

$$\alpha_{SO_4} \equiv \frac{[SO_4^{2-}]}{SO_{4T}} = \{1 + K_{2c}[Al^{3+}]\}^{-1} \approx 1.0 \tag{4.14a}$$

$$\alpha_{Ox} \equiv \frac{[Ox^{2-}]}{Ox_T} = \{1 + K_{3c}[Al^{3+}]\}^{-1} \approx 0.67 \tag{4.14b}$$

The revised value of α_{AlOx} that results from Eq. 4.14b is 0.67. Thus, about two thirds of Al_T is organically complexed and about one third either is complexed with inorganic ligands or is in the free-ion form, which is typical for acidic soil solutions containing dissolved organic ligands at concentrations comparable with Al_T.

This example, despite the approximate nature of the calculations, illustrates all of the salient features of a more exact chemical speciation calculation: *mass balance* (Eq. 4.6), *conditional stability constants* (Eq. 4.7), *distribution coefficients* (eqs. 4.12 and 4.14), and the *iterative refinement* of the distribution coefficients through additional mass balance on the ligands (eqs. 4.13 and 4.14). The approach illustrated can be applied to any soil solution for which the significant aqueous species and their conditional stability constants are known.

4.4 Predicting Chemical Speciation

The distribution of dissolved chemical species in a soil solution can be calculated if three items of information are available: (1) the measured total concentrations of the metals and ligands, along with a pH value; (2) the conditional stability constants for all possible complexes of the metals and H^+ with the ligands; and (3) expressions for the mass balance of each constituent in terms of chemical species (i.e., free ions and complexes). A flowchart outlining the method of calculation given these three items is shown in Figure 4.1.

Total concentration of the metals (M_T) and ligands (L_T), along with a pH value, are the basic input data for the calculation. They are presumed known for all important constituents of a soil solution. The speciation calculation then proceeds on the assumption that mass balance expressions like eqs. 4.6 and 4.13 can be developed for each metal and ligand. The mass balance expressions are converted into a set of coupled algebraic equations with the free-ion

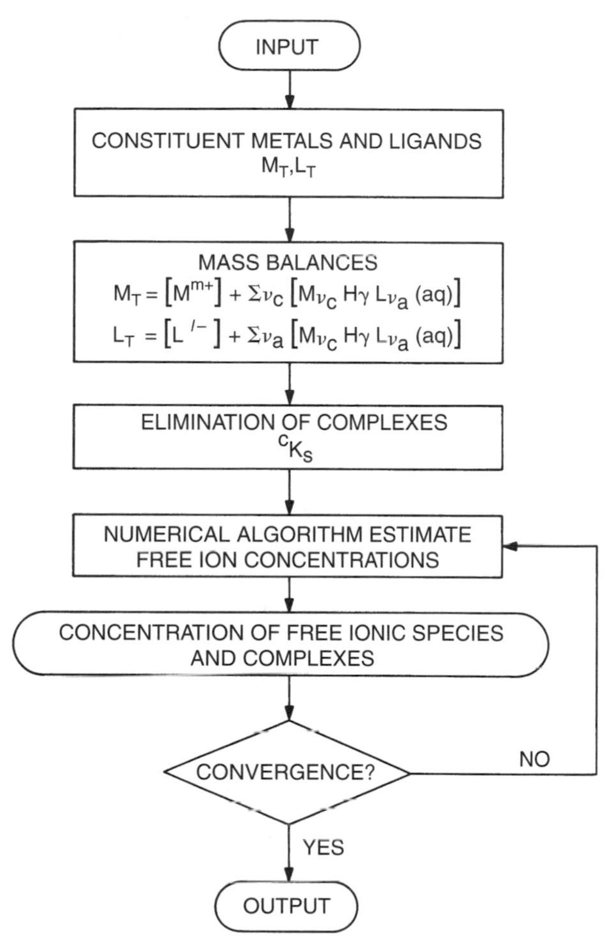

Figure 4.1. Flowchart outlining a chemical speciation calculation based on mass balance and the use of conditional stability constants for complex formation (Eq. 4.16).

concentrations as unknowns by substitution for the complex concentrations, as illustrated in Eq. 4.8. This step requires access to a database containing the values of all relevant conditional stability constants. In general, for the complex formation reaction

$$\nu_c M^{m+} + \gamma H^+ + \nu_\alpha L^{1-} = M_{\nu c} H_\gamma L_{\nu\alpha} \tag{4.15}$$

the conditional stability constant is

$$K_{sc} = \frac{[M_{\nu c} H_\gamma L_{\nu\alpha}]}{[M]^{\nu c} [H]^\gamma [L]^{\nu\alpha}} \tag{4.16}$$

where ν_c, γ, and ν_α are stoichiometric coefficients. Equation 4.16 can always be rearranged to express $[M_{\nu c}H_\gamma L_{\nu\alpha}]$ in terms of K_{sc} and the three free-ion concentrations. For example, the formation of the bicarbonate complex $CaHCO_3^+$ can be expressed by the reaction

$$Ca^+ + H^+ + CO_3^{2-} = CaHCO_3^+ \tag{4.17a}$$

for which $K_{sc} \approx 10^{11.5} dm^6\ mol^{-2}$ at 25 °C in a dilute soil solution. Thus, numerically,

$$10^{11.5} = \frac{[CaHCO_3^+]}{[Ca^{2+}][H^+][CO_3^{2-}]} \tag{4.17b}$$

and the concentration of the complex follows as

$$[CaHCO_3^+] = 10^{11.5}[Ca^{2+}][H^+][CO_3^{2-}] \tag{4.17c}$$

The algebraic equations with the free-ion concentrations as unknowns can be solved numerically by standard algorithms based on estimated or "guessed" values. The resulting free-ion concentrations then are used to calculate the complex concentrations with expressions like Eq. 4.17c. The calculated species concentrations are checked by introducing them into the mass balance equations to determine whether they sum numerically to the input total concentrations. If they do within some acceptable error (say, 0.01% difference from the input M_T or L_T), then the calculation is said to have converged and the speciation results may be accepted. If convergence has not been achieved, then the numerical calculation is repeated using the current speciation results to generate new input estimates for the free-ion concentrations in the numerical algorithm.

As a first example of a full chemical speciation calculation, one may return to the example introduced in Section 4.3, an aqueous solution comprising Al, SO_4, and oxalate at pH 4. The results of a calculation using the program MINEQL+ are shown in Table 4.3. (Note that *percentage speciation* is the same as the set of distribution coefficients for a metal or ligand, after multiplication of the coefficients by 100.) The numerical calculation involved consideration of a total of 13 soluble complexes, including three proton complexes of the two ligands. Table 4.3 indicates that the approximate calculation described in Section 4.3 is qualitatively correct: The complex $AlOx^+$ and the free ions Al^{3+}, SO_4^{2-}, and Ox^{2-} are the most important chemical species, as implied by Eqs. 4.12 and 4.14. However, the complex $AlSO_4^+$ is at a lower concentration than the estimated value because of competition from a second oxalate complex—$Al(Ox)_2^-$—not considered previously, which also has reduced the concentration of $AlOx^+$. The two oxalate complexes of Al^{3+} taken together account for about two thirds of Al_T and of Ox_T, as concluded from the simpler results of the estimation made in Section 4.3.

Table 4.3
Chemical speciation of an aqueous solution containing 10 mmol m^{-3} Al, 10 mmol m^{-3} oxalate, and 50 mmol m^{-3} sulfate at pH 4.[a]

Constituent	Percentage speciation
Al	$AlOx^+$ (50.5), Al^{3+} (29.5), $Al(Ox)_2^-$ (9.6), $AlSO_4^+$ (9.3)
Oxalate	$AlOx^+$ (50.5), Ox^{2-} (22.0), $Al(Ox)_2^-$ (19.1), HOx^- (7.8)
Sulfate	SO_4^{2-} (97.2), $AlSO_4^+$ (1.9)

[a] Speciation computed using MINEQL+ (www.mineql.com).

The species competition noted in connection with the interpretation of Table 4.3 brings to mind the questions of whether other metal cations in a soil solution would compete with Al^{3+} for oxalate ligands and whether other ligands than sulfate would compete with oxalate for Al^{3+}. These questions are addressed in Table 4.4, which shows the results of a speciation calculation performed using MINEQL+ and composition data for a Spodosol soil solution at pH 4.3 sampled by the immiscible fluid displacement technique (Section 4.1). Note that the total concentrations of the four additional metals are much larger than that of Al, and that, except for nitrate, the same is true of the total concentrations of the additional ligands when compared with oxalate. Despite this large difference in concentrations, the percentage speciation of Al and oxalate are rather similar in Table 4.4 to what appears in Table 4.3. The reason for this similarity can be appreciated by considering Eq. 4.7c applied to both Al^{3+} and Ca^{2+}, then noting that the ratio of the concentration of $AlOx^+$ to that of $CaOx^0$ is equal to the ratio of their respective conditional stability constants times the ratio of their respective free-ion concentrations:

$$\frac{[AlOx^+]}{[CaOx^0]} = \frac{K_c^{AlOx}}{K_c^{CaOx}} \frac{[Al^{3+}]}{[Ca^{2+}]} \tag{4.18}$$

Even if the ratio of $[Ca^{2+}]$ to $[Al^{3+}]$ is very large (e.g., something like 10), the ratio of $[AlOx^+]$ to $[CaOx^0]$ will not necessarily be small unless the conditional stability constants for the two complexes are comparable in value. In the current example, the conditional stability constant for the formation of $AlOx^+$ is $10^{6.0}$ mol^{-1} dm^3, whereas that for the formation of $CaOx^0$ is $10^{3.2}$ mol^{-1} dm^3. It follows that the Ca^{2+} concentration would have to be about three orders of magnitude larger than that of Al^{3+} before the concentration of $CaOx^0$ would even begin to approximate that of $AlOx^+$. The important point here is that the concentrations of metal complexes in soils depend not only on the concentration of the free metal ion (a *capacity* factor), but also on the conditional stability constant (an *intensity* factor).

The methodological approach outlined in Figure 4.1 is widely used to estimate the concentrations of metal and ligand species in extracted soil solutions as a basis for understanding the mobility and bioavailability of nutrients

Table 4.4
Composition and speciation of Spodosol soil solution (pH 4.3).[a]

Constituent	C_T (mmol m^{-3})	Percentage speciation
Ca	350	Ca^{2+} (94.5), $CaSO_4^0$ (4.2), $CaOx^0$ (1.1)
Mg	80	Mg^{2+} (96.0), $MgSO_4^0$ (3.4)
K	210	K^+ (100)
Na	130	Na^+ (100)
Al	25	$AlOx^+$ (42.4), $Al(Ox)_2^-$ (36.5), $AlSO_4^+$ (8.8), Al^{3+} (7.2), $Al(Ox)_3^{3-}$ (4.6)
SO_4	310	SO_4^{2-} (93.1), $CaSO_4^0$ (4.7)
Cl	820	Cl^- (100)
C_2O_4	50	$Al(Ox)_2^-$ (36.5), Ox^{2-} (23.1), $AlOx^+$ (21.2), $CaOx^0$ (7.9), $Al(Ox)_3^{3-}$ (6.9), HOx^- (3.7)
NO_3	20	NO_3^- (100)

[a] Speciation computed using MINEQL+ (www.mineql.com).

or toxicants. There are, however, several important limitations on chemical speciation calculations that should not be forgotten:

First, *pertinent chemical reactions and, therefore, important chemical species, may have been unintentionally omitted in formulating mass balance equations. Conditional stability constants for the species included in the mass balance equations may not be accurate, or in some other way may not be appropriate for soil solutions.* The compilation of stability constants by Smith and Martell [Smith, R. M., and A. E. Martell. (2001) *NIST standard reference database 46. Critically selected stability constants of metal complexes database.* U.S. Department of Commerce, Gaithersburg, MD.] is perhaps the most useful source of these parameters available. However, temperature and pressure variations may require attention. Significant temperature gradients exist in nearly all natural soils, but adequate data on the temperature dependence of conditional equilibrium constants may not, because most available databases refer to 25 °C. A major challenge also arises in respect to the suite of chemical species to be considered when metal complexation by dissolved humus must be included in a speciation calculation. Progress in meeting this difficult challenge has been reviewed carefully by Dudal and Gérard [Dudal, Y., and F. Gérard. (2004) Accounting for natural organic matter in aqueous chemical equilibrium models: A review of the theories and applications. *Earth Sci. Rev.* **66**:199.]. Although the biomolecules in humus, such as aliphatic organic acids and siderophores (Section 3.1), play very important roles in the chemical speciation of metal cations in soil solutions, dissolved and particulate humic substances often dominate the suite of organic ligands that influence metal solubility and bioavailability. The current picture of humic substances portrays them as supramolecular associations of many diverse components held together by hydrogen bonding and hydrophobic interactions (see Section 3.2 and Problem 2 in Chapter 3). Despite this molecular-scale complexity, the principal acidic functional groups in humic substances fall into just two classes—carboxyl and phenolic OH groups—and these two classes are likely to be important contributors to the metal-complexing properties of humic and fulvic acids. A key issue to be addressed, therefore, in developing a model of metal speciation that includes humic substances in the spirit of the approach taken in Section 4.2 is how to formulate metal cation interactions with carboxyl and phenolic OH groups to express the concentration of the resulting metal–humic substance complexes in terms of conditional stability constants and free-ion concentrations. When this key issue has been resolved, an appropriate mathematical relationship then can be substituted into a mass balance equation that includes the concentration of metal–humic substance complexes, just as Eq. 4.7c was substituted into the mass balance equation for Al (Eq. 4.6) to express the concentration of the Al–oxalate complex in terms of a conditional stability constant and free-ion concentrations.

Second, *analytical methods for the constituents in a soil solution may be inadequate to distinguish among various physical and chemical forms (e.g., dissolved vs. particulate, oxidized vs. reduced, monomeric vs. polymeric).* Laboratory methods that quantitate total elemental concentrations may inadvertently include particulate forms because of inadequate extraction of a soil solution (i.e., filterable forms of the element are included with truly dissolved forms). Specialized techniques are usually required to distinguish between elements in different oxidation states, as discussed in *Methods of Soil Analysis.* The problem of quantitating free-ion concentrations or the concentrations of specific complexes, which is especially challenging, has been reviewed by Kalis et al. [Kalis, E. J. J., W. Liping, F. Dousma, E. J. M. Temminghoff, and W. H. van Riemsdijk. (2006) Measuring free metal ion concentrations in situ in natural waters using the Donnan membrane technique. *Environ. Sci. Technol.* 40:955.] The studies in which free-ion concentrations of metals have been measured directly and compared with the results of chemical speciation calculations are few in number, but generally report good agreement between the two methodologies—say, within a factor of two over a broad concentration range of relevance to soils.

Third, *the kinetics of certain chemical reactions assumed to be at equilibrium on the basis of studies of simpler aqueous solutions may be retarded in soil solutions by the formation of intermediate species that do not exist in the simpler systems.* Oxidation–reduction reactions and mineral dissolution reactions can exhibit inherently slow kinetics in the absence of catalysis or in the presence of ligands that form exceptionally stable complexes. The situation becomes particularly complicated when the timescale of interest overlaps that of the kinetics of a reaction of interest, as can occur when the uptake of an element in the soil solution by the biota is investigated. Under these conditions, chemical speciation kinetics must be considered carefully, especially in regard to the *lability* of metal–ligand complexes (i.e., the degree to which they do not persist as stable molecular entities on timescales that are long compared with the timescale of interest). Dynamic chemical speciation methodologies have been reviewed carefully by van Leeuwen et al. [van Leeuwen, H. P., R. M. Town, J. Buffle, R. F. M. J. Cleven, W. Davison, J. Puy, W. H. van Riemsdijk, and L. Sigg. (2005) Dynamic speciation analysis and bioavailability of metals in aquatic systems. *Environ. Sci. Technol.* **39**:8545.]

Fourth, and last, *an equilibrium-based approach to chemical speciation may be a poor approximation for a particular soil solution because of flows of matter and energy in natural soils.* The appropriate time-invariant state in a soil solution may not be a state of equilibrium, but instead a steady state. Alternatively, mass balance equations may be affected by flows of matter over the timescales of interest in speciation calculation, transforming them from static to dynamic quantities that require considerations of mass transport. It is important in this respect to emphasize the essentially subjective—but critical—initial decision regarding the "free-body cut" when applying a mass balance approach (i.e., the choice of a closed model system that is to mimic the actual open system in nature).

4.5 Thermodynamic Stability Constants

Conditional stability constants, as the name implies, vary with the composition and total electrolyte concentration of the soil solution. For example, in a very dilute solution, the conditional stability constant for the formation of $CaHCO_3^+$ (Eq. 4.17) has the value 3.4×10^{11} dm^6 mol^{-2}. In a solution of 50 mol m^{-3} NaCl it is 0.70×10^{11} dm^6 mol^{-2}, and in 50 mol m^{-3} $CaCl_2$ it is 0.37×10^{11} dm^6 mol^{-2}. This variability requires the compilation of a different database each time a speciation calculation is performed, which is *not* an efficient approach to the problem!

Instead, concepts in chemical thermodynamics may be called on to define a *thermodynamic stability constant.* This parameter is *by definition* independent of chemical composition at a chosen temperature and pressure, usually 25 °C (298.15 K) and 1 atm. For the complex formation reaction in Eq. 4.15, the thermodynamic stability constant is defined by the equation

$$K_s \equiv (\mathbf{M_{\nu c}H_{\gamma}L_{\nu\alpha}})/(\mathbf{M})^{\nu c}(\mathbf{H})^{\gamma}(\mathbf{L})^{\nu\alpha} \qquad (4.19)$$

where boldface parentheses refer to the thermodynamic *activity* of the chemical species enclosed. Unlike K_{sc} in Eq. 4.16, K_s has a fixed value, regardless of the composition of the soil solution. To make this assertion a reality, the activity of a species is related to its concentration (in moles per cubic decimeter) through an *activity coefficient:*

$$(\mathbf{i}) \equiv \gamma_i[\mathrm{i}] \qquad (4.20)$$

where i is some chemical species, like Ca^{2+} or $MnSO_4^0$, and γ_i is its activity coefficient. By convention, γ_i has the units cubic decimeters per mole such that the activity has *no units* and the thermodynamic stability constant is *dimensionless.*

Conventions and laboratory methods have been developed to measure γ_i, (i), and K_s in electrolyte solutions. All species activity coefficients, for example, are required to approach the value 1.0 $dm^3\ mol^{-1}$ as a solution becomes infinitely dilute. Thus, in the limit of infinite dilution, activities become equal *numerically* to concentrations and K_{sc} becomes equal *numerically* to K_s. With Eqs. 4.16, 4.19, and 4.20, one can derive the relationship

$$\log\ K_s = \log\ K_{sc} + \log\left\{\gamma_{MHL}/\gamma_M^{\nu c}\gamma_H^{\gamma}\gamma_L^{\nu\alpha}\right\} \tag{4.21}$$

The second term on the right side of Eq. 4.21 must vanish in the limit of infinite dilution, so a graph of log K_{sc} against a suitable concentration function must extrapolate to log K_s at zero concentration. Experiment and theory have shown that a useful concentration function for this purpose is the *ionic strength, I:*

$$I = \frac{1}{2}\sum\nolimits_k Z_k^2\ [k] \tag{4.22}$$

where the sum is over all charged *species* (with valence Z_k) in a solution. The effective ionic strength is related closely to the conductivity of a solution. Experimentation with soil solutions has indicated that the *Marion–Babcock equation,*

$$\log I = 1.159 + 1.009 \log \kappa \tag{4.23}$$

is accurate for ionic strengths up to about 0.3 mol dm^{-3}. In Eq. 4.23, I is in units of moles per cubic meter, and κ, the conductivity, is in units of decisiemens per meter ($dS\ m^{-1}$ for a discussion of the units used, see the Appendix.)

Experimental and theoretical studies of electrolyte solutions have led to semiempirical equations that relate species activity coefficients to the effective ionic strength. For charged species (free ions or complexes), one uses the *Davies equation* (at 25 °C):

$$\log \gamma_i = -0.512\ Z_i^2\left[\frac{\sqrt{I}}{1+\sqrt{I}} - 0.3I\right] \tag{4.24}$$

where Z_i is the species valence. The accuracy of Eq. 4.24 can be tested after substituting it into Eq. 4.21:

$$\log\ K_s = \log\ K_{sc} + 0.512\left[\frac{\sqrt{I}}{1+\sqrt{I}} - 0.3I\right]\Delta Z^2 \tag{4.25}$$

where

$$\Delta Z^2 \equiv \nu_c m^2 + \gamma + \nu_\alpha \ell^2 - (\nu_c m + \gamma - \nu_\alpha \ell)^2 \tag{4.26}$$

in terms of the valences of M, H, L, and $M_{\nu c}H_\gamma L_{\nu\alpha}$ in Eq. 4.15. According to the Davies equation, a graph of $\Delta \log K \equiv \log K_s - \log K_{sc}$ against the

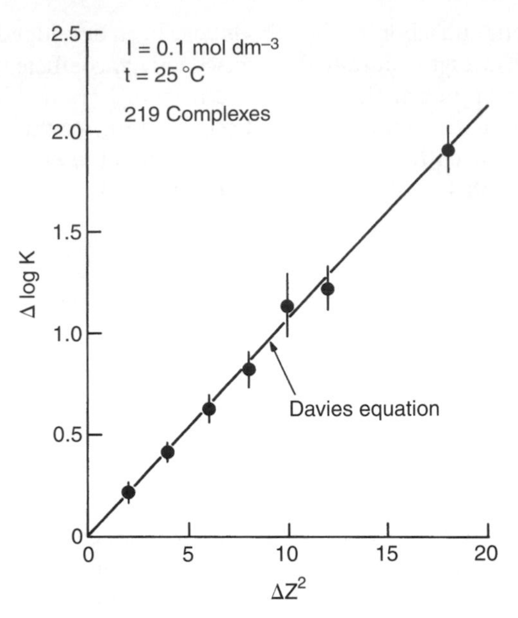

Figure 4.2. A test of Eq. 4.25 (line through the data points) at I = 0.1 mol dm^{-3} and 25 °C.

parameter ΔZ^2 should be a straight line with a positive slope that varies with the value of I, as indicated in Eq. 4.25. Figure 4.2 shows a verification of this result at I = 0.1 mol dm^{-3} for 219 metal complexes for which Δ log K has been measured and the corresponding ΔZ^2 calculated. The line through the data is Eq. 4.25, with I = 0.1 mol dm^{-3}.

For uncharged monovalent metal–ligand complexes, uncharged proton–ligand complexes, and uncharged bivalent metal–ligand complexes, some semiempirical equations for log γ_i are (25 °C)

$$\log\ \gamma_{ML} = \frac{-0.192I}{0.0164 + I}\ (M = Na^+,\ K^+,\ etc.) \quad (4.27a)$$

$$\log\ \gamma_{HL} = 0.1I \quad (4.27b)$$

$$\log\ \gamma_{ML} = -0.3I\ (M = Ca^{2+},\ Mg^{2+},\ etc.) \quad (4.27c)$$

for I < 0.1 mol dm^{-3}. These expressions conform to a theoretical requirement for neutral species that log γ becomes proportional to I in the infinite dilution limit.

With equations for estimating γ_i, it is possible to calculate sets of conditional stability constants under varying composition from a single set of thermodynamic stability constants. For charged complexes, the necessary relationship is given in Eq. 4.25, whereas for uncharged complexes described with Eq. 4.27, one of the three expressions for log γ_i must be added to the right side of Eq. 4.25. For example, in the case of the bicarbonate complex $CaHCO_3^+$, at $I = 0.05$ mol dm^{-3},

$$\log K_{sc} = 11.529 - 0.512\left[\frac{\sqrt{0.05}}{1+\sqrt{0.05}} - 0.3\,(0.05)\right] \times 8 = 11.529 - 0.687 = \mathbf{10.842}$$

according to Eq. 4.25, after rearrangement to calculate log K_{sc}. In the case of $H_2CO_3^0$ at $I = 0.05$ mol dm^{-3}, Eq. 4.27b must be added to Eq. 4.25 and, with $K_s = 7.36 \times 10^{13}$,

$$\begin{aligned}\log K_{sc} &= 13.867 - 0.512[0.1677] \times 4 + 0.1(0.05)\\ &= 13.867 - 0.343 + 0.005 = \mathbf{13.529}\end{aligned}$$

In a speciation calculation based on the flowchart in Figure 4.1, a database of K_s values would be used to create the required database of K_{sc} values, as illustrated earlier. An estimate of I (e.g., based on Eq. 4.23) would be needed to do this, and the K_{sc} database would be refined in each iteration along with the species concentrations and the value of I in Eq. 4.22. Convergence of the calculation then would result in a mutually consistent set of species concentrations, K_{sc} values, and calculated ionic strength.

The conceptual meaning of the activity of a chemical species stems from the formal similarity between Eqs. 4.16 and 4.19. The conditional stability constant is a convenient parameter with which to characterize equilibria, but it is composition dependent, in that it does not correct for the electrostatic interactions among species that must occur as their concentrations change. In the limit of infinite dilution, these interactions die out, and the extrapolated value of K_{sc} represents the chemical equilibrium of an *ideal solution* wherein species interactions other than those involved to form a complex are unimportant. Thus, the activities in Eq. 4.19 play the role of hypothetical concentrations of species in an ideal solution. But the real solution is not ideal as its concentration increases, because species are brought closer together to interact more strongly. When this occurs, K_{sc} must begin to deviate from K_s. The activity coefficient then is introduced to correct the concentration factors in K_{sc} for nonideal species behavior and thereby restore the value of K_s via Eq. 4.21. This correction is expected to be larger for charged species than for neutral complexes (dipoles), and larger as the species valence increases. These trends are indeed apparent in the model expressions in eqs. 4.24 and 4.27.

For Further Reading

Langmuir, D. (1997) *Aqueous environmental geochemistry.* Prentice Hall, Upper Saddle River, NJ. Chapters 2 through 6 of this advanced textbook offer comprehensive discussions of aqueous chemical speciation, including two chapters on carbonate chemistry.

Loeppert, R. H., A. P. Schwab, and S. Goldberg (eds.). (1995) *Chemical equilibrium and reaction models.* Soil Science Society of America, Madison, WI. A useful compendium of applications-oriented articles on chemical speciation, including a discussion of how conditional stability constants are screened for quality, by the creators of the National Institute of Standards and Technology (NIST) database (Section 4.4), and descriptions of several computer programs for performing speciation calculations.

Richens, D. T. (1997) *The chemistry of aqua ions.* Wiley, New York. Chapter 1 of this advanced treatise surveys the experimental methods for characterizing aqueous species. Subsequent chapters provide details of the structure and reactivity of aqueous species organized according to the groups of the Periodic Table of elements.

Schecher, W. D., and D. C. McAvoy. (2001) *MINEQL+: A chemical equilibrium modeling system workbook.* Environmental Research Software, Hallowell, ME. A useful working guide to applying chemical speciation software, based on one of the more popular computer programs.

Schwab, A. P. (2000) The soil solution, pp. B-85–B-122. In: M. E. Sumner (ed.), *Handbook of soil science.* CRC Press, Boca Raton, FL. This chapter surveys the same material that appears in the current chapter, but in more detail and with the explicit use of chemical thermodynamics.

Sparks, D. L. (Ed.). (1996) *Methods of soil analysis: Part 3. Chemical methods.* Soil Science Society of America, Madison, WI. This is the standard reference work on laboratory methods for determining the concentrations and speciation of chemical elements in soils and soil solutions.

Stumm, W., and J. J. Morgan. (1996) *Aquatic chemistry.* Wiley, New York. Chapters 2 through 6 of this classic advanced textbook provide an excellent reference for the technical details of aqueous chemical speciation, including kinetics, with many applications to natural waters.

Problems

The more difficult problems are indicated by an asterisk.

1. In the table presented here are composition data for drainage waters collected at the litter–soil interface and at a point 0.3 m below that interface in a soil supporting a deciduous forest. Discuss possible causes for the differences in pH, and in K and Ca concentrations between the two soil solutions. Calculate the total moles of cation and anion charge per cubic meter, as well as the net charge per cubic meter, for each soil solution. Explain why the net charge in each case is not zero and why it is larger in absolute value for the litter solution than for the subsoil solution.

	Constituent ($mmol\ m^{-3}$)								
	Ca	Mg	Na	K	NH_4	NO_3	Cl	SO_4	pH
Litter	50	37	11	63	5	1	36	62	4.86
Soil	23	33	19	39	8	2	40	50	5.98

2. The temperature dependence of rate coefficients often can be expressed mathematically by the *Arrhenius equation:*

$$\log k = A - B/RT$$

where A and B are constant parameters, R is the molar gas constant (see the Appendix), and T is absolute temperature. The value of B for the rate coefficient k_f^* in Eq. 4.7c is 59 kJ mol^{-1}, whereas that for the rate coefficient k_b is 63 kJ mol^{-1}. Calculate the values of the two rate coefficients, their associated intrinsic timescales, and the conditional stability constant for the reaction in Eq. 4.1 at 15 °C.

3. Develop a rate law to describe the kinetics of the metal complexation reaction

$$M^{2+} + L^{\ell -} = ML^{2-\ell}$$

and apply it to the complexation of Cd^{2+} by the synthetic chelating ligands $EDTA^{4-}$(ethylenedinitrilotetraacetate), $HEDTA^{3-}$ [N-(2-hydroxyethyl)ethylenedinitrilotriacetate], and $CDTA^{4-}$(trans-1,2-cyclohexylenedinitrilotetraacetate), for which the respective log K_{sc} values are 16.5, 13.7, and 19.7 in a 100 mol m^{-3} electrolyte background solution. Measured values of the rate coefficient for complex dissociation are 1.8×10^{-4}, 1.5×10^{-3}, and $9.9 \times 10^{-6} s^{-1}$ respectively. Calculate the second-order rate coefficient for the formation of each complex. What are the intrinsic timescales associated with the two rate coefficients if the initial concentration of Cd^{2+} is 1 μmol m^{-3}? Plant uptake of Cd^{2+} at this initial concentration occurs in timescales on the order of several minutes. Does this fact imply a kinetic influence on uptake could occur from either the formation or dissociation of the Cd complexes?

4. Develop an appropriate rate law like that in Eq. 4.3a for the formation of AlF^{2+} from Al^{3+} and F^-. The value of k_f for this reaction at 25 °C is 110 $dm^3\ mol^{-1}\ s^{-1}$ at pH 3.9, and 726 $dm^3\ mol^{-1}\ s^{-1}$ at pH 4.9. The Arrhenius parameter B = 25 kJ mol^{-1} (see Problem 2). What are the corresponding values of k_f at 10 °C? Calculate the half-life for AlF^{2+} formation at both pH and temperature values, given $[Al^{3+}]_0 = [F^-]_0 =$ 10 mmol m^{-3}.

5. Dissolved CO_2 can react directly with hydroxide ions to form bicarbonate:

$$CO_2 + OH^- = HCO_3^-$$

as an alternative to the reaction in Eq. 4.1. Develop a rate law for CO_2 loss by direct transformation to bicarbonate. Given $k_f = 8500\ dm^3\ mol^{-1}\ s^{-1}$ and $k_b = 2 \times 10^{-4}\ s^{-1}$ for this reaction at 25 °C, show that the pH value above which the rate of loss of CO_2 by reaction with OH^- will exceed that driven by the forward reaction in Eq. 4.1 is about 8.6.

*6. The film diffusion model discussed in Special Topic 3 (Chapter 3) can also be applied to a gas diffusing across an air–water interface such as exists in soil pores (Section 1.4). This interface is characterized by a boundary layer that separates soil air from a bulk soil solution. An intrinsic timescale for diffusion across this boundary layer can be defined by the ratio δ^2 to D, where the two parameters are the same as those that appear in Eq. S.3.1 of Special Topic 3. The quantity $[i]_{surf}$ in this latter equation is now interpreted as a concentration at the soil air–soil solution interface. Its value can be calculated using Henry's law (Eq. 1.1 and Table 1.6) if the partial pressure of a gas in soil air is known. The quantity $[i]_{bulk}$ in Eq. S.3.1 applies to the bulk soil solution and, therefore, is influenced by chemical reactions in this phase. Take $i = CO_2$ and consider the loss of dissolved CO_2 to form the neutral complex $H_2CO_3^0$ as a reaction that could influence $[CO_2]_{bulk}$ (Eq. 4.1). Derive an equation for δ_{crit}, the boundary layer thickness above which diffusion of CO_2 across the boundary layer will be influenced by the kinetics of $H_2CO_3^0$ formation. Given $D = 2 \times 10^{-9}\ m^2\ s^{-1}$ for CO_2 in water, calculate the value of δ_{crit} at 25 °C and interpret your result by comparison with typical soil pore sizes. (*Hint:* Derive an equation for δ_{crit} based on the timescales for diffusion across the boundary layer and loss of CO_2 to form $H_2CO_3^0$, yielding $\delta_{crit} \approx 0.2$ mm.)

7. The mass balance of carbonate in a soil solution, ignoring complexes with metals, can be expressed as

$$CO_{3T} = [H_2CO_3^*] + [HCO_3^-] + [CO_3^{2-}]$$

Given the conditional stability constants (at 25 °C),

$$K_{1c} = [H_2CO_3^*] / [H^+]^2 [CO_3^{2-}] \approx 10^{16.7}$$
$$K_{2c} = [HCO_3^-] / [H^+][CO_3^{2-}] \approx 10^{10.3}$$

derive equations for the distribution coefficients of the three carbonate species. Use the approximation $[H^+] \approx 10^{-pH}$ to estimate the range of pH over which HCO_3^- is dominant. (*Hint:* See Problem 15 in Chapter 1 for the definition of $H_2CO_3^*$.)

8. Combine K_{1c} and K_{2c} in Problem 7 with K_H in Table 1.6 to derive the equation

$$K_{1c}/K_{2c}K_H = P_{CO_2}/\left[H^+\right]\left[HCO_3^-\right] \approx 10^{7.8} \text{ atm dm}^6 \text{ mol}^{-2}$$

where K_H is the equilibrium constant for the formation of carbonic acid, as in Eq. 2.11a. This equation shows that the CO_2 *partial pressure and* $\left[HCO_3^-\right]$ *are sufficient to determine pH.* Calculate the pH value in equilibrium with $\left[HCO_3^-\right] = 1$ mmol m^{-3} and $P_{CO_2} = 10^{-3.5}$ or 10^{-2} atm (the range typical for soils).

*9. The *carbonate alkalinity* of a soil solution is defined by the equation

$$\text{Alk} \equiv \left[HCO_3^-\right] + 2\left[CO_3^{2-}\right]$$

Use the conditional equilibrium constants in problems 7 and 8 to calculate the carbonate alkalinity of the soil solutions described in Problem 1, given $P_{CO_2} = 10^{-3}$ atm. Carbonate alkalinity may be interpreted as the ANC (Section 3.3) of a soil solution that is contributed by carbonate species. Estimate the ANC of each soil solution in Problem 1 that is contributed by dissolved humus. (*Hint:* Reconsider the charge balance calculations in Problem 1 in terms of carbonate alkalinity and humus ANC.)

*10. The base 10 logarithm of the thermodynamic stability constant for the formation of bicarbonate $\left(HCO_3^-\right)$ at 25 °C is 10.329 according to the conventions used in Eq. 4.19 $\left(\nu_c = 0,\ \gamma - 1,\ \nu_a = 1;\ L = CO_3^{2-}\right)$. Calculate the value of pH_{dis} for $H_2CO_3^0$ and compare it with the values listed in Table 3.1 and with the average value of 2.93 for the carboxyls in humic acid (Problem 6 in Chapter 3). (*Hint:* Subtract log K_2 for the formation of bicarbonate from that for the formation of $H_2CO_3^0$ and apply the definition of pH_{dis}.)

11. Given that ΔZ^2 is usually positive, what general conclusion can be drawn from Eq. 4.25 about the effect of increasing salinity on soluble complex formation?

12. The value of log K_s for the formation of $AlSO_4^+$ from Al^{3+} and SO_4^{2-}, a reaction expected when gypsum is added to an acidic soil (see Problem 12 in Chapter 2), is 3.89. Calculate log K_{sc} in a soil solution that has a conductivity of 2.4 dS m^{-1}. Does increasing conductivity enhance or diminish $AlSO_4^+$ formation?

*13. Show that the concentration of the complex $CaHCO_3^+$, in a soil solution is proportional to the concentration of Ca^{2+} times the partial pressure of CO_2 in the soil atmosphere. Calculate the value of the constant of proportionality and then compute values of the ratio $\left[CaHCO_3^+\right]/\left[Ca^{2+}\right]$ over the typical range of P_{CO_2} in soils.

14. The conductivity of a soil solution saline enough to affect salt-sensitive plants is 1.5 dS m^{-1}. Calculate the activities of Ca^{2+} and $CaSO_4^0$ in this solution if the concentrations of Ca^{2+} and SO_4^{2-} are both 2.8 mol m^{-3}, and log K_s for the formation of $CaSO_4^0$ is 2.36.

15. Calculate the effect of increasing the conductivity of a soil solution from 0.5 to 3.0 dS m^{-1} (low to high salinity) on the concentration of Si $(OH)_4^0$ maintained at a constant activity of 10^{-4} by solubility equilibrium with quartz (SiO_2).

5

Mineral Stability and Weathering

5.1 Dissolution Reactions

Soil minerals such as aluminosilicates and metal oxides have strong chemical bonds between their cationic constituents and oxygen. Exchangeable ions on the surfaces of these minerals (e.g., Na^+ and Mg^{2+} on a clay mineral or Cl^- on a metal oxide) can be solvated by water molecules from the soil solution and diffuse away quickly, but the framework ions cannot be dislodged so easily. For their removal, it is necessary to create a strong perturbation of the bonds holding them in the mineral structure, and this can be accomplished only by a highly polarizing species, like the proton or a ligand that forms inner-sphere complexes (Section 4.2).

Proton attack begins with H^+ adsorption by the anionic constituent of a mineral (e.g., OH in a metal oxyhydroxide, CO_3 in a carbonate, or PO_4 in a phosphate). This relatively rapid reaction is followed by the slower process of polarizing the metal–anion bonds near the site of proton adsorption, with subsequent detachment of the metal–anion complex. The two-step mechanism involved is illustrated schematically for the edge surface of the mineral gibbsite [γ-$Al(OH)_3$] in Figure 5.1. A similar process also is shown in Figure 5.1 for ligand attack. In this latter case, a strongly complexing ligand in the soil solution (e.g., oxalate, F^-, or PO_4^{3-}) exchanges for a water molecule bound to Al, as was illustrated in Eq. 3.12:

$$\equiv Al - OH_2^+(s) + F^- = \equiv AlF(s) + H_2O(\ell) \qquad (5.1)$$

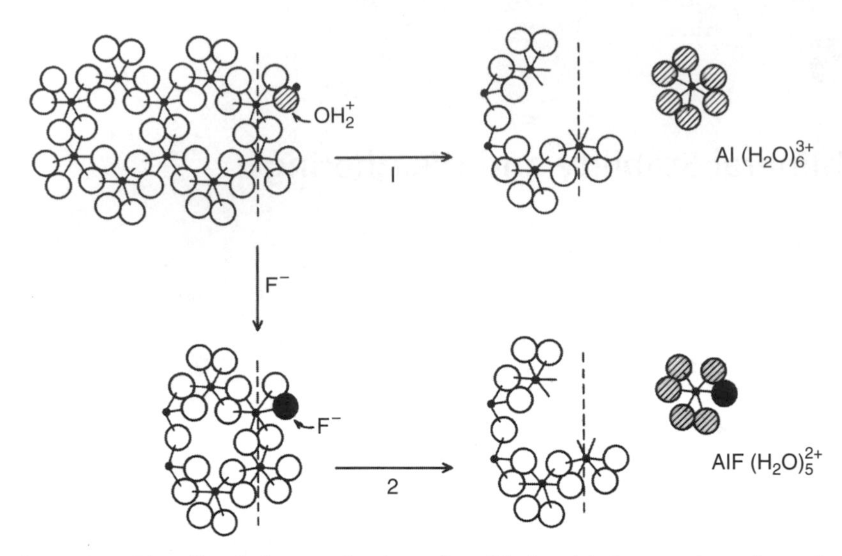

Figure 5.1. Two dissolution mechanisms for gibbsite. (1) Protonation of an edge surface hydroxyl group to form OH_2^+ and detachment of Al^{3+} as a solvation complex (pH < 5). (2) Ligand exchange of OH_2^+ for F^- and detachment of Al^{3+} as the AlF^{2+} complex.

with subsequent detachment of the AlF^{2+} complex, which then equilibrates with F^- in the soil solution, followed by adsorption of H^+ to form the species $\equiv Al - OH_2^+$ once again. Detachment of the metal cation into the soil solution is always the slowest step of a mineral dissolution process.

For soil minerals with ionic constituents that are readily solvated and detached [e.g., NaCl (halite) or $CaSO_4 \cdot 2\,H_2O$ (gypsum)], or for exchangeable ions adsorbed on insoluble minerals, the kinetics of dissolution can be described in terms of the film diffusion mechanism introduced in Special Topic 3. The dissolution reactions of these rather soluble minerals or exchangeable ions are therefore *transport controlled.* For soil minerals like the clay minerals, metal oxides, and most carbonates, however, the rate of dissolution is *surface controlled* and is observed to follow zero-order kinetics, described mathematically in Table 4.2. If [A] is the aqueous-phase concentration of an ionic constituent of a mineral (e.g., Al^{3+}), then the rate law for surface-controlled dissolution is expressed by the equation

$$\frac{d[A]}{dt} = k_d \tag{5.2}$$

where the parameter k_d is a rate coefficient that is independent of [A], but is a function of temperature, pressure, $[H^+]$, and, if appropriate, the concentration of a strongly complexing ligand promoting dissolution via the second mechanism in Figure 5.1. Typically the pH dependence of k_d has the roughly

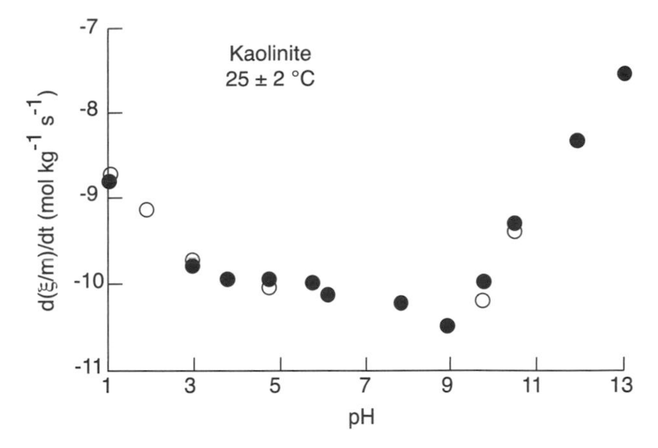

Figure 5.2. Dependence on pH of the logarithm of the mass-normalized rate of dissolution of kaolinite suspended in 1 mol m^{-3} NaCl solution at 25 °C. Open circles are data based on Al release; solid circles are based on Si release. Data from Huertas, F.J., L. Chou, and R. Wollast (1999) Mechanism of kaolinite dissolution at room temperature and pressure. Part II: Kinetic study. *Geochim. Cosmochim. Acta* **63**: 3261–3275.

U-shape form illustrated in Figure 5.2 for kaolinite dissolution at 25 °C. The dissolved species with a concentration that appears in the rate law (Eq. 5.2) is $Si(OH)_4^0$ in this example, but the rate of silica release has been mass normalized (units of moles per kilogram per second) through division by the solids concentration (kilograms per liter).

As introduced in Problem 7 of Chapter 2, when a zero-order rate law applies, an intrinsic timescale can be associated with the kinetics of mineral dissolution:

$$\tau_{\text{dis}} = (M_r \times \text{ dissolution rate})^{-1} \tag{5.3}$$

where M_r is the relative molecular mass of the dissolving mineral and the dissolution rate is in units of moles A per gram of mineral per second, as in Figure 5.2. The value of τ_{dis} characterizes the timescale on which one mole of a mineral will dissolve in water, thus allowing comparisons to be made among minerals of differing composition and density. Figure 5.3 shows a graph of log τ_{dis} plotted against the Si-to-O molar ratio for several of the primary silicates listed in Table 2.3. The values of τ_{dis}, which are expressed in years, pertain to proton-promoted dissolution at pH 5 and 25 °C. Increasing Si-to-O, which implies increasingly strong chemical bonds in a primary silicate (Section 1.3) and, therefore, increasing resistance to weathering, is correlated positively with the intrinsic timescale for dissolution. Note that the timescales for hornblende and quartz, approximately one millennium and a few thousand millennia, respectively, are consistent with the sharp drop in the content of hornblende relative to quartz illustrated in Figure 2.6. The persistence of both minerals in

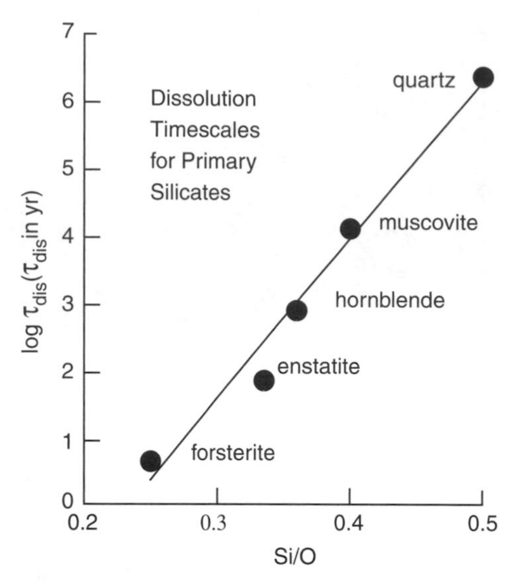

Figure 5.3. Dependence on the Si-to-O molar ratio of the logarithm of the dissolution timescale (Eq. 5.3) at pH 5 and 25 °C for primary silicates.

the chronosequence over timescales that appear to be much longer than τ_{dis} is a reminder that rates of dissolution measured in the laboratory are typically up to three orders of magnitude smaller than those measured in field studies. This well-known discrepancy arises because of the great complexity of mineral dissolution processes in natural soils, where temperature and water content, organic and inorganic coatings on mineral surfaces, and near-equilibrium solubility conditions intervene to obviate the simplicity of Eq. 5.2.

The rate expression in Eq. 5.2 applies to a surface-controlled mineral dissolution reaction after any ion exchange or solvation reactions have occurred, but well before equilibrium between the mineral and the soil solution is reached. The same consideration applies to transport-controlled dissolution reactions governed by an expression like Eq. S.3.10 in Special Topic 3 (Chapter 3). As equilibrium approaches, the rate of dissolution becomes influenced by the stoichiometry of the dissolution reaction. Dissolution reactions for the minerals albite, allophane, anorthite, biotite, calcite, chlorite, orthoclase, and smectite were illustrated in sections 1.5, 2.2, 2.3, and 2.5. (Some of these reactions involved incongruent dissolution.) Two other important examples are the dissolution reactions of gypsum and gibbsite:

$$CaSO_4 \cdot 2\,H_2O(s) = Ca^{2+} + SO_4^{2-} + 2\,H_2O(\ell) \tag{5.4}$$

$$Al(OH)_3(s) = Al^{3+} + 3\,OH^- \tag{5.5}$$

Following the chemical thermodynamics concepts introduced in Section 4.5, one can define a *dissolution equilibrium constant* for the reactions in eqs. 5.4

and 5.5:

$$K_{dis} = (Ca^{2+})(SO_4^{2-})(H_2O)^2/(gypsum) \tag{5.6}$$

$$K_{dis} = (Al^{3+})(OH^-)^3/(gibbsite) \tag{5.7}$$

where the boldface parentheses indicate a thermodynamic activity of the species they enclose. The solid-phase activities of gypsum and gibbsite are defined to have unit value if the minerals exist in pure macrocrystalline form at $T = 298.15$ K and 1 atm pressure. If, as often can be true in soils, the solid phases are "contaminated" with minor elements (Section 1.3) or are not well crystallized (Chapter 2), their activity will differ from 1.0.

The *solubility product constant* for gypsum or gibbsite is defined by the equations

$$K_{so} \equiv K_{dis}\ (gypsum)\ (H_2O)^2 = (Ca^{2+})(SO_4^{2-}) \tag{5.8}$$

$$K_{so} \equiv K_{dis}\ (gibbsite) = (Al^{3+})(OH^-)^3 \tag{5.9}$$

By convention, $K_{so} = K_{dis}$ *numerically* when the solid phase is pure and macrocrystalline (no structural imperfections), and the aqueous solution phase is infinitely dilute. In this case, the solid and water activities are both defined as equal to 1.0. In the current example, $K_{so} = 2.5 \times 10^{-5}$ for gypsum, and $K_{so} = 1.3 \times 10^{-34}$ for gibbsite, according to published compilations of thermodynamic data such as the NIST database mentioned in Section 4.4. Usually K_{so} values for hydroxide solids are reported as $^*K_{so}$, which is K_{dis} for the dissolution reaction that is obtained by replacing OH^- with H^+ using the formation reaction for liquid water. In the case of gibbsite, for example, one adds $3[OH^- + H^+ = H_2O(\ell)]$ to Eq. 5.5 and replaces Eq. 5.9 with the definition

$$^*K_{so} \equiv {}^*K_{dis}(gibbsite)(H_2O)^3 = (Al^{3+})/(H^+)^3 \tag{5.10}$$

Because the equilibrium constant for the water reaction is 10^{14}, $^*K_{so} = 10^{42} \times 1.3 \times 10^{-34} = 1.3 \times 10^8$. The right sides of Eqs. 5.8 through 5.10 contain the *ion activity product* (IAP) corresponding to the solid phases that are dissolving. For the dissolution reaction of a generic solid $M_aL_b(s)$,

$$M_aL_b(s) = M^{m+} + L^{\ell-} \tag{5.11}$$

the IAP is defined by the equation

$$IAP \equiv (M^{m+})(L^{\ell-}) \tag{5.12}$$

Evidently $IAP = (Ca^{2+})(SO_4^{2-})$ for gypsum and $IAP = (Al^{3+})(OH^-)^3$ [or $(Al^{3+})(H^+)^{-3}$] for gibbsite.

According to the method discussed in Section 4.5, the IAP can be calculated solely with data on the chemical speciation of a soil solution. *Thus, Eq. 5.12 can be evaluated regardless of whether the dissolution reaction in Eq. 5.11*

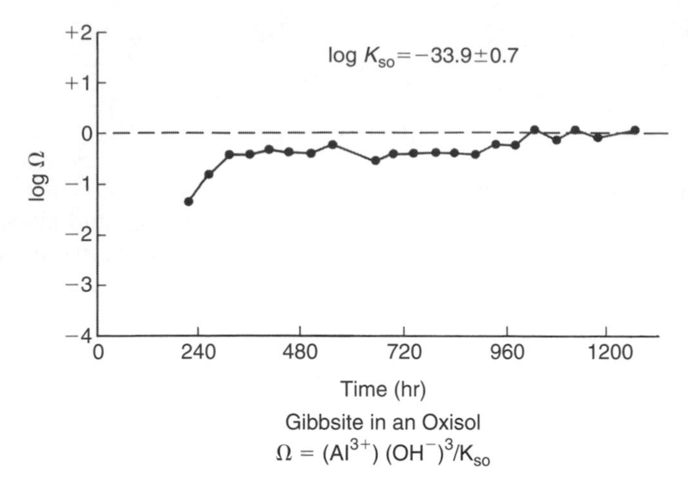

Figure 5.4. Time evolution of the relative saturation (Eq. 5.13) for gibbsite in an Oxisol. Data from Marion, G.M., D.M. Hendricks, G.R. Dutt, and W.H. Fuller (1976) Aluminum and silica solubility in soils. *Soil Sci.* **121**: 76–85.

is actually at equilibrium. Used in this way, the IAP becomes a useful probe for determining whether dissolution equilibrium actually has been achieved. This kind of test is made by examining measured values of the *relative saturation:*

$$\Omega = \text{IAP}/\text{K}_{\text{so}} \tag{5.13}$$

If $\Omega < 1$ within some tolerance interval determined by experimental precision, the soil solution is said to be *undersaturated*; if $\Omega > 1$, it is *supersaturated*; and when a dissolution reaction is at equilibrium, $\Omega = 1$, again within experimental precision. Figure 5.4 shows the approach of Ω from undersaturation to unit value in the soil solution of an Oxisol containing gibbsite (as confirmed by X-ray diffraction analysis). Ion activity products $[(\text{Al}^{3+})\,(\text{OH}^-)^3]$ were determined in aliquots of leachate from the Oxisol during slow elution. After about 40 days of elution, $\Omega \approx 1.0$, and thermodynamic equilibrium between the soil solution and dissolving gibbsite may be assumed to have been achieved. Matters can become complicated, however, in the case of gibbsite precipitation, both because of the formation of metastable Al–hydroxy polymers that transform slowly in the aqueous solution phase and because of structural disorder in the growing solid phase (Section 2.4). Under these obfuscating conditions, the interpretation of measured values of Ω becomes problematic.

The quantitative role of Ω in dissolution–precipitation kinetics can be sharpened by a rate-law analysis of the reaction in Eq. 5.11. As in Section 4.2, the rate of increase of the concentration of M^{m+} can be postulated to be equal to the difference of two functions, R_f and R_b, which depend, respectively, on powers of the concentrations of the reactant and products in Eq. 5.11. This line of reasoning is analogous to that associated with Eqs. 4.2 and 4.3a. The overall rate law is then

$$\frac{d[M^{m+}]}{dt} = k_{dis}[M_aL_b]^{\delta} - k_p[M^{m+}]^{\alpha}[L^{\ell-}]^{\beta} \tag{5.14a}$$

where k_{dis} and k_p are rate coefficients; and α, β, and δ are partial reaction orders. If the solid phase is in great excess during dissolution or precipitation, its concentration factor can be absorbed into the dissolution rate coefficient $k_d \equiv k_{dis}[M_aL_b]^{\delta}$, as was done with H_2O in Eq. 4.3b. If also it is assumed that $\alpha = a$, $\beta = b$, as done in connection with Eq. 4.3c, then the overall rate law

$$\frac{d[M^{m+}]}{dt} = k_d - k_p[M^{m+}]^a[L^{\ell-}]^b \tag{5.14b}$$

can be postulated, where k_d and k_p depend on temperature, pressure, soil solution composition, and the nature of the dissolving solid phase. The rate coefficients, by analogy with Eq. 4.4, are related to a *conditional solubility product constant:*

$$K_{soc} \equiv [M^{m+}]_e^a[L^{\ell-}]_e^b = k_d/k_p \tag{5.15}$$

The combination of Eqs. 5.13 to 5.15 now produces the model rate law

$$\frac{d[M^{m+}]}{dt} = k_d(1 - \Omega) \tag{5.16}$$

where the relationship (cf. Eq. 4.21)

$$\begin{aligned} \Omega_c &\equiv [M^{m+}]^a[L^{\ell-}]^b/K_{soc} \\ &= \gamma_M^a\gamma_L^b[M^{m+}]^a[L^{\ell-}]^b/K_{so} \\ &= (M^{m+})^a(L^{\ell-})^b/K_{so} = \Omega \end{aligned} \tag{5.17}$$

has been applied. Equation 5.16 demonstrates the role of Ω as a discriminant in the kinetics of dissolution–precipitation reactions. If a soil solution is highly undersaturated, $\Omega \ll 1$, and Eq. 5.16 reduces to Eq. 5.2. Near equilibrium, however, $\Omega \approx 1$, and the rate of dissolution becomes very small, a complicating characteristic of mineral weathering in natural soils.

5.2 Predicting Solubility Control: Activity–Ratio Diagrams

Graphical methods based on dissolution equilibria offer a simple and direct approach to the interpretation of soil mineralogy data. The two most common methods are the *activity–ratio diagram* and the *predominance diagram.* Although both methods ultimately tell the same story, each has features appealing to different aspects of the patterns of mineral stability in soils. They are both designed to respond to the questions: Does a dissolving solid phase control the concentration of a given chemical element in a soil solution under given conditions? If so, which solid phase is it likely to be? This query, facile in appearance, turns out to be complex in application, thus the abiding need for qualitative analyses, despite the ever-increasing sophistication of quantitative speciation calculations.

The construction of an activity–ratio diagram can be summarized in four steps:

1. Identify a set of solid phases that contain a chemical element of interest and are likely candidates for controlling its solubility. Write a reaction for the congruent dissolution of each solid, with the *free ionic species* of the element as one of the products. Be sure that the stoichiometric coefficient of the free ion (metal or ligand) in each reaction is 1.0.
2. Compile values of K_{dis} for the solid phases. Write an algebraic equation for log K_{dis} in terms of log [activity] variables for the products and reactants in the corresponding dissolution reaction. Rearrange the equation to have log[(solid phase)/(free ion)]—the log *activity ratio*—on the left side, with all other log[activity] variables on the right side.
3. Choose an *independent* log [activity] variable against which log[(solid)/(free ion)] can be plotted for each solid phase. Select fixed values for all other log[activity] variables, corresponding to an assumed set of soil conditions.
4. Use the fixed activity values and that of log K_{dis} to develop a linear relation between log[(solid)/(free ion)] and the independent log [activity] variable for each solid phase considered. Plot all of these equations on the same graph.

For a chosen value of the log[activity] parameter that has been taken as the independent variable, and under the assumption that all solid phases have activity equal to 1.0, the solid phase that produces the *largest* value of the logarithm of the activity ratio is the one that is most stable, because the activity of the free ion is then smallest. This conclusion follows directly from the fact that a solid phase, which produces the smallest soil solution activity of a free ionic species, will also produce the smallest concentration of that species. The tendency of an ion, if several solids containing it were to be present initially, would be to diffuse to the region of the soil solution where its concentration will be least [recall the discussion of Fick's law in Special Topic 3 (Chapter 3)]. Therefore, the less stable solid phases would continually be dissolving to replenish the ions that diffuse away, leaving as the sole survivor the solid phase capable of producing the smallest soil solution activity of the ion (i.e., the most stable mineral that contains the ion).

As an example, consider an activity–ratio diagram for the control of Al solubility by secondary minerals in an acidic soil. The Jackson–Sherman weathering scenario (Table 1.7) tells us that, when soil profiles are leached free of silica with freshwater, 2:1 layer-type clay minerals are replaced by 1:1 layer-type clay minerals, and ultimately these are replaced by metal oxyhydroxides. This sequence of clay mineral transformations (discussed in Section 2.3) can be represented by the successive *congruent* dissolution reactions of smectite, kaolinite, and gibbsite at 25 °C:

$$\begin{aligned} &Mg_{0.208}[Si_{3.82}Al_{0.18}]Al_{1.29}Fe(III)_{0.335}Mg_{0.445}O_{10}(OH)_2(s) \\ &\quad + 3.28H_2O(\ell) + 6.72H^+ = 1.47Al^{3+} + 0.335Fe^{3+} \\ &\quad + 0.653Mg^{2+} + 3.82Si(OH)_4^0 \quad \log K_{dis} = 3.2 \end{aligned} \tag{5.18a}$$

$$\begin{aligned} &Si_2Al_2O_5(OH)_4(s) + 6\,H^+ = 2Al^{3+} + 2\,Si(OH)_4^0 + H_2O(\ell) \\ &\quad \log K_{dis} = 7.43 \end{aligned} \tag{5.18b}$$

$$Al(OH)_3(s) + 3H^+ = Al^{3+} + 3H_2O(\ell) \quad \log^* K_{dis} = 8.11 \tag{5.18c}$$

The solid-phase reactant in Eq. 5.18a is half a formula unit of montmorillonite, with Mg^{2+} as the interlayer exchangeable cation. Its dissolution reaction is a generalization of that in Eq. 5.11 to the case of a multicomponent solid. The value of K_{dis} for the dissolution of kaolinite (also half a formula unit) reflects a moderately well-crystallized solid phase. Poorly crystallized kaolinite—typical of intensive soil weathering conditions—would yield $\log K_{so} \approx 10.5$. The reaction for gibbsite dissolution differs from that in Eq. 5.7 by subtraction of the water ionization reaction, with a corresponding change in the value of $\log K_{dis}$ (Eq. 5.10). Like kaolinite, gibbsite is assumed to be well crystallized; poorly crystallized gibbsite would yield $\log {}^*K_{so} \approx 9.35$.

Equation 5.18 can be used to construct an activity–ratio diagram for Al solubility as influenced by the leaching of silicic acid $[Si(OH)_4^0]$. The equations for $\log[(\text{solid})\ (Al^{3+})]$ are as follows:

$$\begin{aligned} &\log[(\text{montmorillonite})/(Al^{3+})] = -2.18 + 0.228\log(Fe^{3+}) \\ &\quad + 0.444\log(Mg^{2+}) + 4.57\,pH + 2.6\log(Si(OH)_4^0) - 2.23\log(H_2O) \end{aligned} \tag{5.19a}$$

$$\log[(\text{kaolinite})/(Al^{3+})] = -3.72 + 3\,pH + \log(Si(OH)_4^0) + \frac{1}{2}\log(H_2O) \tag{5.19b}$$

$$\log[(\text{gibbsite})/(Al^{3+})] = -8.11 + 3\,pH + 3\log(H_2O) \tag{5.19c}$$

Note that Eq. 5.18a and its $\log K_{dis}$ value must be divided by 1.47, and that Eq. 5.18b and its $\log K_{dis}$ value must be divided by 2, before Eq. 5.19 can be derived. If $(Si(OH)_4^0)$ is to be the independent activity variable plotted, then pH, (H_2O), and the activities of Fe^{3+} and Mg^{2+} in the soil solution must be prescribed. Useful working values are pH = 5, $(H_2O) = 1.0$, $(Fe^{3+}) = 10^{-13}$, and $(Mg^{2+}) = 6 \times 10^{-3}$. The resulting linear activity–ratio equations are then

$$\log[(\text{montmorillonite})/(\text{Al}^{3+})] = 16.72 + 2.6\log(\text{Si(OH)}_4^0) \quad (5.20a)$$

$$\log[(\text{kaolinite})/(\text{Al}^{3+})] = 11.28 + \log(\text{Si(OH)}_4^0) \quad (5.20b)$$

$$\log[(\text{gibbsite})/(\text{Al}^{3+})] = 6.89 \quad (5.20c)$$

The activity–ratio diagram resulting from plotting Eq. 5.20 is shown in Figure 5.5. The portions of the three straight lines shown in bold depict the largest values of the activity ratio as the activity of silicic acid decreases from left to right in the graph. The range of silicic acid activity typical of all but the most leached soils is indicated by two vertical lines denoting the solubility of quartz $[(\text{Si(OH)}_4^0) = 10^{-4}]$ and poorly crystalline solid silica $[(\text{Si(OH)}_4^0) = 10^{-2.7}]$. Thus, the effect of soil profile leaching is simulated by moving from left to right along the x-axis. At $(\text{Si(OH)}_4^0) \approx 10^{-2.7}$, which reflects conditions during the intensive weathering of primary silicates in an acidic soil, Figure 5.5 indicates that smectite is the most stable solid phase with respect to solubility control on Al. As leaching and loss of silica proceed, the silicic acid activity will decrease, and when (Si(OH)_4^0) falls well below 10^{-4}, gibbsite becomes the most stable Al-bearing solid phase. This progression agrees with the Jackson–Sherman weathering sequence in Table 1.7.

At a given silicic acid activity, the three lines in Figure 5.5 can be pictured as a sequence of (Al^{3+}) "steps," in the sense that (Al^{3+}) decreases as each line is crossed while moving upward in the diagram. If (Si(OH)_4^0) is controlled by poorly crystalline silica, for example, (Al^{3+}) becomes equal successively to $10^{-6.9}$, $10^{-8.6}$, and $10^{-9.7}$, as the lines representing gibbsite, kaolinite, and smectite are crossed. This stepwise decrease in (Al^{3+}) not only tracks the decreasing Al solubility of the minerals at pH 5, but also implies a sequence of solid-phase *precipitation* that can occur in soils if intermediate solid phases form during the intensive weathering of primary silicates or during Si biocycling via phytoliths (the name given to poorly crystalline silica precipitated in plants, particularly grasses). This possibility is formalized in the *Gay–Lussac–Ostwald (GLO) Step Rule:*

> *If the initial composition of a soil solution is such that several solid phases can precipitate a given ion, the solid phase that forms first will be the accessible one for which the activity ratio is nearest above its initial value in the soil solution. Thereafter, the remaining accessible solid phases will form in order of increasing activity ratio, with the rate of formation of a solid phase in the sequence decreasing as its activity ratio increases. In an open system, any one of the solid phases may be maintained "indefinitely."*

The GLO Step Rule is a qualitative empirical guide to the kinetics of precipitation from *supersaturated* solutions. In a closed system, a sequence of

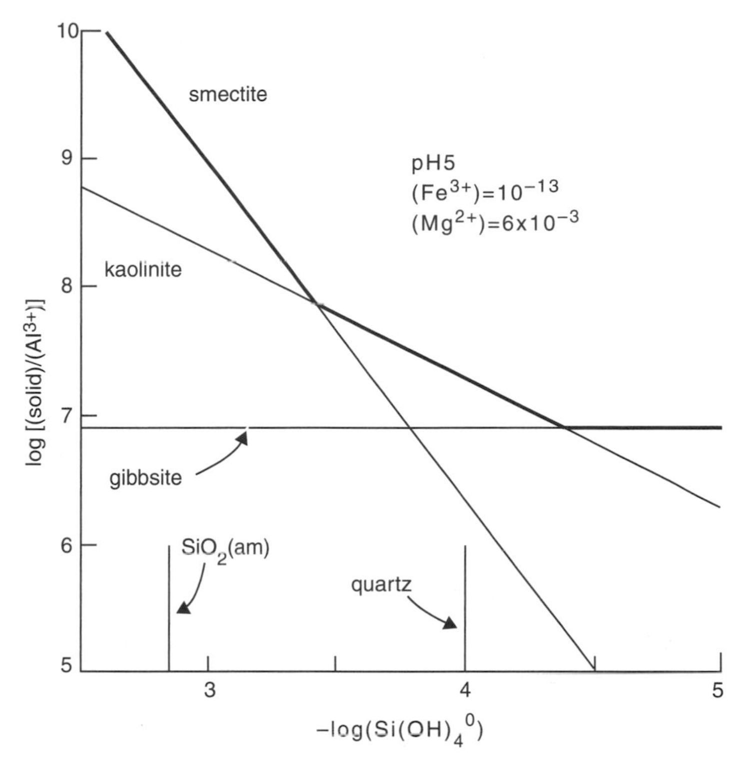

Figure 5.5. Activity–ratio diagram for smectite (montmorillonite), kaolinite, and gibbsite at pH 5 based on Eq. 5.20 with silicic acid activity as the independent variable. A solubility "window" for $SiO_2(s)$ is shown, ranging from that of amorphous silica to that of quartz.

solid-phase intermediates is predicted that depends on the process by which initial conditions of temperature, pressure, and composition result in the formation of a series of increasingly stable states. Each of these states transforms into the one of next greater stability more slowly than it itself came into being (otherwise, intermediate solid phases would not be observed). The mechanistic basis of this sequence of transformations may be related to the fact that solid phases exhibit a larger rate of precipitation from a supersaturated solution as their solubility increases. In an open system, the input of matter can be such as to maintain the initial composition fixed, with the result that the solid phase will be preserved that has least stability consistent with that composition and with the possible reaction pathways, despite its expected dissolution to form more stable phases.

Applied to the activity–ratio diagram in Figure 5.5, the GLO Step Rule implies that, for example, if $(Al^{3+}) > 10^{-7}$ at pH 5 and $(Si(OH)_4^0) = 10^{-2.7}$ in a closed system, the least stable phase, gibbsite, could precipitate before the most stable phase, smectite, is formed. This possibility underlies Problem 9 in

Chapter 2, which proposes kaolinite and gibbsite precipitation from feldspar weathering instead of smectite precipitation (Eq. 2.6). Field observations confirm the existence of all three minerals represented in Figure 5.5 as weathering products of feldspars. Poorly crystalline kaolinite or gibbsite, as mentioned earlier, are associated with larger K_{so} values that would decrease the constant terms in Eqs. 3.19b and 3.19c to the values –5.25 and –10.8 respectively. Thus, the horizontal line in Figure 5.5 would be plotted 1.24 log units lower, and the kaolinite line would be shifted downward by 1.53 log units. These changes create "windows" of gibbsite and kaolinite stability between well-crystallized and poorly crystallized forms that replace the single lines in the diagram and thus enlarge the range of silicic acid activity over which smectite remains the most stable solid phase, within the variability permitted by the windows. This kind of variability and the typical $(Si(OH)_4^0) = 8 \times 10^{-4}$ in acidic soils suggests that smectite, kaolinite, and gibbsite will indeed coexist in active soil weathering environments.

5.3 Coprecipitated Soil Minerals

Coprecipitated soil minerals (Section 1.3) provide ubiquitous evidence for the diverse ionic composition of soil solutions. As discussed in Chapter 2, specific examples of these mixed solid phases include the clay minerals, if metals replace Si in the tetrahedral sheet or Al in the octahedral sheet; calcite, if Mg, Sr, Fe, Mn, or Na replaces Ca; and hydroxyapatite $[Ca_5OH(PO_4)_3]$, if Ca is replaced by Sr or other metals, or if OH is replaced by F or other ligands. In coprecipitation through the formation of a solid solution, the resulting solid phase is a homogeneous mass with its minor substituents distributed uniformly. Thus the basic requirements for this type of coprecipitation are the free diffusion and relatively high structural compatibility of the minor substituents with the precipitate as it is forming. These conditions, incidentally, often are met when minerals precipitate from a silicate melt to form the parent materials of soils: Feldspars and micas are well-known examples of primary-mineral solid solutions (Section 2.2).

The principal effect of coprecipitation is on the solubility of the elements in the solid. If the soil solution is in equilibrium with a solid solution, the activity in the aqueous phase of an ion that is a minor component of the solid may be significantly smaller than what it would be in the presence of a pure solid phase comprising the ion. This effect can be deduced from Eq. 5.11 after noting that $(M_aL_b) \ll 1.0$ could reflect a very small concentration of either the metal M or the ligand L occurring as the compound M_aL_b in a mixed solid phase. The value of K_{so} would thus be much less than that of K_{dis}, with a corresponding reduction in the IAP. Often the dissolution of solid-solution minerals in a soil will be dictated by complicated kinetic considerations, and the prediction of the composition of the soil solution as influenced by these solid phases will be quite difficult. However, if equilibrium exists between the solid and aqueous phases, or even if it is desired only to have a general understanding of reaction pathways, a thermodynamic description of a dissolving solid-solution mineral can be valuable.

Suppose that a solid solution forms because of the coprecipitation of two metal cations, M^{m+} and N^{n+}, with a ligand $L^{\ell-}$. The components (*end members*) of the solid solution are the compounds M_aL_b and $N_cL_d(s)$, where $am = \ell b$ and $cn = \ell d$ to ensure electroneutrality. For each component of the solid, an expression analogous to Eqs. 5.8 and 5.9 can be developed:

$$(M^{m+})^a(L^{\ell-})^b = K_M(M_aL_b) \tag{5.21a}$$

$$(N^{n+})^c(L^{\ell-})^d = K_N(N_cL_d) \tag{5.21b}$$

where K_M and K_N are equilibrium constants for the dissolution of the two pure solid phases. If it is *assumed* that the solid dissolves congruently while retaining a constant composition, its dissolution reaction can be expressed analogously to Eq. 5.11:

$$\begin{aligned} M_{a(1-x)}N_{cx}L_{b(1-x)+dx}(s) &= a(1-x)M^{m+} + cxN^{n+} \\ &+ [b(1-x) + dx]L^{\ell-} \end{aligned} \tag{5.22}$$

where x is the stoichiometric coefficient of N_cL_d (taken to be the minor component) in the solid. This reaction describes a dissolution equilibrium state known as *stoichiometric saturation.* This state is possible if the timescales for changes in the composition of the dissolving solid and for subsequent precipitation of any solid phase (i.e., incongruent dissolution) are much longer than that for the congruent dissolution of the solid. Its existence must be established experimentally. If Eq. 5.22 is applicable, a corresponding dissolution equilibrium constant can be expressed as follows:

$$K_{dis}^{ss} \equiv \frac{(M^{m+})^{a(1-x)}(N^{n+})^{cx}(L^{\ell-})^{b(1-x)+dx}}{(M_{a(1-x)}N_{cx}L_{b(1-x)+dx})} \tag{5.23}$$

and a solubility product constant K_{so}^{ss} can be defined as the product of K_{dis}^{ss} with the activity of the solid, analogous to Eq. 5.8 or Eq. 5.9. This approach treats the solid as if it were a single phase with an activity equal to 1.0.

Solid solutions of diaspore (AlOOH) and goethite (FeOOH) are commonly observed in soils that are subject to flooding, with the value of x in the mixed solid $Fe_{1-x}Al_xOOH$ ranging up to 0.15 (see Problem 6 in Chapter 2). Thus, diaspore is the minor component and the solid-solution mineral is termed *Al-goethite.* By analogy with Eq. 5.9, the dissolution reactions of the two solid-phase components can be expressed in the following form:

$$FeOOH(s) + 3\,H^+ = Fe^{3+} + 2\,H_2O(\ell) \quad \log K_{dis} = -0.36 \tag{5.24a}$$

$$AlOOH(s) + 3\,H^+ = Al^{3+} + 2\,H_2O(\ell) \quad \log K_{dis} = 7.36 \tag{5.24b}$$

These two reactions do not quite fit the format of Eq. 5.21, but, as in the case of gibbsite (Eq. 5.10), they can be adapted to it formally by setting $(H_2O) = 1.0$ and $(L^{\ell-})^b \equiv (H^+)^{-3}$:

$$(Fe^{3+})(H^+)^{-3} = 10^{-0.36}(FeOOH) \qquad (5.25a)$$

$$(Al^{3+})(H^+)^{-3} = 10^{7.36}(AlOOH) \qquad (5.25b)$$

Thus, $(H^+)^{-1}$ plays the role of an aqueous "ligand" activity. Equation 5.22 thus takes the following form:

$$Fe_{1-x}Al_xO(OH)(s) + 3H^+ = (1-x)Fe^{3+} + xAl^{3+} + 2\,H_2O(\ell) \qquad (5.26)$$

with a solubility product constant like that in Eq. 5.10:

$$^*K_{so}^{ss} = (Fe^{3+})^{1-x}(Al^{3+})^x(H^+)^{-3} \qquad (5.27)$$

The effect of Al-goethite on Al solubility can be illustrated through a reconsideration of the activity–ratio diagram in Figure 5.5, but with the system simplified to comprise only kaolinite and gibbsite in addition to Al-goethite. For $(H_2O) = 1.0$, Eqs. 5.19b and 5.19c yield the activity ratios for kaolinite and gibbsite:

$$\log[(\text{kaolinite})/(Al^{3+})] = -3.72 + 3\,\text{pH} + \log(Si(OH)_4^0) \qquad (5.28a)$$

$$\log[(\text{gibbsite})/(Al^{3+})] = -8.11 + 3\,\text{pH} \qquad (5.28b)$$

Equation 5.25b can be developed to have the same form and meaning as these two expressions after dividing both sides by $(AlOOH)^*$, where the asterisk refers to pure single-phase diaspore, and then setting $(AlOOH)/(AlOOH)^*$ equal to x, the stoichiometric coefficient of diaspore in Al-goethite:

$$\log[(\text{diaspore})^*/(Al^{3+})] = -7.36 + 3\text{pH} - \log x \qquad (5.28c)$$

The condition $(AlOOH)/(AlOOH)^* = x$ defines what is known as the *ideal solid solution model.* It states that the change in diaspore activity by virtue of its coprecipitation with goethite is modeled quantitatively simply through equating the activity of a solid in the solid solution to its stoichiometric coefficient in the solid solution. (In general, the solid activity would be expected to be a more complicated function of x.) The effect of this model is to increase the activity ratio, because $x < 1$ and, therefore, $\log x < 0$. Thus, solid solution formation decreases Al solubility.

Inspection of Eq. 5.28 shows that the activity ratios for all three solids have the same dependence on pH. Like gibbsite, the activity ratio for Al-goethite will plot as a horizontal line. Because $\log x \leq 0$, the activity ratio satisfies the inequality $-7.36 - \log x > -8.11$, and ideal Al-goethite exhibits a larger activity ratio than gibbsite at any Al content. Thus, ideal Al-goethite will be more stable than gibbsite, regardless of the activity of silicic acid or the pH value. The same conclusion holds for ideal Al-goethite relative to kaolinite if $\log x - 3.64 - \log(Si(OH)_4^0)$, according to Eqs. 5.28a and c, with $(Si(OH)_4^0) < 10^{-3.4}$ to avoid competition from smectite. At its maximum, $x = 0.33$. Under this condition, $(Si(OH)_4^0) < 10^{-3.15}$ would be sufficient to ensure Al-goethite stability against kaolinite. It follows that *any* x would result in Al-goethite stability over kaolinite. These conclusions, of course, refer to ideal Al-goethite and should be taken as illustrative. Careful studies of synthetic Al-goethite indicate that it is not an ideal solid solution, but instead actually exhibits some immiscibility of its two components.

5.4 Predicting Solubility Control: Predominance Diagrams

A predominance diagram is a two-dimensional field consisting of well-defined regions with coordinate points that are specified by a pH value and the base 10 logarithm of a second relevant activity variable. The boundary lines that define regions in the diagram are specified by equations based on thermodynamic equilibrium constants. In each region of a predominance diagram, either a particular solid phase containing an ion of interest or the free-ion species itself in the aqueous solution contacting the solid will be predominant. Thus, a predominance diagram gives information about changing relative stabilities, at equilibrium, among the solid phases formed by an ion as the pH value and one other activity variable are altered in a soil solution. The construction of this representation of solubility equilibria is summarized as follows:

1. Establish a set of solid-phase species and obtain values of log K for all independent reactions *between the solid-phase species.*
2. Unless other information is available, set the activities of liquid water and all solid phases equal to 1.0. Set all gas-phase pressures at values appropriate to soil conditions.
3. Develop each expression for log K into a relation between a log[activity] variable and pH. In any relation involving aqueous species, choose values for the activities of these species.
4. Plot all the expressions resulting from step 3 as boundary lines on the same graph with pH as the x-axis variable.

These steps will be illustrated with the mineral dissolution reactions in Eq. 5.18 so a comparison can be made between activity–ratio and predominance diagrams. A corresponding set of chemical reactions that relates the solid-phase species to one another is (again with half formula units for the clay minerals):

$$Al_2Si_2O_5(OH)_4(s) + 5\,H_2O(\ell) = 2\,Al(OH)_3(s) + 2\,Si(OH)_4^0 \quad \log K = -8.79 \qquad (5.29a)$$

$$Mg_{0.208}[Si_{3.82}Al_{0.18}]Al_{1.29}Fe(III)_{0.335}Mg_{0.445}O_{10}(OH)_2(s) + 7.69\, H_2O(\ell) + 2.31\, H^+ = 1.47\, Al(OH)_3(s) + 3.82\, Si(OH)_4^0 + 0.653\, Mg^{2+} + 0.335\, Fe^{3+} \quad \log{}^*K = -8.72 \qquad (5.29b)$$

$$Mg_{0.208}[Si_{3.82}Al_{0.18}]Al_{1.29}Fe(III)_{0.335}Mg_{0.445}O_{10}(OH)_2(s) + 4.02\, H_2O(\ell) + 2.13H^+ = 1.47Al_2Si_2O_5(OH)_4(s) + 2.35\, Si(OH)_4^0 + 0.653\, Mg^{2+} + 0.335\, Fe^{3+} \quad \log K = -2.26 \qquad (5.29c)$$

These three chemical equations are algebraic combinations of Eq. 5.18 designed to relate the three minerals one pair at a time. They also represent *incongruent* dissolution reactions (Sections 1.4 and 2.3—note the resemblance between the smectite dissolution reaction in Eq. 2.7a and that in Eq. 5.29c.), by contrast with the preparation of an activity–ratio diagram, which utilizes *congruent* dissolution reactions.

Inspection of Eq. 5.29 suggests that the activities of H_2O, Mg^{2+}, Fe^{3+}, and $Si(OH)_4^0$ are all candidates for the second aqueous-phase variable in a predominance diagram. To preserve comparability with Figure 5.5, choose $(Si(OH)_4^0)$, with the other three activities fixed as before. These choices reduce the general log K equations to the forms:

$$-8.79 = 2\log(Si(OH)_4^0) - 5\log(H_2O) \qquad (5.30a)$$

$$-8.72 = 3.82\log(Si(OH)_4^0) + 0.653\log(Mg^{2+}) + 0.335\log(Fe^{3+}) + 2.31pH - 7.69\log(H_2O) \qquad (5.30b)$$

$$-2.26 = 2.35\log(Si(OH)_4^0) + 0.653\log(Mg^{2+}) + 0.335\log(Fe^{3+}) + 2.31pH - 4.02\log(H_2O) \qquad (5.30c)$$

to the working boundary-line equations:

$$\log(Si(OH)_4^0) = -4.40 \quad \text{(kaolinite–gibbsite)} \qquad (5.31a)$$

$$\log(Si(OH)_4^0) = -0.763 - 0.605\,pH \quad \text{(smectite–gibbsite)} \qquad (5.31b)$$

$$\log(Si(OH)_4^0) = 1.51 - 0.983\,pH \quad \text{(smectite–kaolinite)} \qquad (5.31c)$$

Figure 5.6 shows these boundary lines for a range of pH values common in acidic soils. At pH 5, the sequence of predominant solid phases predicted to occur as the activity of silicic acid changes is in agreement with the sequence predicted in Figure 5.5. Note that if quartz controls the activity of silicic acid [log $(Si(OH)_4^0) = -4$], there is a shift from kaolinite to smectite predominance at pH 5.6 (i.e., for pure water equilibrated with atmospheric CO_2). If the kaolinite and gibbsite solubility windows described in Section 5.2 are incorporated, it is necessary to reconsider Eq. 5.29 with log K = −8.2, −10.5, and −4.5 respectively. The effect of these changes would be to enlarge the field

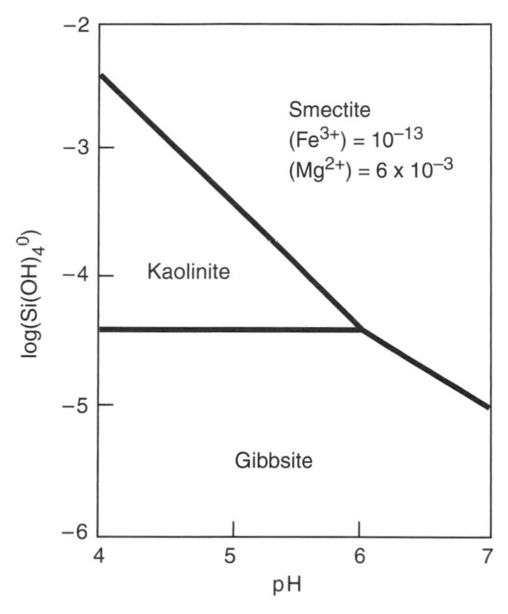

Figure 5.6. Predominance diagram for the same set of secondary minerals and fixed aqueous metal cation activities as in Figure 5.5.

of stability of smectite at the expense of both kaolinite and gibbsite. Thus, poor crystallinity of these two latter minerals makes the persistence of less stable smectite possible in soil profiles. If only gibbsite is assumed to be poorly crystallized, then the stability fields of both kaolinite and smectite grow to push that of gibbsite below a horizontal line at $\log(\mathrm{Si(OH)}_4^0) = -6.35$.

5.5 Phosphate Transformations in Calcareous Soils

Alkaline soils in arid to subhumid environments typically contain significant amounts of calcite (or magnesian calcite), the formation of which is mediated biologically (Section 2.5). The proton-promoted dissolution reaction of calcite is given in Eq. 2.9b:

$$\mathrm{CaCO_3(s)} + \mathrm{H}^+ = \mathrm{Ca}^{2+} + \mathrm{HCO}_3^- \tag{5.32}$$

for which $\log K_{dis} = 1.849$ at 25 °C, if the solid phase is well crystallized, and $\log K_{dis} = 3.939$ if it is very poorly crystallized. In an open system such as a soil profile, CO_2 plays a role in calcite formation that is quantified by combining Eq. 5.32 with the pair of reactions (see Problem 6 in Chapter 4)

$$\mathrm{CO_2(g)} + \mathrm{H_2O}(\ell) = \mathrm{H_2CO_3^*} \qquad \log K_H = -1.466 \tag{5.33a}$$

$$\mathrm{H_2CO_3^*} = \mathrm{H}^+ + \mathrm{HCO}_3^- \qquad \log K_1 = -6.352 \tag{5.33b}$$

to derive the overall dissolution reaction:

$$CaCO_3(s) + 2\,H^+ = Ca^{2+} + H_2O(\ell) + CO_2(g) \qquad (5.34)$$

for which $\log K = 1.849 + 1.466 + 6.352 = 9.667$ if the solid phase is well crystallized, and $\log K = 11.757$ otherwise. Equation 5.34 is convenient for representing the solubility of calcite as controlled by pH and P_{CO_2} following the recipe given in Section 5.2:

$$\begin{aligned}\log[(\text{calcite})/(Ca^{2+})] &= -\log K + 2\,pH + \log P_{CO_2} + \log(H_2O)\\ &= -\log K + 2\,pH + \log P_{CO_2} \qquad (5.35)\end{aligned}$$

if $(H_2O) = 1$ for convenience in applications. Equation 5.35 is suitable either for inclusion in an activity–ratio diagram or for calculating the thermodynamic activity of Ca^{2+} in terms of pH and P_{CO_2}. It shows that solubility control of Ca^{2+} by calcite is favored by high crystallinity (i.e., smaller log *K*), pH, and CO_2 partial pressure. High crystallinity corresponds to a solid phase more stable against dissolution, whereas high pH diminishes the availability of H^+ to promote dissolution (Eq. 5.32), and high P_{CO_2} increases the abundance of bicarbonate ions that promote precipitation (Eq. 5.33). Note that the calcite solubility window ranges over about 2 log units, similar to the windows for kaolinite and gibbsite (Section 5.2).

Suppose now that a "neutral-reaction" phosphate fertilizer containing $CaHPO_4{\cdot}2H_2O$ (dicalcium phosphate dehydrate or *brushite*) is applied to a calcareous soil. What solid phase is likely to control phosphate solubility after equilibration? An answer to this question has been found in experimental studies of the fate of phosphate fertilizers. Depending on soil water content, there is a transformation of brushite to $CaHPO_4$ (dicalcium phosphate or *monetite*), followed by a slow transformation (weeks to months) to $Ca_8H_2(PO_4)_6{\cdot}5H_2O$ (*octacalcium phosphate*). Ultimately, $Ca_{10}(OH)_2(PO_4)_6$ (*hydroxyapatite*) is expected, although octacalcium phosphate may persist for years if phosphate fertilizer is applied continually.

These phosphate transformations can be understood in terms of an activity–ratio diagram involving the four Ca phosphates and calcite. The relevant dissolution reactions for the phosphate solid phases are

$$CaHPO_4 \cdot 2\,H_2O(s) = Ca^+ + HPO_4^{2-} + 2\,H_2O(\ell) \quad \log K_{dis} = -6.62 \qquad (5.36a)$$

$$CaHPO_4(s) = Ca^{2+} + HPO_4^{2-} \quad \log K_{dis} = -6.90 \qquad (5.36b)$$

$$\begin{aligned}\frac{1}{6}\,Ca_8H_2(PO_4)_6 \cdot 5\,H_2O(s) + \frac{2}{3}\,H^+(aq) &= \frac{4}{3}\,Ca^{2+}\\ + HPO_4^{2-} + \frac{5}{6}H_2O(\ell) \quad \log\,K_{dis} &= -3.32 \qquad (5.36c)\end{aligned}$$

$$\frac{1}{6}\,Ca_{10}(OH)_2(PO_4)_6(s) + \frac{4}{3}\,H^+(aq) = \frac{5}{3}\,Ca^{2+} + HPO_4^{2-} + \frac{1}{3}\,H_2O(\ell) \quad \log\,K_{dis} = -2.40 \tag{5.36d}$$

In this soil fertility application, the free-ion activity of interest is (HPO_4^{2-}) and Eq. 5.36 have been arranged so that the stoichiometric coefficient of HPO_4^{2-} is 1.0, following the steps outlined in Section 5.2. The activity of Ca^{2+} in Eq. 5.36 is controlled by calcite. Therefore, Eq. 5.34 with log K_{dis} = 9.667 can be multiplied by the stoichiometric coefficient of Ca^{2+} and subtracted from (i.e., reversed and added to) Eq. 5.36.

The HPO_4^{2-} activity ratios can then be expressed in logarithmic form showing only a dependence on pH and P_{CO_2}. For brushite (DCPDH), the calculation runs as follows:

$$\begin{aligned} -6.62 &= \log(Ca^{2+}) + \log(HPO_4^{2-}) - \log(\mathrm{DCPDH}) \\ &= 9.67 - 2\mathrm{pH} - \log P_{CO_2} - \log[(\mathrm{DCPDH})/(HPO_4^{2-})] \end{aligned}$$

and

$$\log[(\mathrm{DCPDH})/(HPO_4^{2-})] = 16.29 - \log P_{CO_2} - 2\mathrm{pH} \tag{5.37a}$$

Equation 5.35 with ($CaCO_3$) = (H_2O) = 1.0 and log K = 9.667 were used to obtain Eq. 5.37a. In a similar fashion, one can derive expressions for the three other Ca phosphates:

$$\log[(\mathrm{DCP})/(HPO_4^{2-})] = 16.57 - \log P_{CO_2} - 2\,\mathrm{pH} \tag{5.37b}$$

$$\log[(\mathrm{OCP})/(HPO_4^{2-})] = 16.21 - \frac{4}{3}\log P_{CO_2} - 2\,\mathrm{pH} \tag{5.37c}$$

$$\log[(\mathrm{HAP})/(HPO_4^{2-})] = 18.51 - \frac{5}{3}\log P_{CO_2} - 2\,\mathrm{pH} \tag{5.37d}$$

where DCP refers to monetite and obvious abbreviations have been used for the remaining two Ca phosphates. The range of log P_{CO_2} is from approximately –3.52 to –2.5 in coarse-textured calcareous soils, with the larger value representing conditions of high biological activity that tends to occur in the soil rhizosphere.

Figure 5.7 is an activity–ratio diagram for P solubility based on Eq. 5.37 and $P_{CO_2} = 10^{-3.52}$ atm, which is the average value in the atmosphere. At any pH value, the order of decreasing stability of the four Ca phosphates is clearly hydroxyapatite (HAP) $\ll$ octacalcium phosphate (OCP) > monetite (DCP) > brushite (DCPDH), which means that hydroxyapatite should control P solubility at equilibrium. The role of calcite as a mediator of P solubility can be revealed by considering the effects of changing P_{CO_2} or calcite crystallinity on the four parallel lines in Figure 5.7. For example, under rhizosphere conditions, the partial pressure of CO_2 is expected to be larger than its atmospheric value, and the crystallinity of the (biogenic) calcite formed is expected to be less than that precipitated abiotically in a laboratory. Increasing P_{CO_2} at a fixed pH will decrease (Ca^{2+}), according to Eq. 5.34, and accordingly will increase P solubility, as implied by Eq. 5.37 (i.e., the parallel lines in Fig. 5.7 will shift downward, with HAP and OCP shifting more than DCP or DCPDH because of the higher Ca-to-P molar ratio of the former minerals). Decreasing calcite crystallinity, on the other hand, will raise K_{dis} for the reaction in Eq. 5.32, which means that (Ca^{2+}) will increase, if pH and P_{CO_2} are constant, in turn decreasing P solubility and shifting the lines in Figure 5.7 upward by varying amounts. Thus, these two characteristics of biological activity act oppositely on P solubility represented in Figure 5.7.

As discussed in Section 5.2 for the activity–ratio diagram in Figure 5.5, the parallel lines in Figure 5.7 can be viewed as a sequence of HPO_4^{2-} activity "steps" in the sense that, at any fixed pH value, (HPO_4^{2-}) decreases as each line is traversed moving upward in the diagram. For example, at pH 7.5, (HPO_4^{2-}) equals successively $10^{-4.8}$, $10^{-5.1}$, $10^{-5.9}$, and $10^{-9.4}$ as the lines are crossed going from DCPDH to HAP. This monotonic lowering of (HPO_4^{2-}) reflects the decreasing solubility of each phosphate solid and mimics the observed sequence of solid-phase transformations described earlier. Moreover, if the initial pH value and HPO_4^{2-} activity in a calcareous soil solution define a

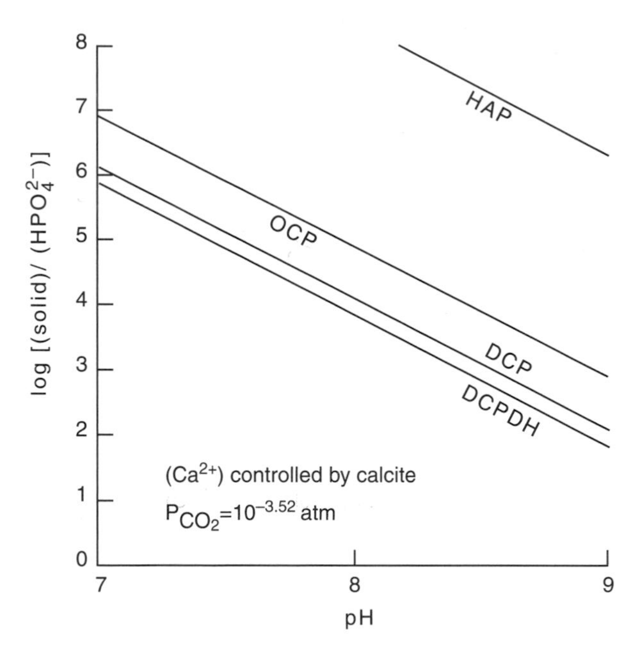

Figure 5.7. Activity–ratio diagram for calcium phosphates in a calcareous soil under atmospheric CO_2 pressure. Abbreviations: DCP, monetite; DCPDH, brushite; HAP, hydroxyapatite; OCP, octacalcium phosphate.

point in the activity–ratio diagram situated between a pair of the lines in the diagram, the solid phase expected to precipitate first is the one with the solubility line closest above the initial point. For example, if $(HPO_4^{2-}) \approx 3 \times 10^{-4}$ at pH 8, then OCP should precipitate, not DCPDH or DCP.

Applied to Figure 5.7, the GLO Step Rule (Section 5.2) indicates that, if DCPDH is added to a calcareous soil, DCP (not HAP) will form first by dissolution of DCPDH. Thereafter, DCP will dissolve and OCP will be formed, with this process occurring more slowly than the DCPDH → DCP transformation. Finally, in a closed system, OCP will slowly dissolve in favor of HAP formation. This overall sequence is what is observed experimentally, and, in laboratory studies with Ca phosphate solutions maintained supersaturated with respect to OCP, but undersaturated with respect to DCPDH or DCP, OCP has been found to precipitate at a rate dependent on $\Omega = (Ca^{2+})^{4/3}(HPO_4^{2-})(H^+)^{-2/3}/K_{dis}$, the appropriate relative saturation variable (Eq. 5.16). In field soils, continual fertilizer applications could maintain supersaturation with respect to OCP and thus stabilize this Ca phosphate for an indefinite period. The GLO Step Rule would predict this stability in an open system. These ideas, however, must be tempered by the possibility that soluble phosphate or calcium complexes, as well as plant uptake of phosphate, could inhibit OCP formation, as could the precipitation of phosphate with cations other than Ca^{2+}.

For Further Reading

Dixon, J. B., and D. G. Schulze (eds.). (2002) *Soil mineralogy with environmental applications.* Soil Science Society of America, Madison, WI. Chapter 4 of this standard reference work gives a brief introduction to solubility equilibria with applications to mineral weathering reactions.

Essington, M. E. (2004) *Soil and water chemistry.* CRC Press, Boca Raton, FL. Chapter 6 of this advanced textbook may be consulted to learn more about the applications of mineral solubility equilibria to contaminant fate and chemical weathering.

Kinniburgh, D. G., and D. M. Cooper. (2004) Predominance and mineral stability diagrams revisited. *Environ. Sci. Technol.* **38**:3641. This useful article describes how to combine the approach in Section 5.4 with chemical speciation calculations to obviate the need to fix any activity values.

Sumner, M. (ed.). (2000) *Handbook of soil science.* CRC Press, Boca Raton, FL. Section F of this advanced treatise contains four chapters giving detailed discussions of the weathering transformations of soil minerals informed by concepts in dissolution equilibria and kinetics.

White, A. F., and S. L. Brantley (eds.). (1995) *Chemical weathering rates of silicate minerals.* Vol. 31. Reviews in mineralogy. Mineralogical Society of America, Washington, DC. This advanced edited monograph offers comprehensive discussions of silicate mineral weathering from microscopic to field scales. Chapter 9, "Chemical Weathering Rates of Silicate Minerals in Soils," is of particular relevance to the current chapter.

Problems

The more difficult problems are indicated by an asterisk.

1. The rate of dissolution of albite ($NaAlSi_3O_8$) at 25 °C at pH < 6 can be described with Eq. 5.2, where A is Al and $k_d = 10^{-10.2} (H^+)^{1/2}$ mol g^{-1} s^{-1}. Calculate the dissolution rates at pH 4.0 ("acid rain") and 5.6. Compare the dissolution timescales of albite at the two pH values. Given the Arrhenius parameter B = 60 kJ mol^{-1} (Problem 2 in Chapter 4), compare the dissolution timescales at the two pH values when the temperature is 12 °C.

2. The rate of dissolution of kaolinite [$Si_4Al_4O_{10}(OH)_8$] as portrayed in Figure 5.2 can be described by Eq. 5.2 with the empirical equation

$$k_d = 10^{-8.28}(H^+)^{0.55} + 10^{-10.45} + 10^{-6.80}(OH^-)^{0.75} \quad (5.37)$$

over the pH range 1 to 13. Show that this rate law exhibits a minimum as a function of pH, either by plotting a graph or by applying differential calculus, and that the minimum value occurs at pH 6.8.

*3. Typically the distribution coefficient for Ca^{2+} in the soil solutions of arid-zone soils is about 0.75. Given this information and the data in Problem 7 of Chapter 4, calculate the IAP for calcite in a soil solution with pH 8, a conductivity of 2.5 dS m^{-1}, an HCO_3^- concentration of 1 mol m^{-3}, and a total Ca concentration of 3.8 mol m^{-3}. (*Answer:* IAP = $\gamma_{Ca^{2+}}[Ca^{2+}]\gamma_{CO_3^{2-}}[CO_3^{2-}] = 3.3 \times 10^{-9}$)

*4. The rate of *precipitation* of calcite ($CaCO_3$) near equilibrium follows Eq. 5.16 ($M^{m+} = Ca^{2+}$) with $k_p = 0.75 \pm 0.08$ L mol^{-1} s^{-1} appearing on the right side. Estimate the value of the *dissolution* rate coefficient k_d.

5. Suppose that dissolved Pb enters an acid soil in runoff water. Lead phosphates are often thought to be the solid phases controlling Pb solubility in acid soils, the two most important minerals being tertiary lead orthophosphate [$Pb_3(PO_4)_2$] and chloropyromorphite [$Pb_5(PO_4)_3Cl$]. The dissolution reactions for these two solid phases can be expressed by the equations

$$Pb(PO_4)_{\frac{2}{3}}(s) + \frac{4}{3}H^+ = Pb^{2+} + \frac{2}{3}H_2PO_4^- \quad \log K_{dis} = -1.80$$

$$Pb(PO_4)_{\frac{3}{5}}Cl_{\frac{2}{5}}(s) + \frac{6}{5}H^+ = Pb^{2+} + \frac{3}{5}H_2PO_4^- + \frac{1}{5}Cl^-$$

$$\log K_{dis} = -5.01$$

Prepare an activity–ratio diagram for Pb solubility control by these two minerals. Use $(H_2PO_4^-) = 10^{-6}$ and $(Cl^-) = 10^{-3}$ as fixed conditions. Which solid phase is expected to control solubility? Does the conclusion change if $(Cl^-) = 10^{-5}$?

6. The weathering of the feldspar anorthite ($CaAl_2Si_2O_8$) to form calcite and montmorillonite in soils (Eq. 2.8) is thought to be limited by unfavorable kinetics of calcite precipitation, which causes the activity of Ca^{2+} to remain larger than what K_{so} for calcite would predict at a given activity of CO_3^{2-}. This hypothesis implies that the activity of Ca^{2+} in equilibrium with anorthite is larger than that in the presence of calcite. Check this assertion by preparing an activity–ratio diagram for Ca solubility control by the two minerals. The congruent dissolution reaction for anorthite is

$$CaAl_2Si_2O_8(s) + 8H^+ = Ca^{2+} + 2Al^{3+} + 2Si(OH)_4^0$$

$$\log K_{dis} = 24.6$$

Assume that Eqs. 5.19c and 5.35 apply, and that the activity of silicic acid is controlled by quartz.

7. The dissolution reaction in Eq. 5.18c has different log $^*K_{dis}$ values depending on the crystallinity of the dissolving gibbsite phase. Prepare an activity–ratio diagram for Al solubility control by gibbsite of differing crystallinity and apply the GLO Step Rule to explain why poorly crystalline gibbsite is likely to be the first solid phase precipitated at pH 5 from soil solutions in which the Al^{3+} concentration exceeds 2 mmol m^{-3}. Estimate the Al^{3+} activity in the soil solution with the chemical speciation described in Table 4.4 and plot it on your activity–ratio diagram. Is gibbsite precipitation expected at this (Al^{3+})? Would your response to this question be different if oxalate were not present? Explain.

8. The transformation of anorthite to montmorillonite and calcite (Eq. 2.8 and Problem 6) is favored by $Si(OH)_4^0$ activities near 10^{-3} and pH values near 8.5. In calcareous soils, however, it is often observed that gibbsite forms instead of smectite when anorthite dissolves incongruently. Use Eqs. 5.19a through 5.19c to construct an activity–ratio diagram at pH 8 like Figure 5.5, then invoke the GLO Step Rule to explain how, when anorthite dissolves, gibbsite may form before montmorillonite.

9. According to the Jackson–Sherman weathering stages (Table 1.7), kaolinite and gibbsite formation are favored by intensive leaching of a soil profile with freshwater. This trend also implies that these two minerals will be disfavored by low levels of soil moisture or by saline waters, both of which are associated with a water activity less than 1.0. Examine this possibility by constructing an activity–ratio diagram like that in Figure 5.5, but with $(H_2O) = 0.5$ instead of 1.0. Take pH and all other log[activity] variables to have the same values as were used in constructing Figure 5.5. Compare

your results with this latter figure. Is smectite favored over a broader range of ($Si(OH)_4^0$) as the water activity decreases?

10. Examine the effect of solid-phase crystallinity on the activity–ratio diagram in Figure 5.5. Prepare activity–ratio diagrams using the alternative values of log K_{dis} for poorly crystalline kaolinite and gibbsite. What is the overall trend in mineral stability among an assembly comprising montmorillonite–kaolinite–gibbsite as crystallinity decreases and the silica concentration diminishes at pH 5?

11. Prepare an activity–ratio diagram for the two lead phosphates described in Problem 5 using log (HPO_4^{2-}) as the x-axis variable. Select pH 8 and (Cl^-) $= 10^{-3}$, noting that

$$H_2PO_4^- = HPO_4^{2-} + H^+ \quad \log K = -7.198$$

Plot lines corresponding to the Ca phosphate dissolution reactions in Eq. 5.36, assuming that calcite controls Ca solubility and $P_{CO_2} = 10^{-3.52}$ atm. Given your results, which Ca phosphate is best to add to a calcareous soil to immobilize Pb as an insoluble phosphate solid?

12. As indicated in Section 1.3 (Table 1.5), Cd may coprecipitate with calcite to form a solid solution of $CdCO_3$ (otavite) and $CaCO_3$. When this happens, the activity of $CdCO_3(s)$ is not 1.0, but instead is equal approximately to the fractional stoichiometric coefficient of Cd (ideal solid solution). Given that log $K_{dis} = -12.1$ for $CdCO_3(s)$, calculate the corresponding log K_{so} for a coprecipitate of otavite and calcite containing 6.3 mol% $CdCO_3$. Show that the activity of Cd^{2+} produced in the soil solution by this mixed solid is 1/16 that which would be produced by pure otavite under the same conditions of temperature, pressure, and soil solution composition.

*13. The clay mineralogy of a forested soil chronosequence developed on volcanic ash parent materials exhibits a transformation from proto-imogolite allophane ($Si_2Al_4O_{10} \cdot 5\,H_2O$) dominance to kaolinite dominance over a period of several thousand millennia. During this time, the silicic acid concentration and pH of the soil solution both decrease, from respective initial values of 0.3 mol m^{-3} and 7.0 to respective final values of 5.6 mmol m^{-3} and 4.6. Given the congruent dissolution reaction

$$Si_2Al_4O_{10} \cdot 5\,H_2O(s) + 12\,H^+ = 4\,Al^{3+} + 2\,Si(OH)_4^0 + 7\,H_2O(\ell)$$
$$\log K_{dis} = 26.0$$

prepare an activity–ratio diagram with log ($Si(OH)_4^0$) as the independent variable to examine solid-phase controls on Al solubility. Use your diagram to discuss the mineralogical transformations observed in the soil chronosequence. (*Hint:* Be sure to consider the effect of kaolinite crystallinity on your calculations.)

14. Prepare a predominance diagram for proto-imogolite allophane and kaolinite based on the dissolution reaction in Problem 13. Use exactly the same coordinate axes as those that appear in Figure 5.7. Plot the soil solution data given in Problem 13 on your diagram and discuss the mineralogical transformations observed in the soil chronosequence. A sharp decline in allophane content and a corresponding increase in kaolinite content is noted in the chronosequence when pH = 5.2 and $(Si(OH)_4^0) = 10^{-4.6}$.

*15. Prepare activity–ratio diagrams analogous to that in Figure 5.7 to verify the conclusions drawn in Section 5.5 concerning the effects of calcite crystallinity and CO_2 partial pressure. At what rhizosphere P_{CO_2} will there be *no* effect of decreasing calcite crystallinity on P solubility as predicted by the activity–ratio diagram?

6

Oxidation–Reduction Reactions

6.1 Flooded Soils

Almost all soils become flooded occasionally by rainwater or runoff, and a significant portion of soils globally underlies highly productive wetlands ecosystems that are intermittently or permanently inundated by water bodies. Peat-producing wetlands (bogs and fens) account for about half of these inundated soils, with swamps and rice fields each accounting for about one sixth more. Wetlands soils hold about one third of the total nonfossil fuel organic C that is stored below the land surface (i.e., about the same amount of C as is found in the atmosphere or in the terrestrial biosphere). This statistic is all the more impressive upon learning that wetlands cover only about 8% of the global land area. On the other hand, they are significant locales for denitrification processes, and they constitute the largest single source of methane entering the atmosphere, emitting half the global total and, therefore, contributing palpably to the stock of greenhouse gases (Section 1.1).

A soil inundated by water is essentially precluded from exchanging gases with the atmosphere, resulting in the depletion of oxygen and the subsequent accumulation of CO_2 because of metabolic processes engaged in by the biota. If sufficient labile humus (i.e., humus readily metabolized by microbes) is available to support respiration (problems 2 and 3 in Chapter 1), then a characteristic sequence of chemical reactions is observed in any submerged soil environment. This sequence is illustrated in Figure 6.1 for two agricultural soils: a German Inceptisol under cereal cultivation and a Philippines Vertisol under paddy rice cultivation. In the former soil, which was maintained in a well-aerated condition prior to inundation, nitrate is observed to disappear first from the soil solution, after which Mn(II) and Fe(II) begin to appear while soluble sulfate is depleted (left side of Fig. 6.1). Methane accumulation increases exponentially in the soil only after sulfate becomes undetectable and the Mn(II) and Fe(II) levels have stabilized. During the incubation time of about 40 days, the pH value in the soil solution increased from 6.3 to 7.5 and acetic acid (Table 3.1) as well as hydrogen gas were produced. These two latter compounds are common products of *fermentation*, a microbial metabolic process that occurs when oxygen levels are very low, resulting in the degradation of humus into simpler organic compounds, especially organic acids, along with the production of H_2 and CO_2. The reported concentrations of acetate (millimolar) and H_2 gas (micromolar in the soil solution) are typical

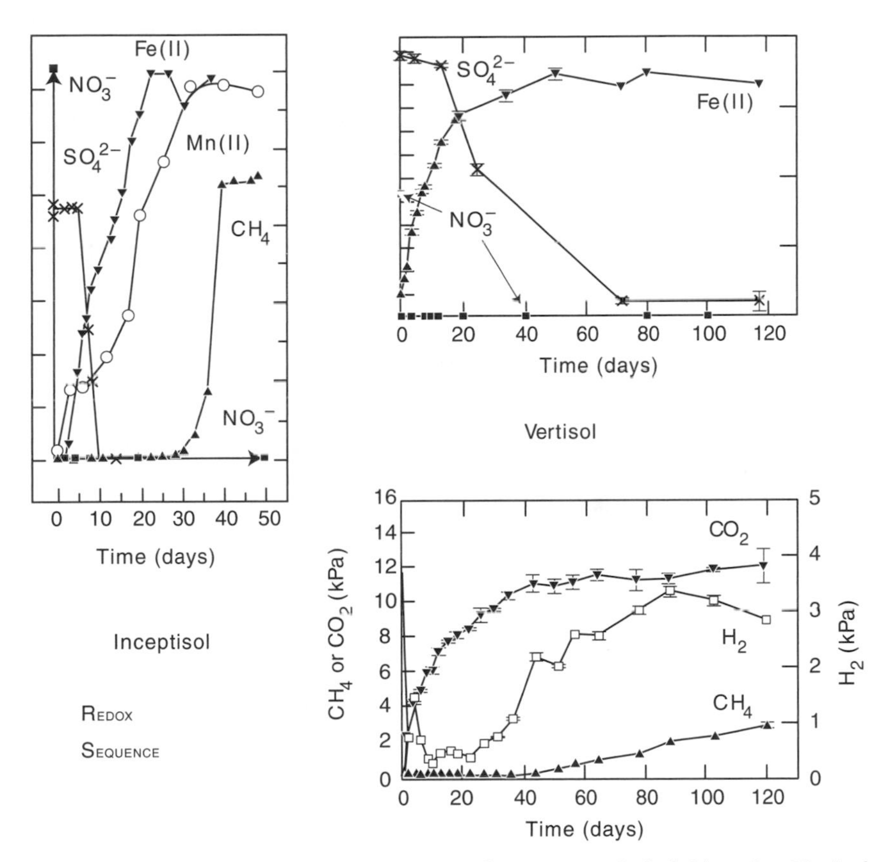

Figure 6.1. Temporal reduction sequences for an Inceptisol (left) and a Vertisol (right). Inceptisol data from Peters, V., and R. Conrad. (1996) Sequential reduction processes and initiation of CH_4 production upon flooding of oxic upland soils. *Soil Biol. Biochem.* 28:371–382. Vertisol data from Yao, H., et al. (1999) Effect of soil characteristics on sequential reduction and methane production in sixteen rice paddy soils from China, the Philippines, and Italy. *Biogeochemistry* 47:269–295.

of active fermentation. These fermentation products accumulate during the early stages of incubation, then are depleted as Mn(II) and Fe(II) levels increase or methane production commences, suggesting consumption by the microbial community during these latter stages.

Similar trends occur in the Vertisol (right side of Fig. 6.1), which was maintained under paddy conditions prior to sampling and inundation. Nitrate disappears quickly, whereas sulfate is depleted gradually over 2 months, after Fe(II) has risen to a plateau value. The characteristic increase in pH noted in the Inceptisol was observed in the Vertisol as well. Acetate levels also followed the same time trend as seen in the Inceptisol. The time trend of net CO_2 production (some of the CO_2 produced microbially is subsequently lost by carbonate precipitation) is remarkably similar to that of Fe(II) production; this strong visual correlation suggests that coupling of some kind is occurring between the two processes. Detailed C balance measurements indicated that the sum total of CO_2 and methane produced results in the loss of just 8% of the initial total organic C in the soil, with 85% of this loss manifest as CO_2. Thus, most of the labile C converted and released was used to produce CO_2 accompanying the accumulation of Fe(II) in the soil.

The temporal sequence of chemical reactions in a flooded soil has a spatial counterpart in sediments that are permanently inundated. Figure 6.2 illustrates this fact with vertical profiles of soluble oxygen, sulfate, methane,

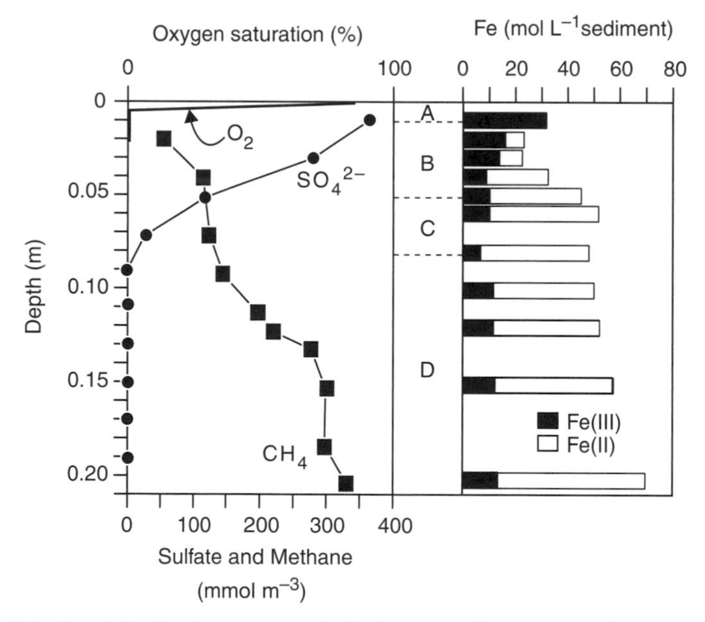

Figure 6.2. Spatial reduction sequence in freshwater sediments. Data from Kappler, A., et al. (2004) Electron shuttling via humic acids in microbial iron(III) reduction in a freshwater sediment. *FEMS Microbiol. Ecol.* 47:85–92.

and Fe observed in uncontaminated freshwater sediments sampled from the bottom of Lake Constance in Germany. Oxygen is depleted over the first few millimeters of zone A, which has a reported rust-brown color that reflects the presence of humus and Fe(III) oxide minerals. A green-brown zone B immediately below zone A is associated with the increase of Fe(II), whereas the reported black color of zone C, defined chemically by the disappearance of soluble sulfate, suggests secondary precipitation of Fe(II) sulfides. Layer D, which has no detectable sulfate, is associated with the increase of significant methane concentrations in the pore water. An expected increase in pH, from 6.8 to 7.3, across the 20-cm depth of the four subsurface zones also was observed. Horizontal spatial zonation akin to the vertical profile in Figure 6.2 can be seen typically in slowly flowing groundwater that has been contaminated by effluent from a contiguous landfill, as illustrated in Figure 6.3 for a study site in Denmark. After the plume of degradable xenobiotic organic compounds invades the sediments below the water table and is advected by ambient groundwater, microbial processes create a sequence of irregular regions with spatial ordering outward from the landfill that reflects the contrast between the incipient aerobic condition of the groundwater and the highly anaerobic conditions that develop near the landfill where the plume is most concentrated. The spatial ordering is, therefore, just the reverse of that observed with increasing depth in sediments lying at the bottom of a river or lake, although the ordering from the tip of the invading plume back toward its landfill source

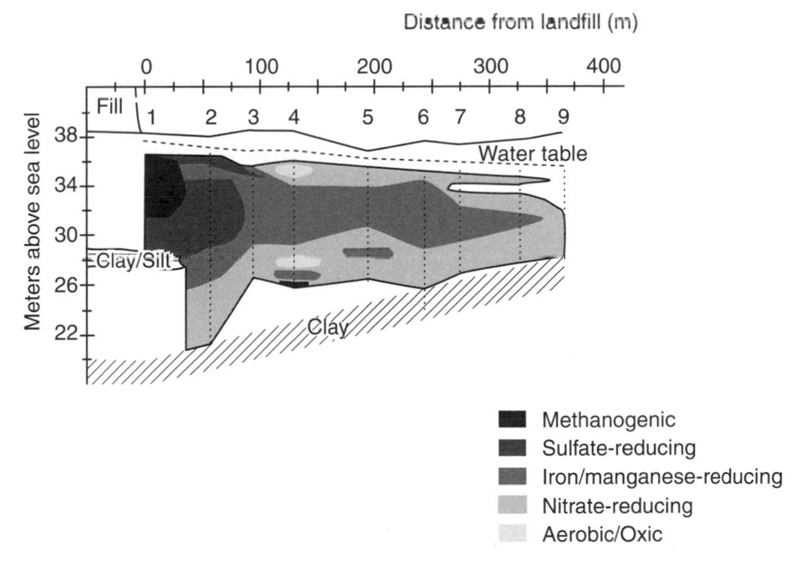

Figure 6.3. Spatial reduction sequence in an organic contaminant plume invading oxic groundwater. Reprinted with permission from Christensen, T. H., et al. (2000) Characterization of redox conditions in groundwater contaminant plumes. *J. Contamin. Hydrol.* 45:165–241.

mimics the spatial sequence in bottom sediments and the temporal sequence in flooded soils.

Detailed microbiological studies of the sequences and profiles depicted variously in figures 6.1 to 6.3 have provided important insights regarding the causes of the characteristic ordering. For example, addition of nitrate to a soil largely depleted of labile humus by a prior long incubation under anaerobic conditions slightly inhibits the production of soluble Fe, but severely inhibits the disappearance of soluble sulfate and the production of methane. These effects, however, are vitiated after fermentation products, such as H_2 gas or acetic acid, are added to the soil. Similarly, addition of ferrihydrite particles (Section 2.4) to a soil low in labile humus suppresses the loss of soluble sulfate and slows the production of methane, and addition of soluble sulfate inhibits methane production—but these effects also can be reversed by supplying fermentation products, especially H_2 gas. Two related overall conclusions can be drawn from these kinds of observations: (1) chemical reactions occurring earlier in the sequence can inhibit those that come later and (2) significant competition for labile humus or microbial fermentation products exists that favors the chemical reactions occurring earlier in the sequence. These conclusions in turn suggest that closer examination of the chemical reactions in the sequence will reveal the operation of general principles underlying the observed biogeochemistry of flooded soils. Evidently, competitive microbial intervention in this biogeochemistry is reflected primarily by the extent to which labile humus or the products of fermentation are depleted as they become consumed. This latter inference is in fact borne out by reports of H_2 gas, with a residence time in soils that is very short (on the order of minutes), being driven to much lower concentrations in the soil solution by the production of soluble Fe than, for example, by methane production, thus indicating that H_2-consuming microbes associated with chemical reactions that occur earlier in the sequence operate much more efficiently than those associated with reactions that occur later on.

6.2 Redox Reactions

An *oxidation–reduction* (or *redox*) reaction is a chemical reaction in which electrons are transferred completely from one species to another. The chemical species that donates electrons in this charge transfer process is called a *reductant*, whereas the one accepting electrons is called an *oxidant*. For example, in the *reductive dissolution* reaction

$$\mathrm{FeOOH(s)} + 3\,\mathrm{H}^+ + \mathrm{e}^- = \mathrm{Fe}^{2+} + 2\,\mathrm{H_2O}(\ell) \qquad (6.1)$$

the solid phase, goethite (Table 2.5 and Fig. 2.11), on the left side is the oxidant that accepts an electron (e^-) and reacts with protons to form the soluble species Fe^{2+} on the right side. As written, Eq. 6.1 is a *reduction half-reaction*, in which an electron in aqueous solution serves as one of the reactants. This latter species, like the proton in aqueous solution, is understood in a formal sense to participate in charge transfer processes. The overall redox reaction always must be the combination of two reduction half-reactions, such that the species e^- does not appear explicitly. Equation 6.1, for example, could be combined (coupled) with the reverse of a half-reaction in which CO_2 is transformed to acetate:

$$\frac{1}{4}\,CO_2\,(g) + \frac{7}{8}\,H^+ + e^- = \frac{1}{8}\,CH_3CO_2^- + \frac{1}{4}\,H_2O\,(\ell) \qquad (6.2)$$

to cancel the aqueous electron and represent the reductive dissolution of goethite coupled to the oxidation of acetate, $CH_3CO_2^-$, which serves as a reductant:

$$\begin{aligned} &FeOOH(s) + \frac{1}{8}\,CH_3CO_2^- + \frac{17}{8}\,H^+ \\ &= Fe^{2+} + \frac{1}{4}\,CO_2(g) + \frac{7}{4}\,H_2O\,(\ell) \end{aligned} \qquad (6.3)$$

Redox reactions can be described in terms of thermodynamic equilibrium constants analogously to the approach used in Chapter 5 for mineral dissolution reactions. The only new feature is the need to account for electron transfer. This is done by associating *oxidation numbers* with oxidants and reductants, while being careful to balance the overall redox reaction in terms of reduction half-reactions, as explained in Special Topic 4 at the end of this chapter. A list of important reduction half-reactions and their thermodynamic equilibrium constants (at 25 °C) is provided in Table 6.1. These equilibrium constants have exactly the same meaning as those discussed in Chapters 4 and 5, even though the reactions to which they refer contain the aqueous electron. The reason for this is the convention by which *the reduction of the proton is defined to have log K* = 0 (the third reaction listed in Table 6.1). Thus, every half-reaction in Table 6.1 may be combined with the reverse of the proton reduction reaction to cancel e^- while leaving log K for the half-reaction completely unchanged numerically. In this sense, each half-reaction in Table 6.1 is equivalent to an overall redox reaction that couples it to the oxidation of H_2 gas serving as the reductant.

The log K data in Table 6.1 can be combined in the usual way to calculate a value of log K for an overall redox reaction. Consider, for example, the combination of Eqs. 6.1 and 6.2 to produce Eq. 6.3. According to Table 6.1, the *reduction* of goethite has log K = 13.34, and the *oxidation* of acetate has log K = 1.20. It follows that the reductive dissolution of goethite by acetate has log K = 13.34 + 1.20 = **14.54**. This equilibrium constant can be expressed in terms of activities related to Eq. 6.3:

$$K = \frac{(Fe^{2+})\,(CO_2)^{\frac{1}{4}}\,(H_2O)^{\frac{7}{4}}}{(FeOOH)\,(H^+)^{\frac{17}{8}}\,(CH_3CO_2^-)^{\frac{1}{8}}} = 10^{14.54} \qquad (6.4)$$

Table 6.1
Some important reduction half-reactions (25 °C).

Reduction half-reaction	log K
$\frac{1}{4}O_2(g) + H^+ + e^- = \frac{1}{2}H_2O(\ell)$	20.75
$\frac{1}{2}O_2(g) + H^+ + e^- = \frac{1}{2}H_2O_2$	11.50
$H^+ + e^- = \frac{1}{2}H_2(g)$	0.00
$\frac{1}{3}NO_3^- + \frac{4}{3}H^+ + e^- = \frac{1}{3}NO(g) + \frac{2}{3}H_2O(\ell)$	16.15
$\frac{1}{2}NO_3^- + H^+ + e^- = \frac{1}{2}NO_2^- + \frac{1}{2}H_2O(\ell)$	14.15
$\frac{1}{4}NO_3^- + \frac{5}{4}H^+ + e^- = \frac{1}{8}N_2O(g) + \frac{5}{8}H_2O(\ell)$	18.90
$\frac{1}{5}NO_3^- + \frac{6}{5}H^+ + e^- = \frac{1}{10}N_2(g) + \frac{3}{5}H_2O(\ell)$	21.05
$\frac{1}{8}NO_3^- + \frac{5}{4}H^+ + e^- = \frac{1}{8}NH_4^+ + \frac{3}{8}H_2O(\ell)$	14.90
$Mn^{3+} + e^- = Mn^{2+}$	25.50
$MnOOH(s) + 3H^+ + e^- = Mn^{2+} + 2H_2O(\ell)$	25.36
$\frac{1}{2}Mn_3O_4(s) + 4H^+ + e^- = \frac{3}{2}Mn^{2+} + 2H_2O(\ell)$	30.68
$\frac{1}{2}MnO_2(s) + 2H^+ + e^- = \frac{1}{2}Mn^{2+} + H_2O(\ell)$	21.82
$\frac{1}{2}MnO_2(s) + \frac{1}{2}CO_2(g) + H^+ + e^- = \frac{1}{2}MnCO_3(s) + \frac{1}{2}H_2O(\ell)$	18.00
$Fe^{3+} + e^- = Fe^{2+}$	13.00
$\frac{1}{2}Fe^{2+} + e^- = \frac{1}{2}Fe(s)$	−7.93
$Fe(OH)_3(s) + 3H^+ + e^- = Fe^{2+} + 3H_2O(\ell)$	17.14
$FeOOH(s) + 3H^+ + e^- = Fe^{2+} + 2H_2O(\ell)$	13.34
$\frac{1}{2}Fe_3O_4(s) + 4H^+ + e^- = \frac{3}{2}Fe^{2+} + 2H_2O(\ell)$	18.16
$\frac{1}{2}Fe_2O_3(s) + 3H^+ + e^- = Fe^{2+} + \frac{3}{2}H_2O(\ell)$	12.96
$\frac{1}{4}SO_4^{2-} + \frac{5}{4}H^+ + e^- = \frac{1}{8}S_2O_3^{2-} + \frac{5}{8}H_2O(\ell)$	4.85
$\frac{1}{8}SO_4^{2-} + \frac{9}{8}H^+ + e^- = \frac{1}{8}HS^- + \frac{1}{2}H_2O(\ell)$	4.25
$\frac{1}{8}SO_4^{2-} + \frac{5}{4}H^+ + e^- = \frac{1}{8}H_2S + \frac{1}{2}H_2O(\ell)$	5.13
$\frac{1}{2}CO_2(g) + \frac{1}{2}H^+ + e^- = \frac{1}{2}CHO_2^-$	−5.22
$\frac{1}{4}CO_2(g) + \frac{7}{8}H^+ + e^- = \frac{1}{8}C_2H_3O_2^- + \frac{1}{4}H_2O(\ell)$	−1.20
$\frac{7}{30}CO_2(g) + \frac{29}{30}H^+ + e^- = \frac{1}{30}C_6H_5COO^- + \frac{2}{5}H_2O(\ell)$	1.76
$\frac{1}{4}CO_2(g) + \frac{1}{12}NH_4^+ + \frac{11}{12}H^+ + e^- = \frac{1}{12}C_3H_4O_2NH_3 + \frac{1}{3}H_2O(\ell)$	0.84
$\frac{1}{4}CO_2(g) + H^+ + e^- = \frac{1}{24}C_6H_{12}O_6 + \frac{1}{4}H_2O(\ell)$	−0.20
$\frac{1}{8}CO_2(g) + H^+ + e^- = \frac{1}{8}CH_4(g) + \frac{1}{4}H_2O(\ell)$	2.86

If the activities of goethite and water are set equal to 1.0, and the usual expressions for the activities of $CO_2(g)$ and H^+ are used (Section 5.5), then Eq. 6.4 can be written in the form

$$(Fe^{2+})\, P_{CO_2}^{\frac{1}{4}}\, 10^{\frac{17}{8}pH} / (CH_3CO_2^-)^{\frac{1}{8}} = 10^{14.54} \quad (6.5)$$

Choosing pH 6 and a CO_2 pressure of $10^{-3.52}$ atm as typical values, one reduces this equation to the simpler expression

$$\frac{(Fe^{2+})}{(CH_3CO_2^-)^{\frac{1}{8}}} = 10^{2.67} \tag{6.6}$$

Equation 6.6 leads to the conclusion that the equilibrium state of the redox reaction in Eq. 6.3 requires the activity of Fe^{2+} in the soil solution to be more than 460 times greater than the eighth root of the activity of acetate in the soil solution. For example, if $(Fe^{2+}) = 10^{-6}$, then Eq. 6.6 predicts $(CH_3CO_2^-) \approx 10^{-69}$. This result shows that, at equilibrium, acetate would be rather well oxidized to CO_2 by the reductive dissolution of goethite.

The reduction half-reactions in Table 6.1 also can be used individually to predict ranges of pH and other log activity variables over which one redox species or another predominates. Nearly all reduction half-reactions are special cases of the generic equation

$$mA_{ox} + nH^+ + e^- = pA_{red} + qH_2O\,(\ell) \tag{6.7}$$

where A is a chemical species in any phase [e.g., $CO_2(g)$ or Fe^{2+}] and "ox" or "red" designates oxidant or reductant respectively. The equilibrium constant for the generic half-reaction is

$$K = \frac{(A_{red})^p\,(H_2O)^q}{(A_{ox})^m\,(H^+)^n\,(e^-)} \tag{6.8}$$

This equation can be rearranged, for example, to provide an expression for pH in terms of other log activity variables. The species A_{ox} and A_{red}, whose activities are related in this way through electron transfer and Eq. 6.8, are termed a *redox couple*. In Eqs. 6.1 and 6.2, for example, the redox couples are goethite/Fe^{2+} and CO_2/acetate respectively.

Application of Eq. 6.8 to soils requires an interpretation of (e^-), the activity of an aqueous electron. This can be accomplished by following the paradigm already well established for the aqueous proton. Soil acidity is expressed quantitatively by the negative common logarithm of the proton activity, the pH value. Similarly, soil "oxidizability" can be expressed by the negative common logarithm of the electron activity, *the pE value:*

$$pE \equiv -\log(e^-) \tag{6.9}$$

Large values of pE favor the existence of electron-poor species (i.e., oxidants), just as large values of pH favor the existence of proton-poor species (i.e., bases). Small values of pE favor electron-rich species, reductants, just as small values of pH favor proton-rich species, acids. Unlike pH, however, pE can take on negative values. This difference results from the separate conventions used to define log K for acid–base and redox reactions (Table 6.2). In soils, pE ranges

Table 6.2
Comparing pE and pH.

Species	Reaction	Predominance	Condition
Acid	Donates H^+	Low pH	Acidic
Base	Accepts H^+	High pH	Basic
Reductant	Donates e^-	Low pE	Reducing
Oxidant	Accepts e^-	High pE	Oxidizing
Reference reactions			
$H_2O(\ell) + H^+ = H_3O^+$		$\log K \equiv 0$	Acid–base
$H^+ + e^- = \frac{1}{2} H_2(g)$		$\log K \equiv 0$	Redox

from around +13.0 to less than −6.0. At circumneutral pH, this range can be partitioned broadly into *oxic* (pE $\gtrsim$ +2), *suboxic* (+12 $\gtrsim$ pE $\gtrsim$ +2), and *anoxic* (pE < +2) zones.

These definitions are motivated after rewriting Eq. 6.8 as an expression for pE [assuming $(H_2O) = 1.0$]:

$$\mathrm{pE} = \log\ K + \log\left[\frac{(A_{ox})^m}{(A_{red})^p}\right] - n\ \mathrm{pH} \tag{6.10}$$

Oxic conditions occur in a soil solution at pH 7 if the partial pressure of oxygen is greater than about 0.01 atm (about 5% of the atmospheric partial pressure). The corresponding pE value can be calculated by introducing the oxygen reduction half-reaction (first reaction listed in Table 6.1) into Eq. 6.10, with $A_{ox} = O_2(g)$ and $A_{red} = H_2O(\ell)$:

$$\mathrm{pE} = 20.75 + \frac{1}{4}\log\ P_{O_2} - 7 = 20.75 - 0.5 - 7 = +\ \mathbf{13.25}$$

Thus, pE values greater than about 12.0 characterize oxic soils. At pE values less than +12.0, the partial pressure of oxygen drops below 0.01 atm and anaerobic conditions obtain.

Suboxic status in a soil at pH 7 can be associated with pE values calculated for nitrate reduction or for the reductive dissolution of $MnO_2(s)$. The latter reaction, listed in the middle of Table 6.1, yields the pE expression [assuming $(MnO_2(s)) = 1.0$]

$$\mathrm{pE} = 21.82 - \frac{1}{2}\log\left(Mn^{2+}\right) - 14 = 7.82 - \frac{1}{2}\log\left(Mn^{2+}\right) \approx +\mathbf{8.8}$$

if $(Mn^{2+}) \approx 10^{-2}$. Similarly, the reduction of nitrate to form ammonium ions (eighth reaction listed in Table 6.1) yields the pE expression

$$\mathrm{pE} = 14.90 + \frac{1}{8}\ \log\left[\frac{\left(NO_3^-\right)}{\left(NH_4^+\right)}\right] - 8.75 \approx +\ \mathbf{6.15},$$

if the pE value for $\left(NO_3^-\right) = \left(NH_4^+\right)$ is taken as a threshold.

Anoxic soils are characterized by the reduction of ferric iron and sulfate along with the production of methane. Returning to Eq. 6.1 as an example, one finds the pE expression

$$pE = 13.34 - \log(Fe^{2+}) - 21.0 = -7.66 - \log(Fe^{2+}) \approx \mathbf{-4.66}$$

if $(Fe^{2+}) \approx 10^{-3}$, an activity typical of a flooded soil. Note that the reductive dissolution of $Fe(OH)_3(s)$, a poorly crystalline solid phase, would yield a pE value near −2 under the same conditions, thus illustrating the need for broad ranges of pE to delineate oxic, suboxic, and anoxic conditions.

Nitrate reduction to form ammonium ions is an example that is useful for emphasizing another important concept about redox reactions. For a fixed pE value, an increase in ammonium ion activity relative to that of nitrate requires lowering the pH value, a trend that also can be deduced directly from Eq. 6.10 by considering decreases in $[(A_{ox})^m/(A_{red})^p]$. *The formation of reductants almost always results in proton consumption and, therefore, an increase in pH.* Thus each reduction half-reaction in Table 6.1 represents a mechanism by which free protons can be removed from the soil solution. Reduction is therefore an important way by which soil acidity can be decreased. Conversely, oxidation can create free protons and increase soil acidity.

It is also very important to keep in mind that the data in Table 6.1 imply that certain redox reactions *can* occur in soils, but not that they *will* occur. A chemical reaction that is favored by a large value of log K is not necessarily favored kinetically. This fact is especially applicable to redox reactions because they are often extremely slow, and because reduction and oxidation half-reactions often do not couple well to each other. For example, the coupling of the half-reaction for $O_2(g)$ reduction with that for acetate oxidation leads to a log K value of 22.0 for the overall redox reaction:

$$\frac{1}{4}\,O_2\,(g) + \frac{1}{8}\,C_2H_3O_2^- + \frac{1}{8}\,H^+ = \frac{1}{4}\,CO_2\,(g) + \frac{1}{4}\,H_2O\,(\ell) \qquad (6.11)$$

For a soil solution that is in equilibrium with the atmosphere ($P_{O_2} = 0.21$ atm), the value of log K just given predicts complete oxidation of acetate at any pH value. But this prediction is contradicted by the observed persistence of dissolved acetate and other components of humus in soil solutions under surface terrestrial conditions. A rather similar example can be developed by considering $N_2(g)$ oxidation coupled to $O_2(g)$ reduction, leading to the conclusion that, under the current oxic conditions at the earth's surface, the oceans should have become nitrate solutions.

The typically sluggish nature of redox kinetics implies that catalysis is required if redox reactions are to equilibrate on timescales comparable with the life cycles of the biota. In soils, the catalysis of redox reactions is effected by microbial organisms and, to a lesser extent, mineral surfaces. In the presence of the appropriate microbial species, a reduction half-reaction can proceed quickly enough in a soil to produce activity values of the reactants and products that largely agree with equilibrium predictions. If the reductant thus produced by the half-reaction accumulates outside the microbial cell, catalysis is termed *dissimilatory*; otherwise, it is *assimilatory*. For example, nitrate reduction by bacteria to yield ammonium ions that are metabolized to form amino acids, such as glutamic and aspartic acid (Table 3.2), is assimilatory, whereas denitrification is dissimilatory. Of course, these possibilities are dependent entirely on the growth and ecological interactions of the soil microbial community and the degree to which the products of biochemical reactions can diffuse readily in the soil solution. In some cases, redox reactions will be controlled by the highly localized and variable dynamics of an open biological system, with the result that redox speciation at best will correspond to local conditions of partial equilibrium. In other cases, including often the important one of the flooded soil, redox reactions will be controlled by the behavior of a closed chemical system that is catalyzed effectively by bacteria and mineral surfaces, for which an equilibrium description is apt. Regardless of which of these two extremes is the more appropriate to characterize redox reactions, the role of organisms (and mineral surfaces) deals only with the kinetics aspect of redox. If a redox reaction is not favored by a positive log K, microbial intervention cannot change that fact.

6.3 The Redox Ladder

A *redox ladder* is a vertical line marked off with "rungs" that are occupied by redox couples, with the oxidant on the left and the reductant on the right. This vertical line is a coordinate axis labeled by pE values calculated using Eq. 6.10 (or an equivalent expression) for a fixed pH value, usually pH 7.0. Construction of the "ladder" is based on three conventional rules.

Rule 1: Each redox couple on the ladder must be related by a reduction half-reaction in which the stoichiometric coefficient of e^- is 1.0. If this half-reaction has the generic form in Eq. 6.7, then pE values are calculated with Eq. 6.10 after fixing the pH value and setting $(H_2O(\ell)) = 1.0$.

Rule 2: If the oxidant and reductant are in the same phase, then (A_{ox}) and (A_{red}) are each set equal to 1.0, yielding a simplified equation for pE at a given pH:

$$\text{pE} = \log K - n\,\text{pH} \qquad (6.12)$$

where K is the thermodynamic equilibrium constant for the reduction half-reaction transforming the oxidant into the reductant in a redox couple and n is the stoichiometric coefficient of H^+ in this half-reaction.

Example: The reduction of sulfate to form bisulfide is described by the half-reaction (Table 6.1)

$$\frac{1}{8}\,\mathrm{SO_4^{2-}} + \frac{9}{8}\,\mathrm{H^+} + \mathrm{e^-} = \frac{1}{8}\,\mathrm{HS^-} + \frac{1}{2}\,\mathrm{H_2O}\,(\ell) \qquad (6.13)$$

for which $\log K = 4.25$ at 25 °C. Placement of the redox couple SO_4^{2-}/HS^-, on the ladder is therefore at $pE = -3.63$, if $pH = 7$ (Fig. 6.4).

Comment: If the activities of the oxidant and reductant are known, they may be used to calculate pE according to Eq. 6.10. For example, $(SO_4^{2-}) = 10^{-3}$ and $(HS^-) = 10^{-4}$ could occur in a fresh groundwater sample, leading to $pE = -3.50$ at pH 7. Note the typical rather small effect of this correction on the pE value.

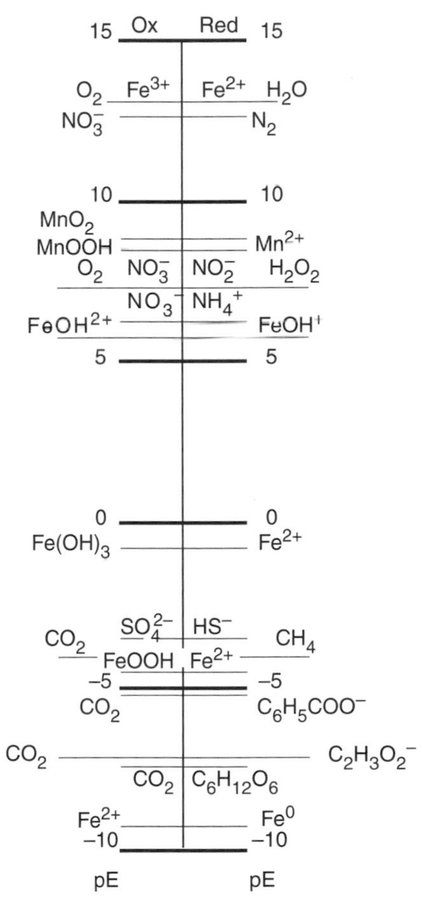

Figure 6.4. A redox ladder constructed for pH 7. Auxiliary conditions imposed on redox species activities are discussed in the text.

Rule 3: If either the oxidant or the reductant is in a gaseous or a solid phase, the gas-phase species activity is equated to the partial pressure in units of atmospheres and the solid-phase species activity is set equal to 1.0. The activity of the remaining, aqueous-phase species in the redox couple is equated to its concentration in moles per cubic decimeter (Section 4.5). Suitable values of the partial pressure or aqueous concentration are used in calculating pE with Eq. 6.10.

Example: Calculations illustrating Rule 3 were presented in Section 6.2 for the reduction of $O_2(g)$ and the reductive dissolution of $MnO_2(s)$. The resulting pE values are also depicted in Figure 6.4. Another example is provided by the reduction of $CO_2(g)$ to form glucose, as occurs in photosynthesis (Table 6.1):

$$\frac{1}{4}\, CO_2\,(g) + H^+ + e^- = \frac{1}{24}\, C_6H_{12}O_6 + \frac{1}{4}\, H_2O\,(\ell)$$

$$\log K = -0.20 \qquad (6.14a)$$

If $P_{CO_2} = 10^{-2}$ atm and $(C_6H_{12}O_6) \approx 5 \times 10^{-4}$ (based on a glucose concentration of 0.5 mol m^{-3}), then at pH 7.0,

$$\begin{aligned} pE &= -\,0.20 + \frac{1}{4}\, P_{CO_2} - \frac{1}{24}\, \log\,(C_6H_{12}O_6) - pH \\ &= -\,0.20 - 0.50 + 0.14 - 7 = \mathbf{-7.56} \end{aligned}$$

This very low pE value is typical of reduction half-reactions involving biomolecules. Note again the small effect of the redox couple activities on the pE value.

Perhaps the most important application of the redox ladder is its use to establish which member of a redox couple is favored (i.e., thermodynamically stable) under given conditions in a soil. This application is initiated by determining where a soil is *poised* with respect to pE. Poising is to pE what buffering is to pH (Section 3.3 and Problem 8 in Chapter 3). Thus, a well-poised soil resists changes in pE, just as a well-buffered soil resists changes in pH. Indeed, pE "poisers" are available with which to calibrate pE electrodes, just as pH buffers are available with which to calibrate pH electrodes. For example, the half-reaction

$$p\text{-benzoquinone} + H^+ + e^- = \frac{1}{2}\ \text{hydroquinone} \qquad \log K = 11.83 \qquad (6.15)$$

is often used for calibration, with the poised suboxic pE value being given by pE = 11.83 – pH, according to Eq. 6.12. [*Benzoquinone* comprises a benzene ring with a pair of *carbonyl* (C = O) substituents. If one of the carbonyls is converted to a C—OH group, the resultant compound is termed *semiquinone* and, if both carbonyls are converted, it is termed *hydroquinone.* This latter compound, a powerful reductant, differs from catechol (Section 3.1) in having its two OH substituents lie along a single axis of symmetry of the benzene ring instead of being adjacent to one another on the ring.]

In soils, the most important redox-active elements are H, C, N, O, S, Mn, and Fe, with the addition of Cr, Cu, As, Se, Ag, Pb, U, and Pu for contaminated environments. Poising by a reduction half-reaction involving one of these chemical elements depends on its relative abundance as an oxidant species in a soil. For example, abundant $O_2(g)$ in a soil atmosphere implies poising of the soil by the first reduction half-reaction listed in Table 6.1, with the poised pE value then given by

$$pE = 20.75 + \tfrac{1}{4}\log P_{O_2} - pH \tag{6.16a}$$

according to Eq. 6.10. If P_{O_2} drops well below its nominal atmospheric value (0.21 atm), $O_2(g)$ no longer will be sufficient to poise pE and the reduction half-reactions of nitrate become potential candidates for poising pE. For example, nitrate reduction, described by the two reactions in Table 6.1 with aqueous ions as products, might poise pE in the suboxic range (Fig. 6.4) if nitrate is abundant. Otherwise, poising by the reductive dissolution of $MnO_2(s)$ would be expected, because Mn is a relatively abundant metal element in soils (Table 1.2), with the poised pE value given by

$$pE = 21.82 - \frac{1}{2}\log\left(Mn^{2+}\right) - 2\,pH \tag{6.16b}$$

Note that the other solid-phase oxidant species of Mn listed in Table 6.1 are thermodynamically unfavored relative to $MnO_2(s)$. If the reverse of the reduction half-reaction for $MnO_2(s)$ is added to those for the other two oxidant solid phases, the resulting $\log K > 0$. Indeed, manganite (γ-MnOOH), a typical product of abiotic air oxidation of soluble Mn(II), *disproportionates* into $MnO_2(s)$ and Mn^{2+}:

$$\begin{aligned} MnOOH(s) + H^+ &= \frac{1}{2}MnO_2(s) + \frac{1}{2}Mn^{2+} \\ &+ H_2O(\ell) \quad \log K = 3.54 \end{aligned} \tag{6.17}$$

Nonetheless, either of the solid phases, manganite or hausmannite (Mn_3O_4), may be found in soils as metastable species. [Note also that various species of $MnO_2(s)$ exist (*polymorphs*). Equation 6.16b applies to that most resembling birnessite (Section 2.4). If the most stable species (pyrolusite, β-MnO_2) were considered instead, log K would be changed to 20.56.]

Under anoxic conditions, reduction half-reactions involving oxidant species of Fe, S, or C (if the reductant product is a biomolecule) can poise pE in a soil:

$$pE = 17.14 - \log(Fe^{2+}) - 3pH \tag{6.16c}$$

if the oxidant is $Fe(OH)_3(s)$,

$$pE = 4.25 + \frac{1}{8} \log \left[\frac{(SO_4^{2-})}{(HS^-)} \right] - \frac{9}{8} pH \tag{6.16d}$$

if the oxidant is soluble sulfate, or

$$pE = 2.86 + \frac{1}{8} \log \left[\frac{P_{CO_2}}{P_{CH_4}} \right] - pH \tag{6.16e}$$

if methanogenic bacteria are active. At a given pH value, the pE values for the reductive dissolution of $Fe(OH)_3(s)$, sulfate reduction, and methane production lie successively lower on the redox ladder (Fig. 6.4). To the extent that the pE values are well-separated on the ladder, they are characteristic of the reduction half-reactions from which they are derived. In recognition of this possibility and the ubiquity of dissimilatory microbial catalysis of soil redox reactions, the half-reactions represented by Eq. 6.16 are called *terminal electron-accepting processes* (TEAPs). In microbiological terms, one portrays TEAPs as key chemical reactions governing microbial respiration and portrays the bacteria involved as oxygen-, nitrate-, or iron-respiring, and so on. Thus, pE in soils is pictured as poised by TEAPs involving the abundant oxidant species of the elements O, N, Mn, Fe, S, or C. In polluted soils, TEAPs involving the eight potentially hazardous elements mentioned earlier also may poise pE if abundant oxidant species of them are present.

How is the poising of a soil pE value quantified? The corresponding question of how the buffering of a soil pH value is quantified has a simple answer in terms of pH measurement using a glass electrode—a technological advance perfected by Arnold Beckman more than 75 years ago (see Special Topic 5 at the end of this chapter). Unfortunately, an equivalent success has not occurred in the development of an electrochemical method to measure pE. To be sure, an *electrode potential* (E_H, in units of volts) can be defined formally in terms of pE:

$$E_H \equiv \frac{RT}{F} \ln 10 \; pE = 0.05916 \; pE \quad (25\,°C) \tag{6.18}$$

where R, the molar gas constant; T, the absolute temperature (298.15 K at 25 °C); and F, the Faraday constant; are defined in the Appendix. Equation 6.18 is a purely formal relationship amounting to a transformation of units. In practice, electrochemical measurements of E_H are subject to numerous interferences, notably the lack of thermodynamic equilibrium between oxidant and reductant in a redox couple (i.e., the ambiguity inherent to interpreting a voltage read at zero net current as the unique signature of a single redox couple at equilibrium) and an anomalous selectivity for Fe(III)/Fe(II) redox couples. Measured values of E_H obtained by a suitably calibrated electrode thus have only a qualitative significance in soil solutions. A similar conclusion can be drawn concerning the use of redox indicators, which, like pH indicators, change color at certain pE values. These compounds can be adsorbed by soil particles or complexed by metal cations, and their colors are pH sensitive.

The most common method used to measure pE in soils and aquifers is quantitation of redox couples. For example, $O_2(aq)$ concentrations can be measured to determine whether poising by O_2 reduction is occurring. If these concentrations are below about 15 mmol m^{-3} (corresponding to a partial pressure of about 0.01 atm), then O_2 reduction cannot be the TEAP that poises soil pE. Similarly, nitrate concentrations below about 3 mmol m^{-3} would eliminate nitrate reduction as a candidate for the TEAP-poising pE. On the other hand, Mn(II) or Fe(II) concentrations exceeding about 100 mmol m^{-3} may signal pE poising by the reductive dissolution of Mn(IV) or Fe(III) oxy(hydr)oxides respectively. Depletion of soluble sulfate below 100 mmol m^{-3} would tend to rule out sulfate reduction as the pE-poising TEAP, whereas methane concentrations above about 50 mmol m^{-3} point to methane production as a poiser. Supporting microbiological evidence for large numbers of the bacteria utilizing a proposed pE-poising TEAP is a helpful adjunct to quantitation. As implied in Figures 6.1 to 6.3, however, overlapping TEAPs identified by quantitation can obfuscate this approach.

6.4 Exploring the Redox Ladder

The redox ladder in Figure 6.4 shows the O_2/H_2O couple on the top "rung" and the $CO_2/C_6H_{12}O_6$ couple on a very low rung. Oxygen gas in the atmosphere constitutes an enormous oxidant reservoir, whereas humus and the biota, loosely represented by glucose, constitute an equally important reservoir of organic reductants. As noted in Section 6.2, large values of pE favor oxidants like $O_2(g)$, whereas small (e.g., negative) values of pE favor reductants like glucose and other organic molecules. In a redox reaction, two reduction half-reactions are combined after one of them is reversed, such that the resulting overall reaction does not exhibit e^- as a participant. How does one determine which of the two half-reactions to reverse? When two half-reactions are coupled, *electron transfer always must be from low pE (electron rich) on the redox ladder to high pE (electron poor) on the ladder.* Thus, in the current example, the half-reaction involving glucose is the one to be reversed, making glucose a reactant and yielding the overall redox reaction

$$\frac{1}{4}\,O_2\,(g) + \frac{1}{24}\,C_6H_{12}O_6 = \frac{1}{4}\,CO_2\,(g) + \frac{1}{4}\,H_2O\,(\ell) \qquad (6.19)$$

which loosely depicts the aerobic oxidation of humus, akin to the reaction in Eq. 6.11. Equation 6.19 may be interpreted as an electron titration of humus, analogous to the proton titration of humus described in Section 3.3. Oxidant [$O_2(g)$] reacts with reductant C to yield oxidant C and water, just as base reacts with acidic C to yield basic C and water (Table 6.2). Out of this analogy, humus emerges as an important terrestrial reservoir of both reactive protons and reactive electrons, with the capability, therefore, of both buffering soil pH and poising soil pE. Whether this potential is realized, of course, depends on the relative abundance of competing inorganic buffers and poisers, and the ability of the soil microbial community to catalyze the relevant electron transfer reactions.

Suppose, for example, that soil pE is poised by the reductive dissolution of $MnO_2(s)$ coupled with the oxidation of humus (the pE value for the MnO_2/Mn^{2+} couple lies above those for CO_2/organic molecule couples in Fig. 6.4). Thus, if $(Mn^{2+}) = 10^{-2}$ and pH = 7, pE is poised at +8.8, according to Eq. 6.16b (Fig. 6.4). Above this rung on the redox ladder are redox couples with oxidant members that are sustained in equilibrium with the reductant members only at higher pE values. If pE drops below the rung for a given redox couple, the oxidant member is destabilized and the reductant member becomes highly favored. For example, if pE = 8.8 is introduced into Eq. 6.16a at pH 7, $P_{O_2} \approx 10^{-5}$ atm and $O_2(g)$ has effectively disappeared in favor of $H_2O(\ell)$. On the other hand, just the opposite trend applies to redox couples perched on rungs below that at pE = +8.8. For them, it is the reductant member that is destabilized, because these redox couples are sustainable at equilibrium only when pE drops to lower values. If pE = 8.8 is introduced into Eq. 6.16c, for example, the resulting Fe^{2+} activity is only about 2×10^{-13}. The general conclusion to be drawn here is the following:

If pE is poised at a certain value on the redox ladder, the favored species in all redox couples perched at higher (lower) pE values than the poised pE is the reductant (oxidant) species in the couples.

It is in this context that the reduction sequences for flooded soils shown in Figures 6.1 to 6.3 can be understood and interpreted in terms of pE descending the redox ladder. Electrons are produced in copious amounts by the microbially mediated oxidation of both humus (e.g., the reverse of the reaction in Eq. 6.14) and the reductants produced in fermentation processes [e.g., organic acids and $H_2(g)$]. As electrons accumulate and the pE value of the soil solution drops below +12.0, enough e^- become available to reduce $O_2(g)$ to $H_2O(\ell)$. Below pE 5, oxygen is not stable in neutral soils. Above pE 5, it is consumed in the respiration processes of aerobic microorganisms. As the pE value decreases further, electrons become available to reduce NO_3^-. This reduction is catalyzed by nitrate respiration (i.e., NO_3^- serving as a biochemical electron acceptor like O_2) involving bacteria that ultimately excrete NO_2^-, N_2, N_2O, NO, or NH_4^+.

As soil pE value drops into the range 9 to 5, electrons become plentiful enough to support the reduction of Mn(IV) in solid phases. The reductive dissolution of Fe(III) minerals does not occur until O_2 and NO_3^- are depleted, but Mn reduction can be initiated in the presence of nitrate. As the pE value decreases below +2, a neutral soil becomes anoxic and, when pE < 0, electrons are available for sulfate reduction catalyzed by a variety of anaerobic bacteria.

Typical products in aqueous solution are H_2S, bisulfide (HS^-), or thiosulfate ($S_2O_3^-$) ions. Methane production ensues for $pE < -4$, a value characteristic of fermentation processes.

The chemical sequence for the reduction of O, N, Mn, Fe, and S or for methane production induced by changes in pE is also an ecological sequence for the biological catalysts that mediate these reactions. Aerobic microorganisms that utilize O_2 to oxidize organic matter do not function below pE 5. Nitrate reducing bacteria thrive in the pE range between +10 and 0, for the most part. Sulfate-reducing bacteria do not do well at pE values above +2. These examples show that the redox ladder portrays domains of stability for both chemical and microbial species in soils.

It is noteworthy that Fe(III)/Fe(II) redox couples span the entire length of the redox ladder in Figure 6.4–some 22 orders of magnitude in electron activity! Five rungs are occupied by these couples, beginning with the free-ion species at $pE = +13.0$ and ending with the Fe^{2+}/Fe^0 couple at $pE = -7.93 + \frac{1}{2}\log(Fe^{2+})$. The redox couples perched between these two extremes comprise complexed species of the two free cations Fe^{3+} and Fe^{2+}. These couples are associated with log K values in Table 6.1 that reflect the influence of complex formation on the two free-ion species. If $L^{\ell-}$ is a ligand that forms a complex with Fe^{3+} and Fe^{2+}, then the reduction half-reaction that relates the oxidant $FeL^{(3-\ell)}$ to the reductant $FeL^{(2-\ell)}$ is the sum of three component reactions:

$$
\begin{array}{ll}
Fe^{3+} + e^- = Fe^{2+} & \log K = +13.0 \\
FeL^{(3-\ell)} = Fe^{3+} + L^{\ell-} & -\log K_L''' \\
Fe2^+ + L^{\ell-} = FeL^{(2-\ell)} & \log K_L'' \\
FeL^{(3-\ell)} + e^- = FeL^{(2-\ell)} & \log K
\end{array}
$$

where K_L is the thermodynamic equilibrium constant describing the *formation* of an FeL complex. The overall equilibrium constant for the $FeL^{(3-\ell)}/FeL^{(2-\ell)}$ couple is then $\log K = 13.0 - \log K_L''' + \log K_L''$. It is apparent that the associated pE value

$$pE_L \equiv 13.0 - \log K_L''' + \log K_L'' \tag{6.20}$$

will *decrease* if, as is almost always true, the stronger complex is formed by Fe^{3+} (i.e., $K_L''' > K_L''$). Thus, for example, if $L^{\ell-} = OH^-$, $\log K_{OH}''' = 11.8$, $\log K_{OH}'' = 4.6$, and $pE_{OH} = +5.8$—a drop by more than seven orders of magnitude in electron activity (Fig. 6.4). A very similar argument can be constructed to account for the placement of the $Fe(OH)_3/Fe^{2+}$ and $FeOOH/Fe^{2+}$ couples, because the two Fe(III) solid phases are formed by reacting Fe^{3+} with H_2O as a ligand that hydrolyzes to yield the solid-phase product and protons, thus producing an overall reaction of the form in Eq. 6.7. Clearly, these solid-phase products are stronger "complexes" than the solvation complex of Fe^{2+}. The line of reasoning presented is quite general, applying as well to the Mn(IV)/Mn(II) and Mn(III)/Mn(II) couples. Indeed, Table 6.1 shows that pE = 25.5 for the Mn^{3+}/Mn^{2+} couple, whereas $MnOOH/Mn^{2+}$ is perched at pE = 4.40 – log(Mn^{2+}) at pH 7—a drop of about 19 orders of magnitude in (e^-) if (Mn^{2+}) $\approx 10^{-2}$.

The Fe^{2+}/Fe^0 couple perched at the bottom of the redox ladder cannot be interpreted as an effect of complexation. This couple involves "zero-valent iron," Fe(s), which, from the point of view of redox reactions, is quite analogous to "zero-valent carbon," as represented, for example, by glucose. This analogy is more transparent if the reduction half-reaction for goethite/Fe^{2+} is added to that for Fe^{2+}/Fe^0 so as to cancel Fe^{2+} and maintain the stoichiometric coefficient of e^- as 1.0:

$$\frac{1}{3}\ \mathrm{FeOOH\ (s)} + \mathrm{H^+} + \mathrm{e^-} = \frac{1}{3}\ \mathrm{Fe\ (s)} + \frac{2}{3}\ \mathrm{H_2O}\ (\ell) \quad \log K = -0.84 \tag{6.14b}$$

which should be compared with Eq. 6.14a. Both half-reactions now span the full range of positive oxidation numbers for C and Fe with remarkably similar log K values, indicating the oxidant to be the favored species thermodynamically.

The reverse of the reaction in Eq. 6.14a is respiration, but might be termed *carbon corrosion* in the spirit of the reverse of the reaction in Eq. 6.14b. Both carbon corrosion and the corrosion of iron are spontaneous processes, thermodynamically speaking. The two couples $CO_2/C_6H_{12}O_6$ and $FeOOH/Fe^0$ occupy nearly the same place on the redox ladder at any pH value. The very low pE values at which they are perched ensures that poising a system with their half-reactions will favor reduced species of virtually every redox-active element (i.e., the reverse reactions will provide a flood of electrons to transform oxidants into reductants if suitable catalysis is available). That this contingency in the case of C has been exploited in the life cycles of soil microorganisms is well-known. That in the case of Fe it provides a means to convert any hazardous element into a reduced species that may be innocuous has, however, only recently been exploited in the design of soil remediation schemes.

6.5 pE–pH Diagrams

A pE–pH diagram is a predominance diagram (Section 5.4) in which electron activity is the dependent activity variable chosen to plot against pH. Thus the pE value plays the same role as the value of log $\left(\mathrm{Si\ (OH)_4^0}\right)$ in Figure 5.5. The construction of a pE–pH diagram is, accordingly, another example of the construction of a predominance diagram. Differences come because of redox reactions involving only aqueous species and because of the interpretation of the diagram, which is in terms of redox species instead of solid phases alone. The steps in constructing a pE–pH diagram are summarized as follows:

1. Establish a set of redox species and obtain values of log K for all possible reactions between the species.
2. Unless other information is available, set the activities of liquid water and all solid phases equal to 1.0. Set all gas-phase pressures at values appropriate to soil conditions.
3. Develop each expression for log K into a pE–pH relation. In *one* relation involving an aqueous species and a solid phase wherein a change in oxidation number is involved, choose a value for the activity of the aqueous species.
4. In each reaction involving two aqueous species, set the activities of the two species equal.

The resulting pE–pH diagram is divided into geometric regions with interiors that are domains of stability of either an aqueous species or a solid phase, and with boundary lines that are generated by transforming Eq. 6.8 (or another suitable expression for an equilibrium constant) into pE–pH relationships like Eq. 6.10. By examining a pE–pH diagram for a chemical element (e.g., Mn or S), one can predict the redox species expected at equilibrium under oxic, suboxic, or anoxic conditions in a soil at a given pH value.

To illustrate these concepts, consider a pE–pH diagram for Fe based on conditions in the Philippines Vertisol with the reduction sequence depicted on the right in Figure 6.1. First, a suite of redox species is chosen: $Fe(OH)_3(s)$, a poorly crystalline Fe(III) mineral similar to ferrihydrite (Section 2.4) that the GLO Step Rule (Section 5.2) would favor in flooded soils; $FeCO_3(s)$, a carbonate mineral observed in anoxic soils (Section 2.5); and Fe^{2+}. The reductive dissolution of $Fe(OH)_3(s)$ is listed in Table 6.1 and its associated pE–pH relationship following the steps just listed appears in Eq. 6.16c. The dissolution reaction for siderite,

$$FeCO_3(s) = Fe^{2+} + CO_3^{2-} \quad \log K_{so} = -10.8 \tag{6.21}$$

is expressed more conveniently after combining it with Eq. 5.33 and the bicarbonate dissociation reaction

$$HCO_3^- = H^+ + CO_3^{2-} \quad \log K_2 = -10.329$$

similar to what was done with the calcite dissolution reaction in Section 5.5. The resulting overall dissolution reaction is

$$FeCO_3(s) + 2H^+ = Fe^{2+} + H_2O(\ell) + CO_2(g) \tag{6.22}$$

for which $\log K = -10.8 + 1.466 + 6.352 + 10.329 = 7.35$ at 25 °C. Note the similarity to Eq. 5.34, although siderite is less soluble than calcite.

The final reaction needed to construct a pE–pH diagram is found by combining the two dissolution reactions just considered:

$$Fe(OH)_3(s) + CO_2(g) + H^+ + e^- = FeCO_3(s) + 2H_2O(\ell) \qquad (6.23)$$

for which log K = 17.14 – 7.35 = 9.79 at 25 °C. Note the similarity to the reaction in Table 6.1 relating $MnO_2(s)$ to $MnCO_3(s)$.

The pE–pH relationships that define the boundary lines in a pE–pH diagram describing the three Fe redox species are then, following Step 2 presented earlier,

$$pE = 17.14 - \log(Fe^{2+}) - 3pH \qquad (6.24a)$$

$$0 = 7.35 - \log(Fe^{2+}) - \log P_{CO_2} - 2pH \qquad (6.24b)$$

$$pE = 9.79 + \log P_{CO_2} - pH \qquad (6.24c)$$

Because Eq. 6.22 is not a reduction half-reaction, it does not involve pE when its log K is expressed in terms of log activity variables. It will plot as a vertical line in a pE–pH diagram.

Equation 6.24 cannot be implemented in a pE–pH diagram until fixed values for (Fe^{2+}) and P_{CO_2} are selected. Reference to Figure 6.1 indicates that $P_{CO_2} \approx 0.12$ atm after about 100 days of incubation of the Philippines soil. A measured value of (Fe^{2+}) is not available, but $(Fe^{2+}) \approx 2 \times 10^{-4}$ is reasonable for a flooded acidic soil. With these data incorporated, Eq. 6.24 becomes the set of working equations

$$pE = 20.84 - 3pH \qquad (6.25a)$$

$$pH = 6.0 \qquad (6.25b)$$

$$pE = 8.86 - pH \qquad (6.25c)$$

The boundary lines based on Eq. 6.25 are drawn in Figure 6.5. Equation 6.25a is the boundary between $Fe(OH)_3(s)$ and Fe^{2+} in respect to predominance of one redox species or the other. Above the line, pE increases and $Fe(OH)_3(s)$ predominates; below the line, the solid phase dissolves to form Fe^{2+} as the predominant species under the given conditions of Fe^{2+} activity and CO_2 partial pressure. This change in predominance as pE decreases has important consequences for the soil solution concentrations of metals like Cu, Zn, or Cd, and of ligands like $H_2PO_4^-$ or $HAsO_4^{2-}$. The principal cause of this secondary phenomenon is the desorption of metals and ligands that occurs when the adsorbents to which they are bound become unstable and dissolve. Typically, the metals that are released in this fashion, including Fe, are soon readsorbed by solids that are stable at low pE (e.g., clay minerals or soil organic matter) and become exchangeable surface species. Redox-driven surface speciation changes have an obvious influence on the bioavailability of the chemical elements involved, particularly phosphorus, arsenic, and selenium.

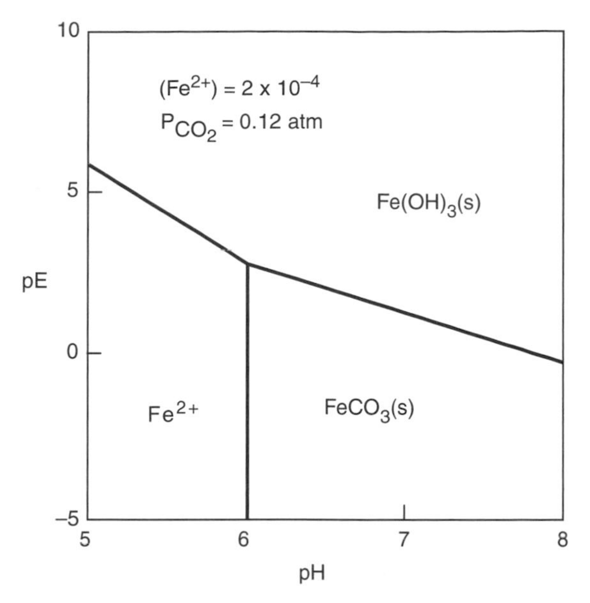

Figure 6.5. A pE–pH diagram for the system $Fe(OH)_3(s)$, $FeCO_3(s)$, and Fe^{2+}.

At pH 6, however, siderite becomes the predominant Fe(II) species at low pE under the fixed conditions assumed. The vertical boundary line signaling this transition in Figure 6.5 is remarkably robust under shifts in the values of (Fe^{2+}) or P_{CO_2}. For example, if (Fe^{2+}) decreases to 10^{-6}, or if P_{CO_2} decreases to $10^{-3.52}$ atm, its atmospheric value, the pH value for siderite precipitation is increased to about 7.0. Increasing (Fe^{2+}) to 10^{-3} (the value assumed in Figure 6.4) decreases the pH value to 5.6. Thus, siderite precipitation can be expected in a flooded soil as its pH increases from above 5 to above 7 during the typical reduction sequence, according to Eq. 6.24. The initial pH value in the Philippines Vertisol was 5.8, reaching 7.0 when the CO_2 partial pressure achieved its plateau (Fig. 6.1). Thus, siderite precipitation in this soil may have occurred. The initial pE value in the soil was estimated (by Pt electrode) to be +8, dropping rapidly to and remaining around –0.6 after only a few days. The initial state of the soil plots comfortably within the $Fe(OH)_3(s)$ field in Figure 6.5, whereas the final state does the same for the $FeCO_3(s)$ field, irrespective of whether (Fe^{2+}) or P_{CO_2} may differ somewhat from their assumed fixed values.

For Further Reading

Brookins, D. G. (1988) *Eh–pH diagrams for geochemistry.* Springer-Verlag, New York. After a careful presentation of the method for their construction, this book presents typical pE–pH diagrams for all the chemical elements of interest in environmental geochemistry. Brookins prefers to use electrode potential (Eq. 6.18) as a surrogate for pE.

Christensen, T. H., P. L. Berg, S. A. Banwart, R. Jakobsen, G. Heron, and H.- J. Albrechtsen. (2000) Characterization of redox conditions in groundwater contaminant plumes. *J. Contamin. Hydrol.* **45**:165. This masterful review article describes a broad variety of techniques for measuring pE and characterizing the redox-related microbial community in groundwater–an excellent read to round out the current chapter.

Kirk, G. (2004) *The biogeochemistry of submerged soils.* Wiley, Chichester, UK. The eight chapters of this outstanding monograph can be read to gain a thorough introduction to the chemistry of wetlands soils.

Langmuir, D. (1997) *Aqueous environmental geochemistry.* Prentice Hall, Upper Saddle River, NJ. Chapters 11 and 12 of this advanced textbook offer useful applications of the concepts developed in the current chapter to environmental geochemistry, with special emphasis on iron and sulfur redox reactions.

Stumm, W., and J. J. Morgan. (1996) *Aquatic chemistry.* Wiley, New York. Chapters 8 and 11 of this classic advanced textbook provide the most thorough discussion available of redox reactions in natural water systems, including issues surrounding the measurement of pE and the mechanistic underpinnings of redox kinetics.

Problems

The more difficult problems are indicated by an asterisk.

1. Gao et al. [Gao, S., K. K. Tanji, S.C. Scardaci, and A. T. Chow (2002) Comparison of redox indicators in a paddy soil during rice-growing season *Soil Sci. Soc. Am. J.* **66**:805.] have investigated the TEAPs that occur over a 90-day period after flooding in a California Vertisol under paddy rice cultivation. The principal results obtained are presented in their Figure 4 and their Table 2. Use these data to prepare graphs similar to those in Figure 6.1 for the "SB-WF" treatment (filled squares in their Figure 4; columns 1 and 2 in their Table 2). (*Hint:* Use Table 1.6 to assist in plotting data for H_2 and CH_4.)

2. Develop a balanced redox reaction for sulfate reduction to bisulfide coupled to glucose oxidation to bicarbonate.

3. Bacteria of the genus *Nitrobacter* catalyze the oxidation of nitrite to nitrate using $O_2(g)$ as an electron acceptor. Write a balanced overall redox reaction for this process and calculate log K. What will be the concentration ratio of NO_3^- to NO_2^- when pE = 7 at pH 6?

*4. Hydrogenotrophic anaerobes like *Paracoccus denitrificans* oxidize H_2 gas while catalyzing denitrification to produce N_2 gas (penultimate N half-reaction in Table 6.1).

a. Show that the oxidation of, say, glucose or acetate to yield H_2 gas is a favorable reaction.

b. Show that the reduction of nitrate to N_2 gas by the oxidation of H_2 gas is a favorable reaction.

c. Show that the oxidation of zero-valent iron to yield H_2 gas is a favorable reaction.

Perform a literature search to determine whether the redox reaction in (c) can be used to generate the H_2 gas required by *P. denitrificans* in catalyzing the reaction in (b).

5. The chloride-bearing variety of green rust (Section 2.4) undergoes the reduction half-reaction

$$Fe_4(OH)_8Cl(s) + 8H^+ + e^- = 4\,Fe^{2+} + Cl^- + 8\,H_2O(\ell)$$

with log K = 42.7 at 25 °C. Place the redox couple, green rust/Fe^{2+}, on the redox ladder in Figure 6.4 using $(Cl^-) = 2 \times 10^{-3}$. Compare its "rung" with those for the other Fe(II)-containing oxide minerals listed in Table 6.1.

6. The hydroxamate siderophore desferrioxamine B (DFOB) complexes Fe^{3+} and Fe^{2+} according to the reactions

$$Fe^{3+} + DFOB = Fe(III)DFOB \qquad \log K_s = 30.6$$

$$Fe^{2+} + DFOB = Fe(II)DFOB \qquad \log K_s = 10.0$$

Use this information to place the redox couple Fe(III)DFOB/Fe(II)DFOB on the redox ladder in Figure 6.4. (See Section 3.1 for a discussion of microbial siderophores.) Explain how complexation by DFOB can be interpreted as a strategy to stabilize Fe in the +III oxidation state down to very low pE levels. This strategy on the part of microorganisms prevents Fe(III) release at sites other than the intended target cell receptor.

7. Develop an expression analogous to Eq. 6.20 to show that precipitation of Fe^{3+} as hematite, goethite, or $Fe(OH)_3(s)$ will always decrease pE on the redox ladder relative to that for the Fe^{3+}/Fe^{2+} couple.

8. Perchloroethene (PCE; $Cl_2C{=}CCl_2$, top line in Fig. 3.4) is a relatively water-soluble dry-cleaning solvent that has become a ubiquitous groundwater contaminant because of improper waste disposal practices. This chlorinated ethene undergoes reductive dechlorination catalyzed by bacteria to form trichloroethene (TCE; $ClCH{=}CCl_2$), which is also a

hazardous organic chemical used industrially as a degreasing solvent:

$$\frac{1}{2}\,\text{PCE} + \frac{1}{2}\,\text{H}^+ + \text{e}^- = \frac{1}{2}\,\text{TCE} + \frac{1}{2}\,\text{Cl}^- \quad \log K = 12.18$$

The product species TCE then undergoes reductive dechlorination to form *cis*1,2-dichloroethene (cDCE),

$$\frac{1}{2}\,\text{TCE} + \frac{1}{2}\,\text{H}^+ + \text{e}^- = \frac{1}{2}\,\text{cDCE} + \frac{1}{2}\,\text{Cl}^- \quad \log K = 11.35$$

which itself can be reductively dechlorinated to form monochloroethene (or VC, vinyl chloride):

$$\frac{1}{2}\,\text{cCDE} + \frac{1}{2}\,\text{H}^+ + \text{e}^- = \frac{1}{2}\,\text{VC} + \frac{1}{2}\,\text{Cl}^- \quad \log K = 9.05$$

Vinyl chloride is, in turn, transformed into relatively harmless ethene (ETH, $H_2C = CH_2$) by this same microbially catalyzed process:

$$\frac{1}{2}\,\text{VC} + \frac{1}{2}\,\text{H}^+ + \text{e}^- = \frac{1}{2}\,\text{ETH} + \frac{1}{2}\,\text{Cl}^- \quad \log K = 8.82$$

Prepare a redox ladder for this set of reduction half-reactions under the assumption that $(Cl^-) = 2 \times 10^{-3}$. Compare your results with the principal redox couples depicted in Figure 6.4, then describe the temporal sequence of chlorinated ethenes expected as a plume of PCE is biodegraded below a water table.

9. Prepare a redox ladder for the hazardous chemical elements Cr, As, and Se based on the following reduction half-reactions:

$$\frac{1}{3}\,\text{CrO}_4^{2-} + \frac{5}{3}\,\text{H}^+ + \text{e}^- = \frac{1}{3}\,\text{Cr(OH)}_3\,(\text{s}) + \frac{1}{3}\,\text{H}_2\text{O}\,(\ell) \quad \log K = 21.06$$

$$\frac{1}{2}\,\text{H}_2\text{AsO}_4^- + \frac{3}{2}\,\text{H}^+ + \text{e}^- = \frac{1}{2}\,\text{As(OH)}_3^0 + \frac{1}{2}\,\text{H}_2\text{O}\,(\ell) \quad \log K = 10.84$$

$$\frac{1}{2}\,\text{HAsO}_4^{2-} + 2\,\text{H}^+ + \text{e}^- = \frac{1}{2}\,\text{As(OH)}_3^0 + \text{H}_2\text{O}\,(\ell) \quad \log K = 14.32$$

$$\frac{1}{2}\,\text{SeO}_4^{2-} + \frac{3}{2}\,\text{H}^+ + \text{e}^- = \frac{1}{2}\,\text{HSeO}_3^- + \frac{1}{2}\,\text{H}_2\text{O}\,(\ell) \quad \log K = 18.19$$

$$\frac{1}{4}\,\text{HSeO}_3^- + \frac{5}{4}\,\text{H}^+ + \text{e}^- = \frac{1}{4}\,\text{Se}\,(\text{s}) + \frac{3}{4}\,\text{H}_2\text{O}\,(\ell) \quad \log K = 13.14$$

Assume pH 7.0 and aqueous species concentrations of 1.0 mmol m^{-3}.

*10. In alkaline, suboxic soils, the important aqueous species of Se are SeO_4^{2-} and SeO_3^{2-}. Given the half-reaction

$$\frac{1}{2}\,\text{SeO}_4^{2-} + \text{H}^+ + \text{e}^- = \frac{1}{2}\,\text{SeO}_3^{2-} + \frac{1}{2}\,\text{H}_2\text{O}\,(\ell) \quad \log K = 14.55$$

determine whether SeO_3^{2-} can be oxidized to SeO_4^{2-} (the more toxic, mobile species) through the reduction of Mn(IV). [*Hint:* Consider poising of soil pE by the reductive dissolution of $MnO_2(s)$ and develop a relationship between (Mn^{2+}) and pH based on pE values for the MnO_2/Mn^{2+} and SeO_4^{2-}/SeO_3^{2-} redox couples.]

11. Over what range of pH will poising by the reductive dissolution of $MnO_2(s)$ be favorable to the complete reductive dechlorination of PCE to ETH as described in Problem 8? Take $(Cl^-) = 2 \times 10^{-3}$ and $(Mn^{2+}) = 10^{-2}$.

*12. An amended soil containing gypsum ($CaSO_4 \cdot 2H_2O$, Section 5.1) and siderite ($FeCO_3$) has a CO_2 pressure of 10^{-3} atm and $(Ca^{2+}) = 10^{-3.65}$ in the soil solution. Calculate the pE value at which FeS ($K_{so} = 10^{-18}$) will precipitate if pH = 8.2. (Log K = 13.92 for the formation of HS^- from S^{2-} and a proton.)

13. Discuss the changes in Figure 6.5 that would occur if goethite is the Fe(III) mineral instead of $Fe(OH)_3(s)$, or the activity of Fe^{2+} imposed is increased to 10^{-3}.

14. Prepare a pE–pH diagram for the three redox species $MnO_2(s)$, $MnCO_3(s)$, and Mn^{2+} given $(Mn^{2+}) = 10^{-3}$ and $P_{CO_2} = 0.12$ atm. Figure 6.5 can be used to guide your work, but shift the pE axis upward by 5 log units, and shift the pH axis to the right by 1 log unit, to acknowledge the difference between anoxic and suboxic conditions. Follow the steps in Section 5.5 to introduce P_{CO_2} as an activity variable.

15. Use the appropriate reactions in Table 6.1 to prepare a pE-pH diagram for S based on the *aqueous* species SO_4^{2-}, H_2S, and HS^-.

Special Topic 4: Balancing Redox Reactions

Redox species differ from other chemical species in that their status as *oxidized* or *reduced* molecular entities must be noted along with their other chemical properties. The redox status of the atoms in a redox species is quantified through the concept of *oxidation number*, a hypothetical valence, denoted by a positive or negative roman numeral, that is assigned to an atom according to the following three rules:

1. For a monoatomic species, the oxidation number equals the valence.
2. For a molecule, the sum of oxidation numbers of the constituent atoms equals the net charge on the molecule expressed in units of the protonic charge.

3. For a chemical bond in a molecule, the shareable, bonding electrons are assigned entirely to the more electron-attracting atom participating in the bond. If no such difference exists, each atom receives half the bonding electrons.

These rules can be illustrated by working out the oxidation numbers for the atoms in the redox species FeOOH, CHO_2^-, N_2, SO_4^{2-}, and $C_6H_{12}O_6$. In FeOOH, oxygen is more electron attracting than Fe or H and is conventionally designated O(–II). Thus, oxygen has oxidation number –2. The hydrogen atom in OH is designated H(I) (oxidation number +1). By Rule 2, the iron atom is designated Fe(III), because FeOOH has a zero net charge and $3+2(-2)+1 = 0$. (Recall that this notation was used already in Chapter 2 to distinguish ferric iron from ferrous iron in soil minerals.) A similar computation can be done for CHO_2^-, in which oxygen and hydrogen are designated as above and, therefore, carbon must be designated C(II), because $2 + 2(-2) + 1 = -1$ = the net number of protonic charges on the formate anion.

In the case of N_2, there is no difference between the two identical atoms in the molecule, and, by Rule 3, neither one can be assigned all of the bonding electrons. Because the molecule is neutral, Rule 2 then leads to the designation N(0) for each constituent N atom. For sulfate, oxygen is again O(–II) and S must be S(VI), according to Rule 2. Finally, in glucose, C on average must be C(0) because the designations O(–II) and H(I) lead by themselves to a neutral $C_6H_{12}O_6$ molecule.

Redox reactions obey the same mass and charge balance laws as described for other chemical reactions in Special Topic 1 at the end of Chapter 1. The only new feature is the need to account for changes in oxidation number when charge balance is imposed. Consider, for example, the aerobic weathering of an olivine to form goethite (Section 2.2). The essential characteristic of this reaction in the current context is the oxidation of Fe(II) to Fe(III). The reduced Fe species is olivine, $Mg_{1.63}Fe_{0.37}SiO_4$, and the oxidized Fe species is goethite, with $O_2(g)$ as a reactant (Eq. 1.3). This latter species must have been reduced to water in order that the electrons released by Fe oxidation be absorbed in the weathering process. Thus the redox aspect is captured by considering how to balance the *postulated* weathering reaction:

$$Mg_{1.63}Fe_{0.37}SiO_4(s) + O_2(g) \rightarrow FeOOH(s) + H_2O(\ell) \qquad \text{(S4.1)}$$

The schematic reaction in Eq. S4.1 can be balanced by first dividing it into reduction and oxidation half-reactions. The Fe oxidation half-reaction is the reverse of the reduction half-reaction,

$$FeOOH(s) + e^- \rightarrow Mg_{1.63}Fe_{0.37}SiO_4(s) \qquad \text{(S4.2)}$$

which is analogous to Eq. 6.1. Mass balance for Fe is obtained by giving olivine the stoichiometric coefficient $1/0.37 = 2.7$, after which mass balance can be imposed as well on Mg and Si:

$$\begin{aligned} & FeOOH(s) + 4.4\,Mg^{2+} + 2.7\,Si(OH)_4^0 + e^- \\ & \quad \rightarrow 2.7\,Mg_{1.63}Fe_{0.37}SiO_4(s) \end{aligned} \tag{S4.3}$$

Mass balance for oxygen is achieved by adding two water molecules to the right side of Eq. S4.3. Proton balance then would require $1 + 2.7(4) - 4 = 7.8\,H^+$ to be added to the right side:

$$\begin{aligned} & FeOOH(s) + 4.4\,Mg^{2+} + 2.7\,Si(OH)_4^0 + e^- = 2.7\,Mg_{1.63} \\ & \quad Fe_{0.37}SiO_4(s) + 2\,H_2O(\ell) + 7.8\,H^+ \end{aligned} \tag{S4.4}$$

This reaction can be shown to meet the requirement of overall charge balance. Note that e^- is essential for this charge balance.

To develop a redox reaction without the aqueous electron, one need only add to the inverse of Eq. S4.4 the reduction half-reaction for $O_2(g)$ in Table 6.1:

$$\begin{aligned} & 2.7\,Mg_{1.63}Fe_{0.37}SiO_4(s) + 0.25\,O_2(g) + 8.8\,H^+ + 1.5\,H_2O(\ell) \\ & \quad = FeOOH(s) + 4.4\,Mg^{2+} + 2.7\,Si(OH)_4^0 \end{aligned} \tag{S4.5}$$

where $0.5H_2O(\ell)$ has been canceled from both sides of the result. Equation S4.5 is a balanced redox reaction showing the weathering of olivine in an oxic environment to form goethite, aqueous magnesium ions, and silicic acid. The procedure by which it was developed can be described as follows:

1. Identify the two redox couples participating in the overall redox reaction.
2. For each redox couple, develop a balanced reduction half-reaction in which 1 mol aqueous electrons is transferred.
3. Combine the two half-reactions developed in Step 2 to cancel the aqueous electron and produce the required reactant and product redox species in the overall reaction.

Special Topic 5: The Invention of the pH Meter

While Linus Pauling was beginning his academic career at the California Institute of Technology, another promising physical chemist came under the tutelage of Roscoe Gilkey Dickinson, Pauling's mentor as a graduate student (see Special Topic 2 in Chapter 2). Arnold Beckman had returned to graduate study after a two-year hiatus provoked by his affection for the woman who was to become his spouse for more than 60 years. What drew Beckman back to Caltech was its strong commitment to developing new technology as a force for improving the lives of ordinary people. Like Pauling, he also was asked to join the faculty of his alma mater upon receiving his Ph.D. degree in 1928. Unlike Pauling, however, Beckman devoted his research not to the foundations but to the applications of chemical science. This focus eventually led him to work on a problem that made him as well-known as Pauling: the development of a reliable, robust, electrode-based instrument to measure pH. The story of this invention was summarized nicely in an article by Elizabeth Wilson commemorating Beckman's 100th birthday. (Reprinted with permission from Chemical & Engineering News, April 10, 2000, 78(15), p.19. Copyright 2000 American Chemical Society.)

In the 1930s, Arnold Beckman was a chemistry professor at California Institute of Technology with a reputation for solving practical problems. Though Beckman prided himself on his considerable and creative teaching skills, his colleague Robert A. Millikan encouraged his inventive side by sending him jobs from people who needed technical and scientific help.

Soon, Beckman was augmenting his modest professorial salary with income from a sideline consulting business, which was particularly welcome during The Depression.

Among his numerous projects were: methods of testing rock samples for elusive "colloidal gold"—flecks of the precious metal that eager gold-hunters believed impregnated otherwise ordinary rock samples; and an inking device for National Postal Meter that was free of clogging problems that had plagued previous devices. Beckman even gained a reputation as an expert scientific witness in court trials.

But the invention that made Beckman a household name was the pH meter. From a device constructed to help a friend measure the acidity of citrus juice came a paradigm shift in the way researchers did science.

Beckman's former undergraduate classmate Glen Joseph, a chemist who worked in the citrus industry in California, desperately needed a consistent, reliable method of testing citrus juice acidity. Such a measurement was vital to the industry, because it would help determine if the juice met legal standards that would allow the fruit to be sold to consumers. If the standards weren't met, the fruit would then be used to make citric acid or pectin.

To be sure, some methods already existed for testing acidity. Litmus paper, familiar to anyone who has taken a chemistry lab course, turns red if a sample's pH is low, and blue if the pH is high. But the sulfur dioxide used by the citrus industry to preserve the juice bleached litmus paper, rendering it useless.

There were also electrochemical methods for measuring acidity, as the current generated from available hydrogen ions in a solution corresponds to its acidity. These methods were quantitatively precise, but caused Joseph innumerable headaches.

Hydrogen electrodes, his first choice, were "poisoned" by sulfur dioxide. Another option was the glass electrode, which was impervious to sulfur dioxide. But the electrical current generated by cells with glass electrodes was very weak, and couldn't be read reliably by the galvanometers Joseph used. He could increase the current by using very large, thin-walled glass electrodes. Unfortunately, these electrodes were so delicate that they routinely broke, creating numerous frustrating delays.

In desperation, Joseph consulted Beckman. The young inventor almost immediately came up with a solution that combined the hardiness of a thick glass electrode with enough electrical sensitivity to produce meaningful measurement.

The key to Beckman's device was the vacuum tube amplifier. A galvanometer, the measurement instrument of choice, had magnetic needles that deflected in proportion to the amount of current. But Beckman realized that by using a vacuum tube-amplifier-based meter instead of a galvanometer, he would be able to boost the signal to readable levels with a robust but relatively insensitive glass electrode.

Perhaps Beckman's greatest insight, however, was to put the whole shebang—electrodes, amplifiers, circuitry—together in one box. This was a revolutionary concept. Until then, chemists had largely built their instruments from scratch, spending much time assembling and tweaking numerous components.

This new ready-made, compact device, which Beckman dubbed the acidimeter, was a time- and sanity-saving godsend. The acidimeter was an instant hit in Joseph's lab, and it soon became apparent to Beckman that many others would be clamoring for it—and they did.

Before Beckman's pH meter, "you could measure pH using a platinum or calomel electrode and a galvanometer and a whole benchful of lab equipment," notes Beckman's longtime friend and Beckman Foundation board member Gerald E. Gallwas. "He made it commercially viable—he put it in a little box you could carry into an orchard."

7

Soil Particle Surface Charge

7.1 Surface Complexes

Given the variety of functional groups present in the organic compounds that form soil humus (sections 3.1 and 3.2), it can be expected that some will come to reside on the interface between particulate humus and the aqueous phase in soil. These molecular units protruding from a solid particle surface into the soil solution are *surface functional groups.* In the case of soil humus, the surface functional groups are necessarily organic molecular units, but in general they can be bound into either organic or inorganic solids, and they can have any molecular structure that is possible through chemical interactions. Because of the variety of possible functional group compositions, a broad spectrum of surface functional group reactivity is also likely. Superimposed on this intrinsic variability is that created by the wide range of stereochemical and charge distribution characteristics possible in a heterogeneous solid matrix. For this reason, no organic surface functional group (e.g., carboxyl) has single-value quantitative chemical properties (e.g., the proton dissociation equilibrium constant), but instead can be characterized only by ranges of values for these properties (Section 3.3). This "smearing out" of chemical behavior is an important feature that distinguishes surface functional groups from those bound to small molecules (e.g., oxalic acid).

Figure 5.1 shows a water molecule bearing a positive charge that is bound to an Al^{3+} ion at the periphery of the mineral gibbsite (upper left). This highly reactive combination of metal cation and water molecule at an interface is called a *Lewis acid site*, with the metal cation identified as the Lewis acid. (*Lewis acid* is the name given to metal cations and protons when their reactions are considered from the perspective of the electron orbitals in ions. A Lewis acid initiates a chemical reaction with empty electron orbitals.) Lewis acid sites can exist on the surface of goethite (Fig. 2.2), if peripheral Fe^{3+} ions are bound to water molecules there, and on the edge surfaces of clay minerals like kaolinite (Fig. 2.8, left side). These surface functional groups are very reactive because a positively charged water molecule is quite unstable and, therefore, is exchanged readily for an organic or inorganic anion in the soil solution, which then can form a more stable bond with the metal cation. This *ligand exchange reaction* is described by Eq. 3.12 for the example of carboxylate and a Lewis acid site.

Lewis acid sites result from the protonation of surface hydroxyl groups, as indicated in Eq. 3.12a and discussed in sections 2.3 and 2.4 for the minerals kaolinite and gibbsite. The protonation reaction, in turn, is provoked by unsatisfied bond valences with an origin that can be understood in terms of Pauling Rule 2 (Section 2.1). The surface of goethite, for example, can expose OH groups that are bound to one, two, or three Fe^{3+} (Fig. 2.2). A single Fe—O bond in goethite has a bond valence of 0.47 vu, according to Eq. 2.3, with the required data given in Table 2.2 and Problem 2 of Chapter 2. An OH group bound to a single Fe^{3+} evidently would bear a net negative charge and would require protonation analogously to the same type of OH group in gibbsite or kaolinite, whereas an OH group bound to two Fe^{3+} should be stable, and one bound to three Fe^{3+} can be stabilized by hydrogen bonding to water molecules, by analogy with what occurs in the bulk goethite structure (Section 2.1). Thus, the three types of OH group exhibit very different reactivity with respect to protonation and subsequent ligand exchange.

The plane of oxygen atoms on the cleavage surface of a layer-type aluminosilicate is called a *siloxane surface.* This plane is characterized by a distorted hexagonal symmetry among its constituent oxygen atoms (Section 2.3). The functional group associated with the siloxane surface is the roughly hexagonal (strictly speaking, ditrigonal) cavity formed by six corner-sharing silica tetrahedra, shown on the left in Figure 2.3. This cavity has a diameter of about 0.26 nm and is bordered by six sets of electron orbitals emanating from the surrounding ring of oxygen atoms.

The reactivity of the siloxane cavity depends on the nature of the electronic charge distribution in the layer silicate structure. If there are no nearby isomorphic cation substitutions to create a negative charge in the underlying layer, the O atoms bordering the siloxane cavity will function as an electron donor that can bind neutral molecules through van der Waals interactions (Section 3.4). These interactions are akin to those underlying the hydrophobic interaction because the planar structure of the siloxane surface is not particularly compatible with that in bulk liquid water. Therefore, uncharged patches on siloxane surfaces may be considered hydrophobic regions to a certain degree, with a relatively strong local attraction for hydrophobic moieties in soil humus or hydrophobic organic molecules in the soil solution. However, if isomorphic substitution of Al^{3+} by Fe^{2+} or Mg^{2+} occurs in the octahedral sheet (Table 2.4

and the right side of Fig. 2.3), a structural charge is created (Eq. 2.8) that can attract cations and polar molecules (e.g., phenols) or moieties in humus. If isomorphic substitution of Si^{4+} by Al^{3+} occurs in the tetrahedral sheet, excess negative charge is created much nearer to the siloxane surface, and a strong attraction for cations and polar molecules is generated. Structural charge of this kind vitiates the otherwise mildly hydrophobic character of the siloxane surface. Thus 2:1 layer-type clay minerals present a heterogeneous basal surface comprising hydrophobic patches interspersed among charged hydrophilic sites.

The complexes that form between surface functional groups and constituents of the soil solution are classified analogously to the complexes that form among aqueous species (Section 4.1). If a surface functional group reacts with an ion or a molecule dissolved in the soil solution to form a stable molecular unit, a *surface complex* is said to exist and the formation reaction is termed *surface complexation*. Two broad categories of surface complex are distinguished on structural grounds. If no water molecule is interposed between the surface functional group and the ion or molecule it binds, the complex is *inner-sphere*. If at least one water molecule is interposed between the functional group and the bound ion or molecule, the complex is *outer-sphere*. As a general rule, outer-sphere surface complexes involve electrostatic bonding mechanisms and, therefore, are less stable than inner-sphere surface complexes, which necessarily involve either ionic or covalent bonding, or some combination of the two. These concepts are quite parallel with those developed in Section 4.1 for aqueous species and in Section 3.5 for bridging complexes. Seen in this light, Figure 3.5 shows Ca^{2+} in an inner-sphere surface complex with a charged site on a siloxane surface while at the same time forming an outer-sphere complex with carboxylate. Figure 3.6 shows the reverse arrangement of the complexes: inner-sphere with carboxylate and outer-sphere with a charged site on the siloxane surface. Bridging complexes thus may also be considered as *ternary surface complexes* involving an organic ligand, a metal cation, and a charged surface site.

Figure 7.1 illustrates the structure of the outer-sphere surface complex formed between Na^+ and a surface site on the siloxane surface of montmorillonite. At about 50% relative humidity, a stable hydrate forms in which there are two layers of water molecules. The Na^+ tends to adsorb as solvated species on the siloxane surface near negative charge sites originating in the octahedral sheet from isomorphous substitution of a bivalent cation for Al^{3+}. This outer-sphere complex occurs as a result of both the strong solvating characteristics of Na^+ and the physical impediment to direct contact between Na^+ and the site of negative charge posed by the layer structure itself. The way in which this negative charge is distributed on the siloxane surface is not well-known, but if the charge tends to be delocalized ("smeared out" over several oxygen ions), that would lend itself to outer-sphere surface complexation. It is pertinent to note that the molecular structure of the solvation complex in Figure 7.1 is very similar to that observed for solvated Na^+ in concentrated aqueous solutions.

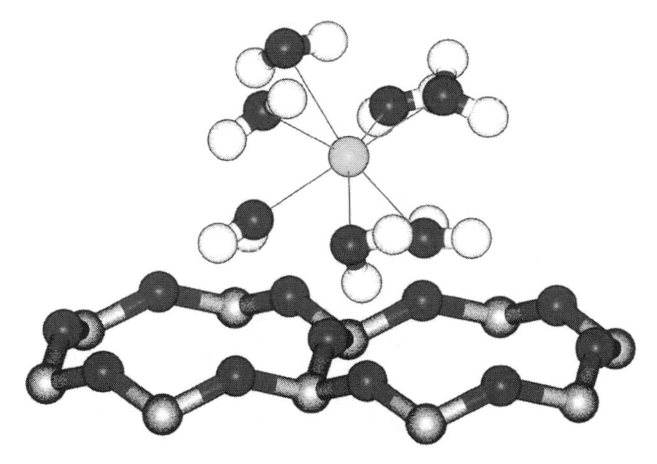

Figure 7.1. An outer-sphere surface complex formed between Na^+ and a charged surface site on montmorillonite. Visualization courtesy of Dr. Sung-Ho Park.

Figure 7.2 illustrates an outer-sphere surface complex formed between Pb^{2+} and a hydrated surface site on the Al oxide, corundum (Section 2.1). The exposed surface comprises triangular rings of six oxygen ions in AlO_6^{3-} octahedra, with each O^{2-} bonded to a pair of neighboring Al^{3+}. Like the basal planes in gibbsite, this surface is protonated when in contact with an aqueous solution. The outer-sphere surface complex has Pb^{2+} coordinated to three water molecules that are also hydrogen bonded to surface hydroxyls on the border of an octahedral cavity in the center of one of the triangular rings of protonated oxygen ions. Two other water molecules solvate the adsorbed Pb^{2+} to give a total solvation shell coordination number of 5, similar to what is observed for Pb^{2+} in concentrated aqueous solutions.

Inner-sphere surface complexes between K^+ or Cs^+ and the siloxane surface of a 2:1 clay mineral with extensive Al^{3+} substitutions in the tetrahedral sheets are especially stable. This type of surface complex requires coordination of the monovalent cation with 12 oxygen atoms bordering two opposing siloxane cavities. The layer charge in the clay mineral vermiculite is large enough (Table 2.4) that each siloxane cavity in a basal plane of the mineral can complex one monovalent cation. Moreover, the ionic radius of K^+ (Table 2.1) is essentially equal to that of a cavity. This combination of charge distribution and stereochemical factors gives K-vermiculite surface complexes great stability and is the molecular basis for the term *potassium fixation. colloquial*

As mentioned earlier, the hydroxyl group coordinated to one Fe^{3+} in goethite can be protonated to form a Lewis acid site. The water molecule can then be exchanged as in Eq. 3.12 to allow formation of an inner-sphere surface complex with the oxyanion HPO_4^{2-}. This surface complex is illustrated in Figure 7.3. It consists of an HPO_4^{2-} bound through its oxygen ions to a pair of

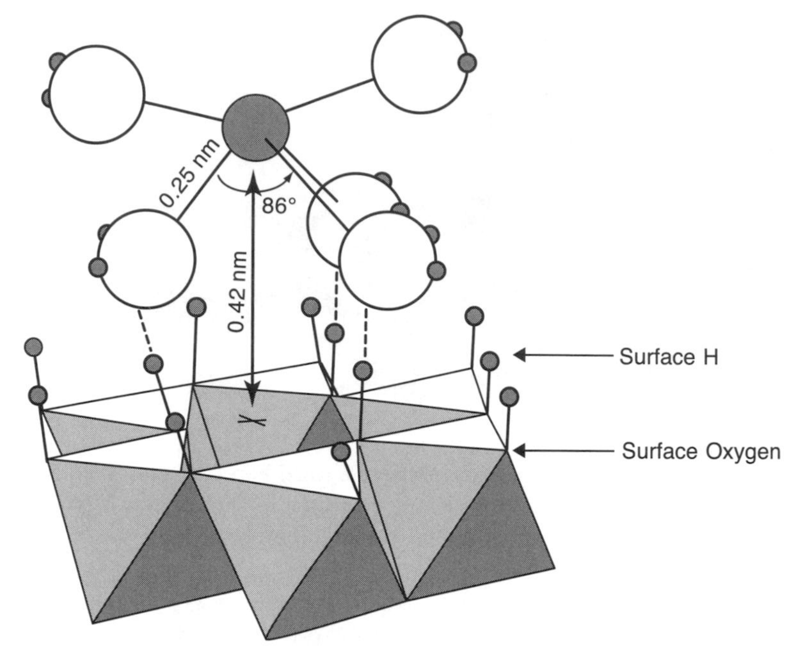

Figure 7.2. An outer-sphere surface complex formed between Pb2+ and the hydroxylated surface of α–Al_2O_3 (corundum). After Bargar, J., S.N. Towle, G.E. Brown, and G.A. Parks (1997) XAFS and bond-valence determination of the structures and compositions of surface functional groups and Pb(II) and Co(II) sorption products on single-crystal α-Al_2O_3. *J. Colloid Interface Sci.* **185**: 473–492.

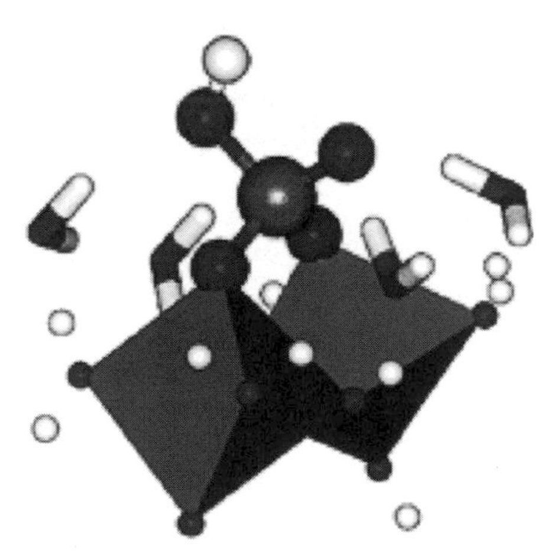

Figure 7.3. An inner-sphere surface complex formed between a biphosphate anion (HPO_4^{2-}) and two adjacent Fe^{3+} in goethite. Visualization courtesy of Dr. Kideok Kwon.

adjacent Fe^{3+} cations (*binuclear surface complex*). The configuration of the o-phosphate unit is compatible with the grooved structure of the goethite surface, thus providing stereochemical enhancement of the stability of the inner-sphere complex. Inner-sphere complexes can also form through the ligand exchange of other oxyanions (e.g., selenite, arsenate, borate) with protonated OH groups on goethite and other metal oxyhydroxides.

7.2 Adsorption

Adsorption is the process through which a chemical substance reacts at the common boundary of two contiguous phases. If the reaction produces enrichment of the substance in an interfacial layer, the process is termed *positive adsorption*. If, instead, a depletion of the substance is produced, the process is termed *negative adsorption*. If one of the contiguous phases involved is solid and the other is fluid, the solid phase is termed the *adsorbent* and the matter that accumulates at its surface is an *adsorbate*. A chemical species in the fluid phase that potentially can be adsorbed is termed an *adsorptive*. As indicated in Section 7.1, if an adsorbate is immobilized on the adsorbent surface over a timescale that is long, say, when compared with that for diffusive motions of the adsorptive, then the adsorbate and the site on the adsorbent surface to which it is bound are termed a *surface complex*.

Adsorption experiments involving solid particles typically are performed in a sequence of three steps: (1) *reaction* of an adsorptive with an adsorbent contacting a fluid phase of known composition under controlled temperature and applied pressure for a prescribed period of time, (2) *separation* of the adsorbent from the fluid phase after reaction, and (3) *quantitation* of the chemical substance undergoing adsorption, both in the supernatant fluid phase and in the separated adsorbent slurry that includes any entrained fluid phase. The reaction step can be performed in either a closed system (batch reactor) or an open system (flow-through reactor), and can proceed over a time period that is either quite short (adsorption kinetics) or very long (adsorption equilibration) compared with the natural timescale for achieving a steady composition in the reacting fluid phase. The separation step is similarly open to choice, with centrifugation, filtration, or gravitational settling being convenient methods to achieve it. The quantitation step, in principle, should be designed not only to determine the moles of adsorbate and unreacted adsorptive, but also to verify whether unwanted side reactions, such as precipitation of the adsorptive or dissolution of the adsorbent, have unduly influenced the adsorption experiment.

Ion adsorption on soil particle surfaces can take place via the three mechanisms illustrated in Figure 7.4 for a monovalent cation on the siloxane surface of a 2:1 clay mineral like montmorillonite. The inner-sphere surface complex shown involves the siloxane cavity, as described in Section 7.1, whereas the outer-sphere surface complex shown includes the cation solvation shell and

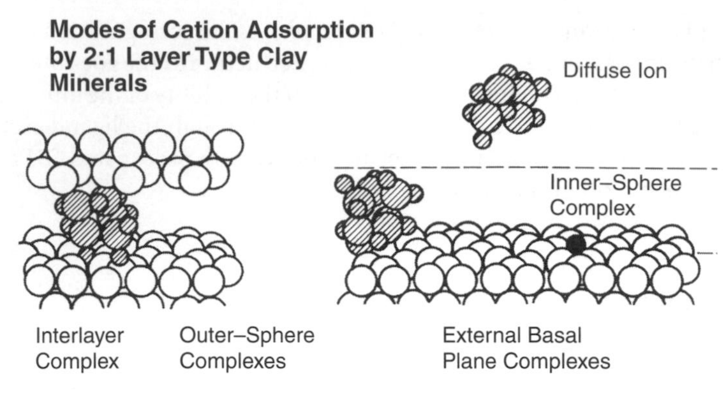

Figure 7.4. The three modes of ion adsorption, illustrated for cations adsorbing on montmorillonite.

is similar to that depicted in Figure 7.2. These two localized surface species constitute the *Stern layer* on an adsorbent. If a solvated ion does not form a complex with a charged surface functional group, but instead screens a surface charge in a delocalized sense, it is said to be adsorbed in the *diffuse-ion swarm*, also shown in Figure 7.4. This last adsorption mechanism involves ions that are fully dissociated from surface functional groups and are, accordingly, free to hover nearby in the soil solution. The diffuse-ion swarm and the outer-sphere surface complex mechanisms of adsorption involve almost exclusively electrostatic bonding, whereas inner-sphere complex mechanisms are likely to involve ionic as well as covalent bonding. Because covalent bonding depends significantly on the particular electron configurations of both the surface group and the complexed ion, it is appropriate to consider inner-sphere surface complexation as the molecular basis of the term *specific adsorption*. Correspondingly, diffuse-ion screening and outer-sphere surface complexation are the molecular basis for the term *nonspecific adsorption*. *Nonspecific* refers to the weak dependence on the detailed electron configurations of the surface functional group and adsorbed ion that is to be expected for the interactions of solvated species.

Readily exchangeable ions in soil are those that can be replaced easily by leaching with an electrolyte solution of prescribed composition, concentration, and pH value. Despite the empirical nature of this concept, there is a consensus that ions adsorbed specifically (like HPO_4^{2-} in Fig. 7.3) are not readily exchangeable. Thus, experimental methods to determine readily exchangeable adsorbed ions must avoid extracting specifically adsorbed ions. From this point of view, *fully solvated ions adsorbed on soils are readily exchangeable ions*, with the molecular definition of *readily exchangeable* thus based on the diffuse-ion swarm and outer-sphere complex mechanisms of adsorption.

More generally, the interactions between adsorptive ions and soil particles can be portrayed as a web of *sorption* reactions mediated by two parameters: timescale and adsorbate surface coverage. Surface complexes are the products of these reactions when timescales are sufficiently short and surface coverage is sufficiently low, with "sufficiently" always being defined operationally in terms of conditions attendant to the sorption process. As timescales are lengthened (e.g., longer than hours) and surface coverage increases, or as chemical conditions are altered (e.g., pH changes) for a fixed reaction time, adsorbate "islands" comprising a small number of ions bound closely together may form. These reaction products are termed *multinuclear surface complexes* by analogy with their counterpart in aqueous solution chemistry. They are the more likely for adsorptive ions that readily form polymeric structures in aqueous solution. Multinuclear surface complexes may in turn grow with time to become colloidal structures that are precursors of either surface polymers or, if they are well organized on a three-dimensional lattice, surface precipitates. Thus, sorption processes need not exhibit the inherently two-dimensional character of positive adsorption processes, although both involve the accumulation of a substance at an interface.

7.3 Surface Charge

Solid particle surfaces in soils develop an electrical charge in two principal ways: either from isomorphic substitutions in soil minerals among ions of differing valence, or from the reactions of surface functional groups with ions in the soil solution. The electrical charge developed by these two mechanisms is expressed conventionally in moles of charge per kilogram ($mol_c\ kg^{-1}$, see the Appendix). Four different types of surface charge contribute to the *net total particle charge* in soils, denoted σ_P. Each of these components can be positive, zero, or negative, depending on soil chemical conditions.

Two components of σ_P have been described in previous chapters. *Structural charge*, σ_0, defined in Eq. 2.7, arises from isomorphic substitutions in 2:1 clay minerals (Section 2.3) and from cation vacancy defects and in manganese oxides (Section 2.4). Although σ_0 can be calculated with chemical composition data for a mineral specimen, in a soil sample it is measured conventionally as *Cs-accessible surface charge* following a reaction of the sample with 50 mol m^{-3} CsCl at pH 5.5 to 6.0. Briefly, the soil is saturated with Cs by repeated washing in CsCl, with a final supernatant solution ionic strength of 50 mol m^{-3}. After centrifugation, the supernatant solution is discarded and the remaining entrained CsCl solution is removed by washing with ethanol. The samples are then dried at 65 °C for 48 hours to enhance formation of inner-sphere Cs surface complexes. Next, the samples are washed in 10 mol m^{-3} LiCl solution to eliminate outer-sphere surface complexes of Cs. The suspension is centrifuged, and the supernatant LiCl solution is removed for analysis, leaving only a slurry containing the soil sample and entrained LiCl solution. Finally, Cs is extracted with 1 mol dm^{-3} ammonium acetate (NH_4OAc) solution, and the LiCl and NH_4OAc solutions are analyzed for Cs. Permanent structural charge is then calculated as minus the difference between moles Cs in the NH_4OAc extract and moles Cs in the entrained LiCl solution, per kilogram of dry soil. This method is reliable even for highly heterogeneous samples that comprise both crystalline and amorphous minerals, organic matter, and biota. Its sensitivity is such that $|\sigma_0|$ values < 1 $mmol_c\ kg^{-1}$ are detectable.

The *net proton charge*, σ_H, discussed at length in Section 3.3 as an attribute of soil humus, is defined for unit mass of any charged particle as the difference between the moles of protons and the moles of hydroxide ions complexed by surface functional groups (cf. Eq. 3.5). Thus, protons and hydroxide ions adsorbed in the diffuse-ion swarm are *not* included in the definition of σ_H. As noted in Section 3.3, the measurement of σ_H remains as an experimental challenge, but consensus exists that it makes a very important contribution to σ_p over a broad range of pH. It receives contributions from all acidic surface functional groups in a soil, including those exposed on humus, on the edges of clay mineral crystallites, and on oxyhydroxide minerals. The sum of structural and net adsorbed proton charge defines the *intrinsic charge*, σ_{in},

$$\sigma_{in} \equiv \sigma_0 + \sigma_H \tag{7.1}$$

which is intended to represent components of surface charge that develop solely from the adsorbent structure.

The *net adsorbed ion charge* is defined formally by the equation

$$\Delta q \equiv \sigma_{IS} + \sigma_{OS} + \sigma_d \tag{7.2}$$

which refers, specifically, to the net charge of ions adsorbed in inner-sphere surface complexes (IS), in outer-sphere surface complexes (OS), or in the diffuse-ion swarm (d). The utility of Eq. 7.2 depends on the extent to which experimental detection and quantitation of these surface species is possible. The partitioning of surface complexes into inner-sphere and outer-sphere is not always possible (or required), however, and Eq. 7.2 can alternatively be written in the simpler form

$$\Delta q \equiv \sigma_S + \sigma_d \tag{7.3}$$

where σ_S denotes the *Stern layer charge* (cf. Fig. 7.4) representing all adsorbed ions not in the diffuse-ion swarm. This latter conceptual distinction, based largely on adsorbed ion mobility, is epitomized by defining the *net total particle charge*, σ_p:

$$\sigma_p \equiv \sigma_{in} + \sigma_S \tag{7.4}$$

which is the surface charge contributed by the adsorbent structure and by adsorbed ions that are immobilized into surface complexes (i.e., adsorbed ions that do not engage in translational motions that may be likened to the diffusive motions of a free ion in aqueous solution). The adsorbed ions that do engage in more or less free diffusive motions must nonetheless contribute a net charge that balances the net total particle charge:

$$\sigma_p + \sigma_d = 0 \tag{7.5}$$

Equation 7.5 is the *condition of surface charge balance* for soil particles. It states simply that any electrical charge these particles may bear is always balanced by a counterion charge in the diffuse swarm of electrolyte ions near their surfaces. An alternative form of Eq. 7.5 can be written down at once after combining Eq. 7.1 with eqs. 7.3. to 7.5:

$$\sigma_0 + \sigma_H + \Delta q = 0 \tag{7.6}$$

Equation 7.6, which does not require molecular-scale concepts, mandates the overall electroneutrality of any soil sample that has been equilibrated with an aqueous electrolyte solution. Structural charge and the portion of net particle charge attributable to surface-complexed protons or hydroxide ions must be balanced with the net surface charge that is contributed by all other adsorbed ions and by H^+ or OH^- in the diffuse-ion swarm.

Equation 7.6 can be used to test experimental surface charge data for self-consistency. A convenient approach is to plot Δq against σ_H over a range of pH values for which these two surface charge components have been measured. A simple rearrangement of Eq. 7.6,

$$\Delta q = -\sigma_H - \sigma_0 \tag{7.7}$$

shows that the slope of this *Chorover plot* must be equal to –1, with both its y- and x-intercepts equal to – σ_0. Figure 7.5 illustrates the application of Eq. 7.7 to an Oxisol, comprising kaolinite, metal oxides, and quartz intermixed with humus, that was equilibrated with LiCl solution at three different ionic strengths over the pH range 2 to 6. The line through the data is based on a linear regression equation,

$$\Delta q = -1.01 \pm 0.07\sigma_H + 12.5 \pm 0.8 \qquad (R^2 = 0.92)$$

with both Δq and σ_H expressed in units of millimoles of charge per kilogram and with 95% confidence intervals following the values of the slope and intercept. The value of σ_0 measured independently by the Cs^+ method is –12.5 $\pm$ 0.04 $mmol_c\ kg^{-1}$, which is in excellent agreement with both the y- and x-intercepts.

7.4 Points of Zero Charge

Points of zero charge are *pH values* at which one of the surface charge components in Eqs. 7.5 and 7.6 becomes equal to zero under given conditions of temperature, applied pressure, and soil solution composition. Three standard definitions are given in Table 7.1.

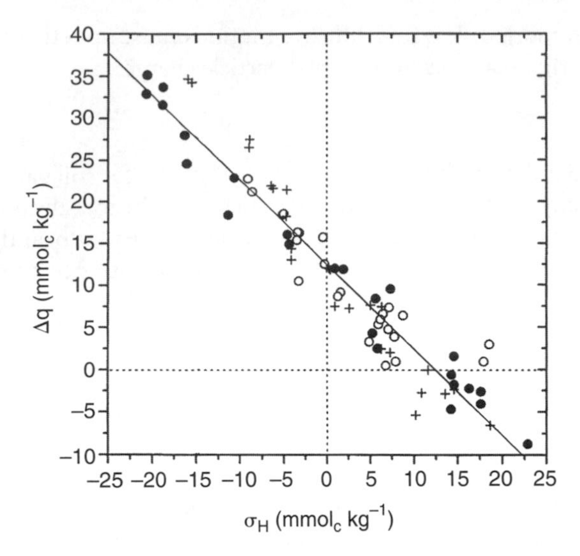

Figure 7.5. A Chorover plot for a cultivated kaolinitic Oxisol suspended in LiCl solutions of varying concentration (open circles = 1 m*M*, crosses = 5 mM, and filled circles = 10 mM) and pH 2 to 6. The vertical and horizontal dashed lines are the coordinate axes. Their intersections with the linear plot are required to be equal if charge balance is confirmed. Original graph courtesy of Dr. Jon Chorover.

The p.z.n.p.c. is the pH value at which the net adsorbed proton charge is equal to zero. A straightforward method to determine this pH value is to measure Δq as a function of pH and then locate the pH value at which $\Delta q = -\sigma_0$, thus taking direct advantage of Eq. 7.6 (Fig. 7.6), given that a separate measurement of σ_0 has been made. Most published reports of p.z.n.p.c. values based on the use of titration measurements to determine σ_H resort to the device of choosing $\sigma_H \equiv 0$ at the crossover point of two $\delta\sigma_{H,titr}$ versus pH curves that have been determined at different ionic strengths. Unfortunately, as mentioned in Section 3.3, each such curve is, in principle, offset differently from a true σ_H curve by an *unknown* $\delta\sigma_{H,titr}$ that corresponds to the particular initial state of the titrated system, thus making the crossover point illusory.

Equation 7.6 imposes a constraint on changes in the net adsorbed proton charge and/or net adsorbed ion charge that may occur in response to controlled

Table 7.1
Some points of zero charge.

Symbol	Name	Definition
p.z.n.p.c.	Point of zero net proton charge	$\sigma_H = 0$
p.z.n.c.	Point of zero net charge	$\Delta q = 0$
p.z.c.	Point of zero charge	$\sigma_p = 0$

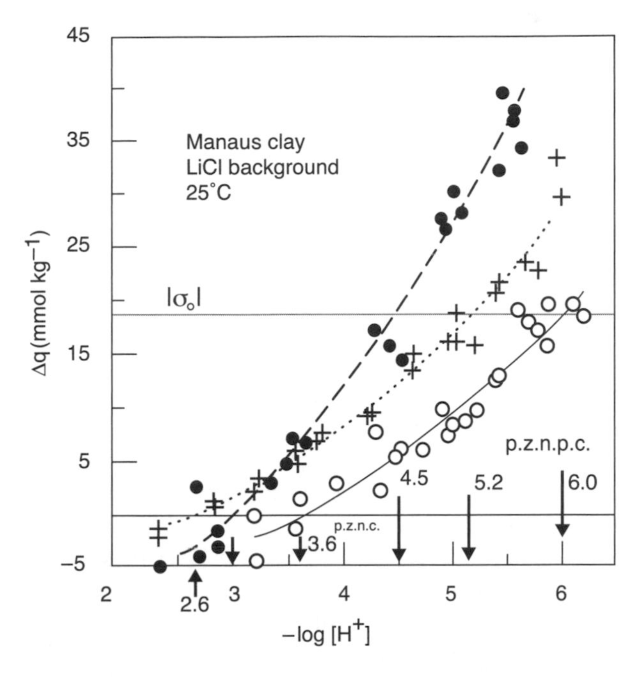

Figure 7.6. Plots of net adsorbed ion charge against pH for an uncultivated kaolinitic Oxisol suspended in LiCl solutions with the same varying concentration and pH values as in Figure 7.5. The upper horizontal line intersects each graph at the p.z.n.p.c., whereas the lower horizontal line intersects each graph at the p.z.n.c. Data from Chorover, J., and G. Sposito (1995) Surface charge characteristics of kaolinitic tropical soils. *Geochim. Cosmochim. Acta* **59**:875–884.

changes in adsorbate or adsorptive composition at fixed temperature (T) and applied pressure (P):

$$\delta\sigma_{II} + \delta\Delta q = 0 \tag{7.8}$$

where δ represents an infinitesimal shift caused by any mechanism that does not alter σ_0. For example, if the ionic strength (I) of the aqueous solution equilibrated with a soil is changed at fixed T and P, Eq. 7.8 can be expressed as

$$\left(\frac{\partial\sigma_H}{\partial I}\right)_{T,P} + \left(\frac{\partial\Delta q}{\partial I}\right)_{T,P} = 0 \tag{7.9}$$

This constraint may be applied to the definition of the *point of zero salt effect* (p.z.s.e.),

$$\left(\frac{\partial\sigma_H}{\partial I}\right)_{T,P} = 0 \quad (\text{pH} = \text{p.z.s.e.}) \tag{7.10}$$

to show that the crossover point of two σ_H versus pH curves must also be that of two Δq versus pH curves. This can be used to verify the accuracy of p.z.s.e. values inferred from the crossover point of $\delta\sigma_{H,\,titr}$ curves.

The p.z.n.c. is the pH value at which the net adsorbed ion charge is equal to zero. A common laboratory method is to utilize *index ions*, such as Li^+ and Cl^-, in the determination of p.z.n.c. from a Δq versus pH curve (Fig. 7.6). Evidently, the value of p.z.n.c. will depend on the choice of index ions, although this dependence tends to be small if the ions are chosen from the following group: Li^+, Na^+, K^+, Cl^-, ClO_4^-, and NO_3^-. As a broad rule, p.z.n.c. values for silica, humus, clay minerals, and most manganese oxides are less than pH 4, whereas those for aluminum and iron oxyhydroxides and for calcite are more than pH 7. Thus, *p.z.n.c. tends to increase as chemical weathering of a soil proceeds if there is an attendant loss of humus and silica* (cf. Table 1.7).

The p.z.c. is the pH value at which the net total particle charge is equal to zero. Thus, by Eq. 7.5, at the p.z.c., there is no surface charge to be neutralized by ions in the diffuse swarm. Therefore, the p.z.c. could be measured by ascertaining the pH value at which a perfect charge balance exists among the ions in an aqueous solution with which soil particles have been equilibrated. More commonly, p.z.c. is inferred from the pH value at which a suspension of particles flocculates rapidly—a condition that is produced by the dominance of attractive van der Waals interactions (Section 3.4) over the coulomb repulsion between particles that is created by a nonzero net total particle charge.

The charge balance conditions in eqs. 7.5 and 7.6 lead to three broad statements about points of zero charge known as *PZC Theorems.* The first of these theorems concerns the relationship between p.z.n.p.c. and p.z.n.c. At the latter point of zero charge, Eq. 7.5 reduces to the condition

$$\sigma_0 + \sigma_H = 0 \quad (\text{pH} = \text{p.z.n.c}). \tag{7.11}$$

If p.z.n.c. > p.z.n.p.c., σ_H must have a negative sign in Eq. 7.11 because σ_H always decreases as pH increases, and $\sigma_H \equiv 0$ at p.z.n.p.c. It follows that the structural charge $\sigma_0 > 0$ if p.z.n.c. > p.z.n.p.c. Similarly, if p.z.n.c. < p.z.n.p.c., $\sigma_0 < 0$. Therefore, we have the first PZC Theorem:

1. *The sign of the difference (p.z.n.c. – p.z.n.p.c.) is the sign of the structural charge.*

For example, p.z.n.c. < p.z.n.p.c. typically for kaolinitic Oxisols (Fig. 7.6) and for specimen kaolinite samples, indicating at once that a negative structural charge exists in these materials, likely from the presence of 2:1 layer-type clay minerals, given the typical lack of isomorphic substitutions in kaolinite. More generally, soil particles with a surface chemistry dominated by 2:1 clay minerals or manganese oxides ($\sigma_0 < 0$) must always have p.z.n.c. values below their p.z.n.p.c. values.

A corollary of PZC Theorem 1 is that, *for soil particles without 2:1 clay minerals* (and, strictly speaking, without oxide minerals having structural charge) *p.z.n.c. = p.z.n.p.c.* Equality of the two points of zero charge means that the pH value at which σ_H is equal to zero can be determined through ion adsorption measurements alone.

The difference between p.z.n.c. and p.z.c. is that a charged diffuse-ion swarm exists at the former pH value, whereas it cannot exist at the latter pH value. The use of suspension flocculation to signal p.z.c. is compromised by the fact that flocculation usually occurs in the presence of a small—but nonzero—electrostatic repulsive force that is not strong enough to preclude van der Waals attraction from inducing flocculation. However, surface charge balance, as expressed by combining eqs. 7.4 and 7.5,

$$\sigma_{in} + \sigma_S + \sigma_d = 0 \tag{7.12}$$

yields a relationship between p.z.n.c. and p.z.c. Suppose that the Stern layer charge $\sigma_S = 0$ at the p.z.n.c. Then σ_d must also vanish because of Eq. 7.12 and the fact that $\sigma_{in} = 0$ at the p.z.n.c. But $\sigma_d = 0$ means pH = p.z.c. Therefore, p.z.c. = p.z.n.c. if $\sigma_S = 0$ at the p.z.n.c. Conversely, if p.z.c. = p.z.n.c., then $\sigma_{in} = 0 = \sigma_d$ and, again by Eq. 7.12, $\sigma_S = 0$ of necessity. The general conclusion to be drawn is in the second PZC theorem:

2. *The p.z.c. is equal to the p.z.n.c. if and only if the Stern layer charge is zero at the p.z.n.c.*

Note that PZC Theorem 2 is trivially true if the only adsorbed species are those in the diffuse-ion swarm. If surface complexes exist, PZC Theorem 2 will not hold unless the ions adsorbed in them (other than H^+ or OH^-) meet a condition of zero net charge at the p.z.n.c. This might occur for monovalent ions adsorbed "indifferently" in outer-sphere surface complexes by largely electrostatic interactions (e.g., Li^+ and Cl^-). Electrolytes for which $\sigma_S = 0$ at the p.z.n.c. are indeed termed *indifferent electrolytes*, in the sense that relatively weak electrostatic interactions cause their more or less equal adsorption. The p.z.c. values of particles suspended in solutions of indifferent electrolytes thus can be determined by ion adsorption measurements.

As originally conceived, the Stern layer comprises both inner-sphere and outer-sphere surface complexes (Fig. 7.4). If these species do not combine to yield zero net charge at the p.z.n.c., then p.z.c. $\neq$ p.z.n.c., according to PZC Theorem 2. The close relationship between p.z.c. and σ_S can be exposed further by applying the charge balance constraint in Eq. 7.8 at the p.z.c.:

$$\delta\sigma_H + \delta\sigma_S = 0 \quad (\text{pH} = \text{p.z.c.}) \tag{7.13}$$

which thus refers to changes under which σ_p remains equal to zero. If the Stern layer charge is made to increase, say, by increasing the amount of surface-complexed cations, then, according to Eq. 7.13, the net proton charge must compensate this change by decreasing, which in turn requires the p.z.c. to increase. The pH value at which $\sigma_H + \sigma_S$ balances σ_0 must be higher, as σ_S becomes higher, in order that σ_H will be negative enough to meet the condition of charge balance. In the same way, the pH value at which $\sigma_d = 0$ must be lower, as σ_S becomes lower through anion adsorption, in order that σ_H will become positive enough to compensate exactly the decrease in σ_S. This line of reasoning is epitomized in the third PZC theorem:

3. *If the Stern layer charge increases, the p.z.c. also increases, and vice versa.*

Theorem 3 indicates the role of cation surface complexation in increasing p.z.c. and that of anion surface complexation in decreasing p.z.c. It does *not* imply, however, that shifts in p.z.c. signal the effect of strong ion adsorption (*specific adsorption*), because changes in the number of outer-sphere surface complexes in the Stern layer are sufficient to change p.z.c.

7.5 Schindler Diagrams

Additional insight into the differences between readily exchangeable and specifically adsorbed ions can be obtained through the use of *Schindler diagrams.* A Schindler diagram is a banded rectangle in which the charge properties of an adsorbent and an adsorptive are compared as a function of pH in the range normally observed for soil particles, say, pH 3 to 9.5. The top band contains a vertical line denoting the p.z.n.c. of the adsorbent. The central band contains vertical lines denoting either the value of –log *K for hydrolysis (based on water as a reactant) of metal cation adsorptives, or the value of log K for protonation (based on the proton as a reactant) of ligand adsorptives:

$$M^{m+} + H_2O(\ell) = MOH^{(m-1)+} + H^+ \quad {}^*K = \frac{(MOH^{(m-1)+})(H^+)}{(M^{m+})} \tag{7.14a}$$

$$L^{\ell-} + H^+ = HL^{(\ell-1)-} \quad K = \frac{(HL^{(\ell-1)-})}{(L^{\ell-})(H^+)} \tag{7.14b}$$

The bottom band shows a horizontal line depicting the range of pH over which adsorption is to be expected when based solely on unlike charge attraction between the adsorbent and the adsorptive. *This pH range, therefore, indicates conditions under which the adsorbent can surely function as a cation or anion exchanger. If adsorption is observed to occur at pH values outside this range, specific adsorption mechanisms are implied.*

As a first example of a Schindler diagram, consider an adsorbent composed primarily of clay minerals and humus, with the adsorptive being an ion of an indifferent electrolyte (e.g., Li^+, Na^+, Cl^-, or NO_3^-). The p.z.n.c. of the adsorbent will not likely exceed 4.0, and the –log *K value for metal hydrolysis as well as log K for anion protonation of indifferent electrolyte adsorptives always will lie outside the pH range between 3 and 9. Therefore, the Schindler diagram will feature a top band with a vertical line at pH 4 (or possibly to its left), a central band that has no vertical lines, and a bottom band with either a horizontal line extending to the right of pH 4 (cations) or one extending to the left of pH 4 (anions). It follows that adsorbents comprising principally humus and clay minerals (e.g., soils from temperate grassland regions) will function

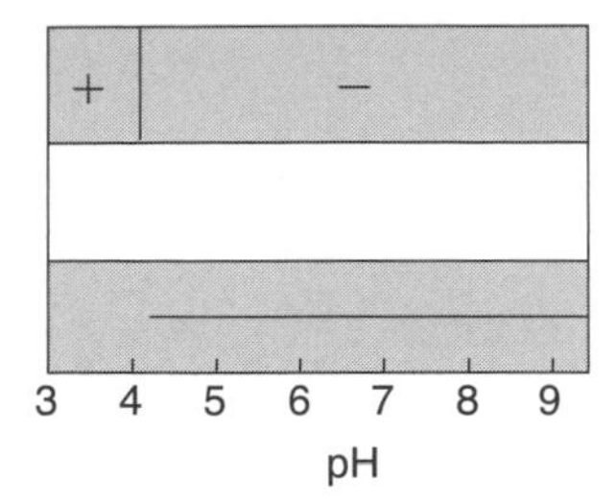

Figure 7.7. Schindler diagram for cations adsorbing on a soil with a clay fraction that is dominated by humus and clay minerals.

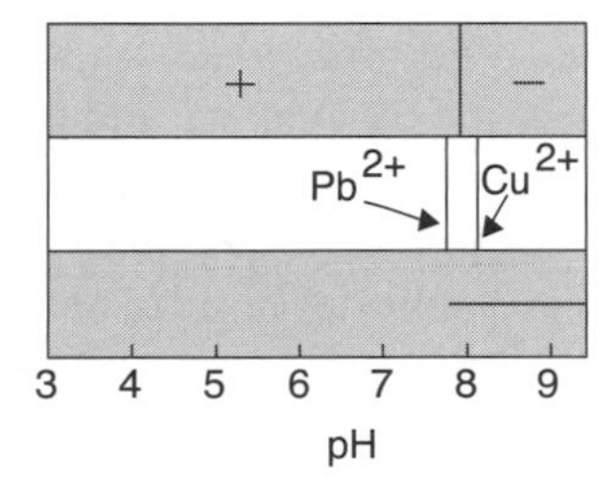

Figure 7.8. Schindler diagram for metal cations adsorbing on $Fe(OH)_3$, a common product of Fe(III) precipitation in soils undergoing alternate flooding and drying conditions.

effectively as cation exchangers under most soil conditions. Conversely, adsorbents comprising principally iron and aluminum oxides (e.g., uncultivated subsoils from the humid tropics), for which p.z.n.c. > 7 typically, will function effectively as anion exchangers. These trends are illustrated for adsorptive cations in Figure 7.7.

A second example can be developed for the adsorbent $Fe(OH)_3$ and the adsorptives Pb^{2+}, Cu^{2+}, and Cd^{2+}. The relevant p.z.n.c. value is 7.9, and the respective –log *K values are 7.7, 8.1, and 10.1. Therefore, the Schindler diagram for this system features a top band with a vertical line at pH 7.9, a central band with vertical lines at pH 7.7 and 8.1, and a bottom band with a horizontal line from pH 7.9 to 9.5 (Fig. 7.8). The rather narrow range of pH over which the adsorbent can function as a cation exchanger is apparent. The adsorption reactions of Pb^{2+} and Cu^{2+} with $Fe(OH)_3$ are in fact typically observed to be very strong at pH $\lesssim$ p.z.n.c. while the adsorbent surface is still positively charged, implying a specific adsorption mechanism. The adsorption of Cd^{2+}, on the other hand, often only commences on $Fe(OH)_3$ for pH > p.z.n.c. and, therefore, is consistent with a cation exchange mechanism.

The same approach can be used to analyze a calcareous soil reacting with borate in solution. The relevant p.z.n.c. value is 9.5, and log K for $B(OH)_4^-$ is 9.23. Therefore, the corresponding Schindler diagram has a top band with

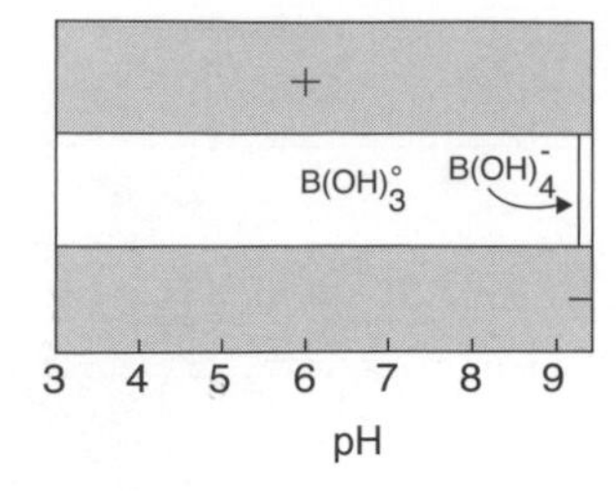

Figure 7.9. Schindler diagram for borate adsorbing on a calcareous Entisol.

a uniformly positive adsorbent surface charge indicated, a central band with a vertical line at pH 9.2, and a bottom band with a horizontal line extending over the very narrow range of pH between 9.2 and 9.5 (Fig. 7.9). Quite clearly, then, specific adsorption mechanisms are involved if the reaction of borate with this soil is significant at pH < 9.2. At pH values less than 9.2, borate anions do exist to some degree and can be attracted to the positively charged adsorbent in increasing numbers as pH increases from 7 to 9. At pH values more than 9.2, the adsorptive is predominantly anionic, but now the adsorbent is also becoming increasingly negatively charged, leading to an expected sharp fall-off in adsorption at pH $>$ p.z.n.c.

For Further Reading

Chorover, J., M. K. Amistadi, and O. A. Chadwick. (2004) Surface charge evolution of mineral–organic complexes during pedogenesis in Hawaiian basalt. *Geochim. Cosmochim. Acta* **68**:4859–4876. This article offers a comprehensive application of surface charge concepts and definitions to soils in a chronosequence for which a variety of chemical, mineralogical, and spectroscopic properties are known.

Johnston, C. T., and E. Tombácz. (2002) Surface chemistry of soil minerals, pp. 37–67. In: J. B. Dixon and D. G. Schulze (eds.), *Soil mineralogy with environmental applications.* Soil Science Society of America, Madison, WI. An excellent survey of surface charge concepts applied to soil minerals and humus that can be read with profit as a companion to the current chapter.

Sposito, G. (1998). On points of zero charge. *Environ. Sci. Technol.* **32**:2815–2819, Sposito, G. (1999). Erratum: On points of zero charge. *Environ. Sci. Technol.* **33**:208. An in-depth treatment of the PZC Theorems together with some of the conceptual issues arising in the measurement of points of zero charge.

Yu, T. R. (1997) *Chemistry of variable charge soils.* Oxford University Press, New York. This research monograph provides a detailed survey of the chemical properties of soils with surface charge characteristics that are highly pH dependent.

Zelazny, L. W., L. He, and A. Vanwormhoudt. (1996) Charge analysis of soils and anion exchange, pp. 1231–1253. In: D. L. Sparks (ed.), *Methods of soil analysis: Part 3. Chemical methods.* Soil Science Society of America, Madison, WI. This book chapter presents a useful discussion of laboratory methods for measuring surface charge components and points of zero charge.

Problems

The more difficult problems are indicated by an asterisk.

1. The table presented here shows the pH dependence of the amounts of Na^+ and Cl^- absorbed by a kaolinitic Brazilian Oxisol at two inoic strengths. Calculate $\Delta q = n_{Na} - n_{Cl}$ as a function of pH, estimating its precision, and determine the p.z.n.c. of the soil at both ionic strengths, also estimating its precision

I = 9 mol m^{-3}			I = 30 mol m^{-3}		
pH	n_{Na} (mmol kg^{-1})	n_{Cl} (mmol kg^{-1})	pH	n_{Na} (mmol kg^{-1})	n_{Cl} (mmol kg^{-1})
2.55	1.02 ± 0.12	6.39 ± 0.41	2.57	1.51 ± 0.52	9.79 ± 0.94
2.76	1.21 ± 0.20	5.73 ± 0.52	2.79	2.03 ± 0.21	8.28 ± 0.62
2.87	1.30 ± 0.47	5.57 ± 0.36	2.92	1.71 ± 0.68	7.64 ± 0.54
3.04	1.51 ± 0.40	4.71 ± 0.41	3.09	2.37 ± 0.16	7.77 ± 0.56
3.46	1.78 ± 0.16	3.88 ± 0.58	3.30	2.54 ± 0.30	6.37 ± 0.70
3.63	2.11 ± 0.28	3.78 ± 0.33	3.57	3.51 ± 0.45	5.72 ± 0.40
3.90	2.05 ± 0.10	2.51 ± 0.91	4.04	4.30 ± 0.67	4.62 ± 0.13
4.16	2.01 ± 0.23	2.60 ± 0.44	4.35	5.27 ± 0.85	3.75 ± 0.65
4.41	3.60 ± 0.33	1.34 ± 0.52	4.63	7.13 ± 0.91	2.46 ± 0.62
4.95	4.49 ± 0.46	1.14 ± 0.45	4.84	8.71 ± 0.75	1.86 ± 0.70

[*Hint:* If σ_{Na} and σ_{Cl} are the respective standard deviations of n_{Na} and n_{Cl}, then $\sigma_{\Delta q} = (\sigma_{Na}^2 + \sigma_{Cl}^2)^{1/2}$.]

2. The amount of Cs^+ adsorbed by gibbsite particles suspended in CsCl solution was found to increase linearly from essentially 0 to 20 mmol kg^{-1} as pH increased from 7.7 to 9.0, whereas the amount of Cl adsorbed decreased linearly from 13 mmol kg^{-1} to essentially 0 mmol kg^{-1} as pH increased from 4 to 9. Calculate p.z.n.p.c. for this mineral.

3. The table presented here shows the pH dependence of σ_H for the kaolinitic Oxisol described in Problem 1. Determine p.z.n.p.c. taking into account the precision of the data.

I = 30 mol m^{-3}		I = 9 mol m^{-3}	
pH	σ_H (mmol$_c$ kg^{-1})	pH	σ_H (mmol$_c$ kg^{-1})
2.57	58.11 ± 1.15	2.55	50.63 ± 6.62
2.79	41.73 ± 2.78	2.76	37.36 ± 3.59
2.92	34.08 ± 1.72	2.87	29.74 ± 6.31
3.09	25.99 ± 1.78	3.04	23.03 ± 2.05
3.30	18.49 ± 2.69	3.46	9.64 ± 1.42
3.57	11.55 ± 1.41	3.63	6.67 ± 0.99
4.04	1.21 ± 1.68	3.90	3.74 ± 1.71
4.35	– 4.06 ± 1.66	4.16	0.09 ± 1.62
4.63	– 7.52 ± 1.01	4.41	–1.81 ± 0.66
4.84	–10.88 ± 2.10	4.95	–4.94 ± 1.04

4. Compare p.z.n.c. and p.z.n.p.c. at each ionic strength for the Oxisol described in Problems 1 and 3. What can be deduced about the existence of structural charge in this soil?

*5. The table presented here shows the pH dependence of Δq and $\delta\sigma_{H,titr}$ at two ionic strengths for a California Alfisol suspended in NaCl solution. For this soil, the Cs^+ method yields $\sigma_0 = -64.5 \pm 0.2$ mmol$_c$ kg^{-1}.

I (mol L^{-1})	pH	Δq (mmol$_c$ kg^{-1})	$\delta\sigma_{H,titr}$ (mmol$_c$ kg^{-1})
0.05	4.20 ± 0.01	63 ± 12	48.1 ± 0.3
	4.59 ± 0.02	72 ± 5	35.4 ± 0.2
	5.53 ± 0.04	94 ± 4	16.20 ± 0.04
	5.84 ± 0.04	91 ± 6	12.40 ± 0.05
	6.52 ± 0.07	102 ± 7	2.50 ± 0.01
	7.04 ± 0.04	104 ± 4	0.03 ± 0.01
	7.30 ± 0.08	117 ± 2	–0.08 ± 0.01
	7.84 ± 0.08	115 ± 6	–1.60 ± 0.08
0.02	4.30 ± 0.04	81 ± 3	56.1 ± 0.6
	4.76 ± 0.03	92 ± 3	37.4 ± 0.2
	5.35 ± 0.02	103 ± 1	25.0 ± 0.1
	5.82 ± 0.02	108 ± 4	17.02 ± 0.06
	6.51 ± 0.01	124 ± 1	7.50 ± 0.01
	7.08 ± 0.07	116 ± 2	0.24 ± 0.01
	7.56 ± 0.09	122 ± 4	–0.17 ± 0.01
	7.91 ± 0.09	128 ± 3	–0.97 ± 0.01

a. Calculate σ_H as a function of pH for each ionic strength, including an estimate of its precision.
b. Plot your results on a single graph with error bars on each data point indicating the imprecision in both σ_H and pH. (*Hint:* The standard deviation of a sum or difference of two quantities is estimated as the square root of the sum of the squares of the standard deviations for the two quantities.)

6. Estimate p.z.n.p.c. for the Alfisol described in Problem 5 at both ionic strengths. Compare your results with p.z.s.e. based on $\delta\sigma_{H,titr}$, taking into account its imprecision. Explain any discrepancy between p.z.n.p.c. and p.z.s.e.

*7. The table presented here gives values of the slope and y-intercept derived from linear regression of σ_H on Δq for the A horizons of four Hawaiian Andisols and an Oxisol that constitute a chronosequence on basaltic parent material.

a. Determine σ_0 for each soil.
b. Interpret the p.z.n.p.c. values for the soils in terms of their properties and the Jackson–Sherman weathering sequence.

Soil age (ky)	Organic C ($g\ kg^{-1}$)	Clay mineralogy[a]	Slope	y-Intercept ($mmol_c\ kg^{-1}$)	p.z.n.p.c.
20	339 ± 5	$F > A \gg Q$	-1.00 ± 0.18	51 ± 6	4.5 ± 0.3
150	390 ± 14	$A > F \gg V > Q$	-0.84 ± 0.07	63 ± 2	4.9 ± 0.2
400	136 ± 6	$A > F \gg V > Q$	-1.02 ± 0.08	92 ± 5	6.4 ± 0.2
1400	125 ± 2	$K > Gi > F \gg Q$	-0.89 ± 0.15	28 ± 3	4.5 ± 0.1
4100	51 ± 1	$K > Gi > Go$	-1.02 ± 0.16	15 ± 4	3.4 ± 0.1

[a]Abbreviations: A, allophane; F, ferrihydrite; Gi, gibbsite; Go, goethite; K, kaolinite; Q, quartz; V, vermiculite, including pedogenic chlorite.

(*Hint:* Review the p.z.n.p.c. values mentioned for soil minerals in Chapter 2 and for humus in Chapter 3. Consider also the reactions of humus with soil minerals discussed in the latter chapter.)

*8. The p.z.c. of a soil low in 2:1 clay minerals is 5.0. After phosphate fertilizer is applied, the soil retains more adsorbed cations at pH 5 than before. Offer an explanation for this observation in terms of particle surface charge concepts.

*9. Potassium fertilizer added to a vermiculitic soils causes the retention of nitrate by the soil at a given pH value to increase. Offer an explanation for this effect in terms of particle surface charge concepts.

10. Discuss the statement: "In soils with low quantities of 2:1 clay minerals, the greater the degree of desilication (silica removal), the higher the

p.z.n.c." Consider both the Jackson–Sherman weathering sequence and the implications for soil fertility (i.e., for adsorbed ion retention).

11. Prepare a Schindler diagram for the Oxisol described in Problem 1.
12. Prepare a Schindler diagram for Hg^{2+} reacting with a soil with a p.z.n.c. = 3.6. Comment on whether the Hg^{2+} adsorption data in the table presented here imply specific adsorption as a likely reaction mechanism.

pH	Adsorbed Hg (mmol kg^{-1})
3.01	50.4
3.20	85.4
3.49	99.0
3.59	115.5
3.70	136.2
3.84	139.5
4.02	144.3
4.37	160.8
4.61	160.4

*13. The fluoroquinolone antibiotic ciprofloxacin (Problem 12 in Chapter 3) has log K values for protonation equal to 6.3 ± 0.1 for its COOH group and 8.6 ± 0.2 for its NH_3^+ group. Prepare a Schindler diagram for ciprofloxacin reacting with a typical temperate-zone soil and use it to predict whether ion exchange mechanisms are likely to be operative.

14. For which of the following diprotic organic acids presented in the table would significant adsorption by the soil described in Figure 7.6 be strong evidence of specific adsorption mechanisms?

Organic acid	log K_1	log K_2
Catechol	9.4	12.8
Phthalic	3.0	5.4
Salicylic	3.0	13.7

15. Why is it not usually appropriate to use p.z.n.p.c. instead of p.z.n.c. to construct a Schindler diagram?

8

Soil Adsorption Phenomena

8.1 Measuring Adsorption

After reaction between an adsorptive i and a soil adsorbent, the moles of i adsorbed per kilogram of dry soil is calculated with the standard equation

$$n_i \equiv n_{iT} - M_w m_i \tag{8.1}$$

where n_{iT} is the total moles of species i per kilogram dry soil in a slurry (batch process) or in a soil column (flow-through process), as described in Section 7.2; M_w is the gravimetric water content of the slurry or soil column (kilograms water per kilogram dry soil); and m_i is the molality (moles per kilogram water) of species i in the supernatant solution (batch process) or effluent solution (flow-through process). (For a discussion of the units of n_{iT}, M_w, and m_i, see the Appendix.) Equation 8.1 defines the *surface excess*, n_i, of a chemical species that has become an adsorbate. Formally, n_i is the excess number of moles of i per kilogram soil relative to its molality in the supernatant solution. As mentioned in Section 7.2, this excess can be positive, zero, or negative.

Consider, for example, a Mollisol containing humus and 2:1 clay minerals that reacts in a batch adsorption process with a $CaCl_2$ solution at pH 7. After the reaction, the soil and supernatant aqueous solution are separated by centrifugation. The resulting soil slurry is found to contain 0.053 mol Ca kg^{-1} and to have a gravimetric water content of 0.45 kg kg^{-1}. The supernatant solution separated from the slurry contains Ca at a molality of 0.01 mol kg^{-1}.

According to Eq. 8.1,

$$n_{Ca} = 0.053 - (0.45)(0.01) = \mathbf{+0.049\ mol\ kg^{-1}}$$

is the *positive* surface excess of Ca adsorbed by the soil. Suppose that the molality of Cl in the supernatant solution is 0.02 mol kg^{-1} and that the soil slurry contains 0.0028 mol Cl kg^{-1}. Then,

$$n_{Cl} = 0.0028 - (0.45)(0.02) = \mathbf{-0.0062\ mol\ kg^{-1}}$$

is the *negative* surface excess of Cl adsorbed by the soil. In both examples, n_i is the relative *excess* moles of species i (per kilogram dry soil) compared with a hypothetical aqueous solution containing M_w kilograms water and species i at the molality m_i. This excess is attributed to the presence of the soil adsorbent.

If the initial molality of species i in the reactant aqueous solution is m_i^0 and the *total* mass of water in this solution that is mixed with 1 kg dry soil, in either a batch or a flow-through process, is M_{Tw}, then the condition of mass balance for species i can be expressed as

$$\underset{\text{(moles added initially)}}{m_i^0 M_{Tw}} = \underset{\text{(moles in slurry)}}{n_{iT}} + \underset{\text{(moles in supernatant solution)}}{m_i\,(M_{Tw} - M_w)} \tag{8.2}$$

Equations 8.1 and 8.2 can be combined to yield

$$n_i = \Delta m_i M_{Tw} \tag{8.3}$$

where $\Delta m_i \equiv m_i^0 - m_i$ is the change in molality, attributed to adsorption. Equation 8.3 is applied frequently to calculate a surface excess as the product of the change in adsorptive concentration times the mass of water *added* per unit mass of dry soil. Note that the right side of Eq. 8.3 refers only to the aqueous solution phase and that Δm_i can be positive, zero, or negative. In practice, the difference between molality and a concentration in moles per liter can be neglected in applying the equation.

As a second, more complicated example of the use of Eq. 8.1, consider a montmorillonitic Entisol that contains both calcite and gypsum (Section 2.5). These soil minerals likely will dissolve to release Ca, as well as bicarbonate and sulfate, when in contact with an aqueous solution. Suppose the soil is equilibrated in batch mode with a NaCl/$CaCl_2$ solution, followed by centrifugation to separate the supernatant solution from a soil slurry with a gravimetric water content that is 0.562 kg kg^{-1}. Quantitation of the electrolyte composition in the slurry and the supernatant solution yields the following data set:

$$n_{NaT} = 13.20\ \text{mmol kg}^{-1} \qquad c_{Na} = 12.67\ \text{mmol L}^{-1}$$
$$n_{CaT} = 79.25\ \text{mmol kg}^{-1} \qquad c_{Ca} = 7.28\ \text{mmol L}^{-1}$$
$$n_{ClT} = 13.90\ \text{mmol kg}^{-1} \qquad c_{Cl} = 25.03\ \text{mmol L}^{-1}$$
$$n_{HCO_3T} = 25.10\ \text{mmol kg}^{-1} \qquad c_{HCO_3} = 0.27\ \text{mmol L}^{-1}$$
$$n_{SO_4T} = 5.00\ \text{mmol kg}^{-1} \qquad c_{SO_4} = 0.98\ \text{mmol L}^{-1}$$

where the difference between molality and moles per liter in the supernatant solution has been neglected. For a montmorillonitic soil, it is reasonable to assume that the concentrations of bicarbonate and sulfate and, therefore, a portion of the Ca present, can be attributed mainly to soil mineral dissolution. Charge balance considerations then would reduce n_{CaT} and c_{Ca} according to the expressions

$$n'_{CaT} \equiv n_{CaT} - \frac{1}{2} n_{HCO_3T} - n_{SO_4T} = 61.70 \text{ mmol kg}^{-1}$$

$$c'_{Ca} \equiv c_{Ca} - \frac{1}{2} c_{HCO_3} - c_{SO_4} = 6.17 \text{ mmol L}^{-1}$$

as a first approximation that neglects the adsorption of the two anions. By Eq. 8.1, again ignoring the minute difference between molar and molal concentrations, the respective surface excesses of Na, Ca, and Cl are

$$n_{Na} = 13.20 - (0.562)12.67 = \mathbf{+6.07\,mmol\,kg^{-1}}$$

$$n_{Ca} = 61.70 - (0.562)6.17 = \mathbf{+58.23\,mmol\,kg^{-1}}$$

$$n_{Cl} = 13.90 - (0.562)25.03 = \mathbf{-0.17\,mmol\,kg^{-1}}$$

The cations are positively adsorbed by the soil under the conditions of measurement, whereas chloride is once again negatively adsorbed, consistent with the montmorillonitic character of the soil.

8.2 Adsorption Kinetics and Equilibria

Experiments have shown that adsorption reactions in soils are typically rapid, operating on timescales of minutes or hours, but that sometimes they exhibit long-time "tails" that extend over days or even weeks. Readily exchangeable ions (Section 7.2) adsorb and desorb very rapidly, with a rate usually governed by a film diffusion mechanism (Section 3.3 and Special Topic 3). Specifically adsorbed ions show much more complicated behavior in that they often adsorb by multiple mechanisms that differ from those involved in their desorption, and their rates of adsorption or desorption are described by more than one equation during the time course of either process. It is usually these ions with adsorption reactions that will have the long-time tails.

Adsorption kinetics for ions are *assumed* to be represented mathematically by the difference of two terms, as in Eq. 4.2:

$$\frac{dn_i}{dt} = R_f - R_b \tag{8.4}$$

where R_f and R_b are forward and backward rate functions respectively. A consensus does not exist regarding which rate laws should be applied to model R_f and R_b. Many different empirical formulations appear in the soil chemistry literature. One popular choice has been a rate law like that in Eq. S.3.7 (Special Topic 3 in Chapter 3):

$$R_f = k_{ads} c_i (n_{imax} - n_i) \qquad R_b = k_{des} n_i \tag{8.5}$$

where n_{imax} is the maximum value of n_i, and k_{ads} and k_{des} are the rate coefficients in Eq. S.3.7. As indicated in Table 4.2, appropriate plotting variables can be identified to determine the rate coefficients under conditions such that either R_f or R_b is negligible. For example, if desorption alone is provoked by placing an equilibrated soil in contact with a very dilute aqueous solution ($R_f = 0$), eqs. 8.4 and 8.5 combine to become a first-order rate law in n_i. The rate coefficient k_{des} is then determined from the slope of a plot of ln n_i versus time (Table 4.2). However, rate laws like those in Eq. 8.5 do not reflect a unique mechanism of adsorption or desorption. They are empirical mathematical models with an underlying mechanistic significance that must be established by independent experiments on the detailed nature of the surface reactions they purport to describe.

A graph of n_i against m_i or c_i at fixed temperature and applied pressure at any time during an adsorption reaction is an *adsorption isotherm*. Adsorption isotherms are convenient for representing the effects of adsorptive concentration on the surface excess, especially if other variables, such as pH and ionic strength, are controlled along with temperature and pressure. Figure 8.1 shows four categories of adsorption isotherm observed commonly in soils.

The *S-curve* isotherm is characterized by an initially small slope that increases with adsorptive concentration. This behavior suggests that the affinity of the soil particles for the adsorbate is less than that of the aqueous solution for the adsorptive. In the example of copper adsorption given in Figure 8.1, the S-curve may result from competition for Cu^{2+} ions between ligands in soluble humus and adsorption sites on soil particles. When the concentration of Cu added exceeds the complexing capacity of the soluble organic ligands, the soil particle surface gains in the competition and begins to adsorb copper ions significantly. In some instances, especially when "hydrolyzable" metals or "polymerizable" organic compounds are adsorbed, the S-curve isotherm is the result of cooperative interactions among the adsorbed molecules. These interactions (e.g., surface polymerization) cause multinuclear surface complexes to grow on a soil particle surface (Fig. 7.5), producing an enhanced affinity for the adsorbate as its surface excess increases.

The *L-curve* isotherm is characterized by an initial slope that does not increase with the concentration of adsorptive in the soil solution. This type of isotherm is the effect of a relatively high affinity of soil particles for the adsorbate at low surface coverage mitigated by a decreasing amount of adsorbing surface remaining available as the surface excess increases. The example of phosphate adsorption in Figure 8.1 illustrates a universal L-curve feature: The isotherm is concave to the concentration axis because of the combination of affinity and steric factors.

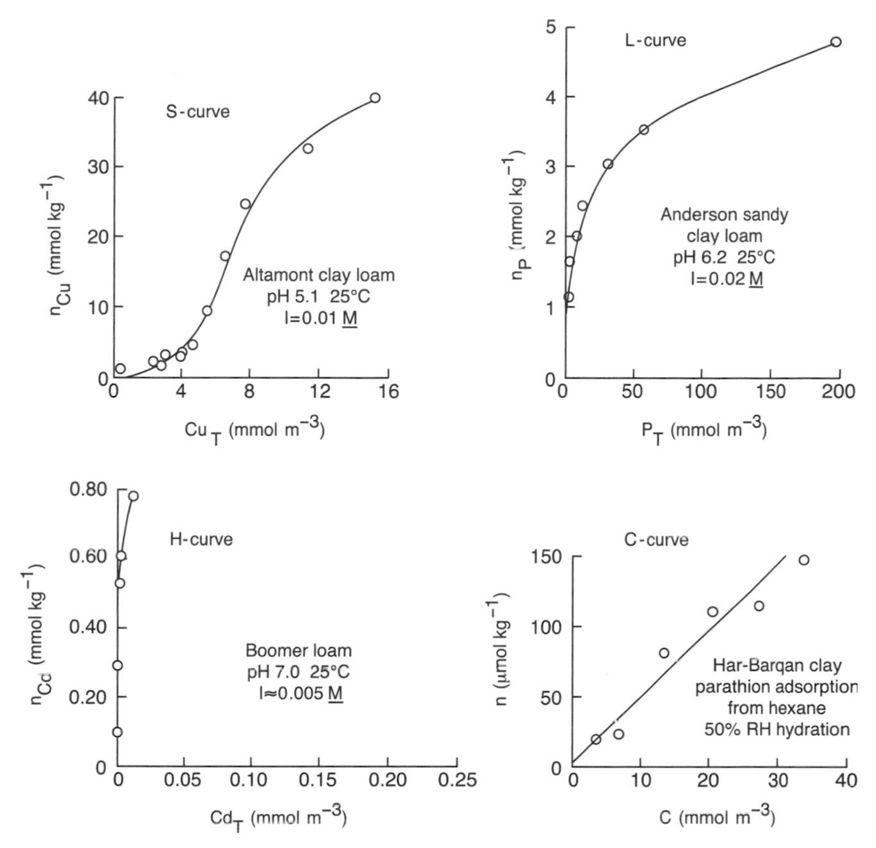

Figure 8.1. The four categories of adsorption isotherm as characterized by their shapes as curves. Abbreviations: I, ionic strength; RH, relative humidity.

The *H-curve* isotherm is an extreme version of the L-curve isotherm (an XL-curve). Its characteristic large initial slope (by comparison with an L-curve isotherm) suggests a very high relative affinity of the soil for an adsorbate. This condition is usually produced either by inner-sphere surface complexation or by significant van der Waals interactions in the adsorption process (sections 3.4 and 3.5). The example of cadmium adsorption shown in Figure 8.1 illustrates an H-curve isotherm evidently caused by specific adsorption. Soil humus and inorganic polymers (e.g., Al-hydroxy polymers) can produce H-curve isotherms resulting from both specific adsorption and van der Waals interactions.

The *C-curve* isotherm is characterized by an initial slope that remains independent of adsorptive concentration until the maximum possible adsorption is achieved. This kind of isotherm can be produced either by a constant partitioning of an adsorptive between the interfacial region and the soil solution, or by a proportionate increase in the amount of adsorbing surface as the surface excess increases. The example of parathion (diethyl *p*-nitrophenyl monothiophosphate) adsorption in Figure 8.1 shows constant partitioning of this compound between hexane, a hydrophobic liquid, and the layers of water on soil particles that accumulate at 50% relative humidity (RH). Similarly, the adsorption of a hydrophobic organic compound by soil humus is often a constant partitioning between the latter solid phase and the soil solution as described by a C-curve isotherm (Section 3.4).

The adsorption isotherm categories illustrated in Figure 8.1 can be quantified by expressing the data in terms of the *distribution coefficient*,

$$K_{di} \equiv n_i / c_i \tag{8.6}$$

where c_i represents a soil solution concentration of an adsorptive species i. Equation 3.10 is the special case of Eq. 8.6 obtained by dividing both sides of the latter expression with the soil organic C content (f_{oc}). Thus the Chiou distribution coefficient is simply a distribution coefficient normalized to the soil organic C content (i.e., $K_{oc} = K_d/f_{oc}$). Comparatively, the C-curve corresponds to a distribution coefficient that is independent of the surface excess, whereas the S-curve corresponds to one that increases initially with the surface excess. The L- and H-curve isotherms, by contrast, correspond to a distribution coefficient that decreases with increasing surface excess.

Equation 8.6 necessarily provides a complete mathematical description of the C-curve isotherm because the left side of the equation is a constant parameter. The L-curve isotherm usually is described mathematically by the *Langmuir equation*:

$$n_i = \frac{bKc_i}{1 + Kc_i} \tag{8.7}$$

where b and K are adjustable parameters. The *capacity parameter* b represents the value of n_i that is approached asymptotically as c_i becomes arbitrarily large. The *affinity parameter* K determines the magnitude of the initial slope of the isotherm. Equation 8.7 can be derived from the rate law obtained by combining eqs. 8.4 and 8.5:

$$\frac{dn_i}{dt} = k_{ads}\, c_i\, (n_{max} - n_i) - k_{des}\, n_i \tag{8.8}$$

Under steady-state conditions, the left side of Eq. 8.8 is zero and Eq. 8.7 follows upon solving for n_i and making the parameter identifications:

$$b \equiv n_{max}, K \equiv k_{ads}/k_{des} \tag{8.9}$$

Equation 8.9 provides a connection between an empirical rate law and the Langmuir equation. Note that the affinity parameter K is large if adsorption is rapid and desorption is slow (i.e., $k_{ads} \gg k_{des}$). After multiplying both sides of

Eq. 8.7 by $(1/c_i + K)$ and solving for K_{di}, one finds that the Langmuir equation is equivalent to the linear expression

$$K_{di} = bK - Kn_i \tag{8.10}$$

Thus, a graph of K_{di} against n_i should be a straight line with a slope equal to $-K$ and an x-intercept equal to b, if the Langmuir equation is applicable.

Adsorption isotherm equations cannot be interpreted to indicate any particular adsorption mechanism *or even if adsorption—as opposed to precipitation—has actually occurred.* On strictly mathematical grounds, it can be shown that a sum of two Langmuir equations with its four adjustable parameters will fit *any* L-curve isotherm, regardless of the underlying adsorption mechanism. Thus, adsorption isotherm equations, like rate laws, should be regarded as curve-fitting models without particular mechanistic significance, but with predictive capability under defined conditions.

To see this latter point in detail, consider the typical situation in which the distribution coefficient decreases with increasing surface excess (Fig. 8.2). If K_d extrapolates to a finite value as the surface excess tends to zero and extrapolates to zero at some finite value of the surface excess, then adsorption isotherm data can *always* be fit to the two-term equation

$$n_i = \frac{b_1 K_1 c_i}{1 + K_1 c_i} + \frac{b_2 K_2 c_i}{1 + K_2 c_i} \tag{8.11}$$

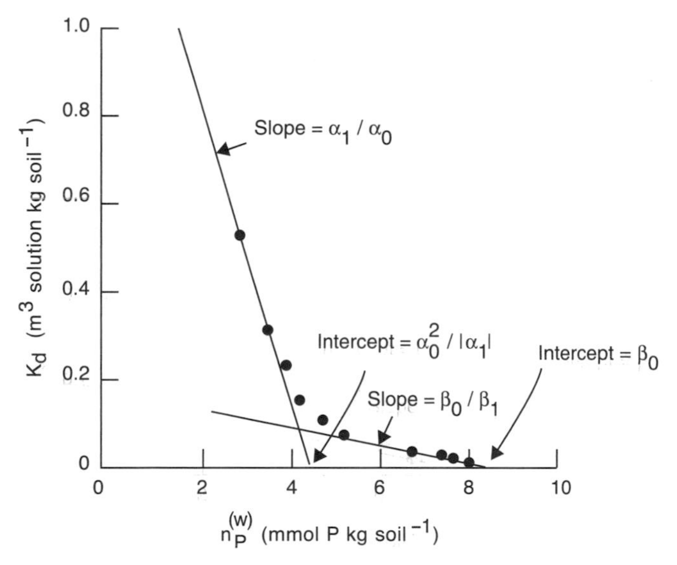

Figure 8.2. Plot of the distribution coefficient (Eq. 8.6) for phosphate adsorption by a clay loam soil, with lines illustrating Eqs. 8.14 and 8.15.

where b_l, b_2, K_1, and K_2 are empirical parameters, and c_i is a soil solution concentration. Equation 8.11 can be derived rigorously, but its correctness as a universal approximation emerges after using Eq. 8.6 to substitute for c_i in terms of K_{di} and n_i to generate a second-degree algebraic equation (dropping the subscript i for convenience):

$$K_d^2 + (K_1 + K_2)K_d n + K_1 K_2 n^2 - (b_1 K_1 + b_2 K_2)K_d - bK_1 K_2 n = 0 \tag{8.12}$$

where $b \equiv b_1 + b_2$. Equation 8.10 is recovered if K_2 and b_2 are set equal to zero. The derivative of K_d with respect to n follows from Eq. 8.12 as

$$\frac{dK_d}{dn} = -\frac{(K_1 + K_2)K_d + 2K_1 K_2 n - bK_1 K_2}{2K_d + (K_1 + K_2)n - (b_1 K_1 + b_2 K_2)} \tag{8.13}$$

As n tends to zero, K_d can be approximated by a linear equation in n, and Eqs. 8.12 and 8.13 combine to show that

$$K_d \approx \alpha_0 + (\alpha_i/\alpha_0)n \quad (n \downarrow 0) \tag{8.14}$$

where

$$\alpha_0 = b_1 K_1 + b_2 K_2, \quad \alpha_1 = -(b_1 K_1^2 + b_2 K_2^2)$$

According to Eq. 8.14, the x-intercept of the linear expression is $\alpha_0^2/|\alpha_1|$, as indicated in Figure 8.2. As n tends to its maximum value *b*, K_d drops to zero, according to Eq. 8.12. Thus, K_d is once again a linear function of n, and Eqs. 8.12 and 8.13 can be used to demonstrate that

$$K_d \approx (\beta_0^2/|\beta_1|) + (\beta_0/\beta_1)n \quad (n \uparrow b) \tag{8.15}$$

where

$$\beta_0 = b = b_1 + b_2, \quad \beta_1 = -\frac{b_1}{K_1} - \frac{b_2}{K_2}$$

The slope of the line now is $\beta_0/\beta_1 < 0$, and its x-intercept is b. If adsorption data are plotted as in Figure 8.2, then the limiting slopes and the two x-intercepts can be determined graphically. The four values found can be used to solve *uniquely* for the four empirical parameters b_1, K_1, b_2, and K_2. These parameters, like those in Eq. 8.7, have no particular *chemical* significance in terms of adsorption reactions.

If the distribution coefficient does not extrapolate to a finite value as the surface excess tends to zero, then the Langmuir equation can be generalized to a power–law expression known as the *Langmuir–Freundlich equation*:

$$n_i = \frac{b(\tilde{K}c_i)^\beta}{1 + (\tilde{K}c_i)^\beta} \quad (0 < \beta \leq 1) \tag{8.16}$$

where b is the maximum value of n_i and $\tilde{K}$ is an affinity parameter analogous to K in the Langmuir equation, to which it reduces if the exponent $\beta = 1$. The "linearized" form of Eq. 8.16 is analogous to Eq. 8.10:

$$\frac{n_i}{c_i^{\beta}} = b\tilde{K}^{\beta} - \tilde{K}^{\beta}\, n_i \tag{8.17}$$

Equation 8.17 can be applied to determine b and $\tilde{K}$ once β is known. This exponent is determined by considering Eq. 8.16 at values of c_i low enough to justify the approximation

$$n_i \approx Ac_i^{\beta} \qquad (c_i \downarrow 0) \tag{8.18}$$

which is termed the *van Bemmelen–Freundlich equation.* The value of β is then found by a log–log plot of surface excess against soil solution concentration that, if it is linear, yields β as its slope. Then n_i/c_i^{β} can be calculated with adsorption data and plotted according to Eq. 8.17 to estimate b and $\tilde{K}$. A generalization of Eq. 8.16 analogous to that of the Langmuir equation in Eq. 8.11 also can be made. The resulting six-parameter equation has been applied successfully to model metal cation adsorption reactions (Section 8.5).

However, no mechanistic interpretation of these adsorption isotherm models can be had on the basis of goodness-of-fit criteria alone—a conclusion that extends even to determining whether an adsorption reaction has occurred, as opposed to a precipitation reaction. Not only do the data sets for this latter reaction yield plots similar to that in Figure 8.2 under a broad variety of experimental settings, but they also are often consistent with undersaturation conditions because of coprecipitation phenomena (Section 5.3), making identification of the reaction mechanism even more problematic. When no molecular-scale data on which to base a decision regarding mechanism are available, the loss of an adsorptive from aqueous solution to the solid phase can be termed *sorption* (Section 7.2) to avoid the implication that either adsorption or precipitation has definitely taken place. As a general rule, a surface precipitation mechanism is favored by high soil solution concentrations and long reaction times in sorption processes.

8.3 Metal Cation Adsorption

Metal cations adsorb onto soil particle surfaces via the three mechanisms illustrated in Figure 7.4. The relative affinity a metal cation has for a soil adsorbent depends in a complicated way on soil solution composition, but, to a first approximation, adsorptive metal cation affinities can be rationalized in terms of inner-sphere and outer-sphere surface complexation and diffuse-ion swarm concepts. As discussed in Section 7.2, the relative order of decreasing interaction strength among the three adsorption mechanisms is inner-sphere complex > outer-sphere complex >> diffuse-ion swarm. In an inner-sphere surface complex, the electronic structures of the metal cation and surface functional group are important, whereas for the diffuse-ion swarm only metal cation valence and surface charge should determine adsorption affinity. The outer-sphere surface complex is intermediate, in that valence is probably the most important factor, but the stereochemical effect of immobilizing a cation in a well-defined complex must also play a role in determining affinity (e.g., Fig. 7.2).

As a rule of thumb, the relative affinity of a *free* metal cation for a soil adsorbent will increase with the tendency of the cation to form inner-sphere surface complexes. This tendency is correlated positively with ionic radius (Table 2.1). For a given valence Z, the ionic potential Z/R (Section 1.2) decreases with increasing ionic radius R. This trend implies that metal cations with larger ionic radii will create a smaller electrical field and will be less likely to remain solvated during complexation by a surface functional group. Second, larger R implies a more labile electron configuration and a greater tendency for a metal cation to polarize in response to the electrical field of a charged surface functional group. This polarization is necessary for distortion of the electron configuration leading to covalent bonding. It follows that relative adsorption affinity series (*selectivity sequences*) can be established solely on the basis of ionic radius (Table 2.1):

$$Cs^+ > Rb^+ > K^+ > Na^+ > Li^+ \text{(Group IA)}$$

$$Ba^{2+} > Sr^{2+} > Ca^{2+} > Mg^{2+} \text{(Group IIA)}$$

$$Hg^{2+} > Cd^{2+} > Zn^{2+} \text{(Group IIB)}$$

These selectivity sequences, which encompass both Class A and Class B character (Section 1.2), have been observed often in soil sorption experiments. For borderline metals (i.e., bivalent transition metal cations), however, ionic radius is not adequate as a predictor of adsorption affinity, because electron configuration also plays a very important role in the complexes of these cations (e.g., Mn^{2+}, Fe^{2+}, Ni^{2+}). Their relative affinities tend to follow the *Irving–Williams* order:

$$Zn^{2+} < Cu^{2+} > Ni^{2+} > Co^{2+} > Fe^{2+} > Mn^{2+}$$

If a soil adsorbent is dominated by humus, either in particulate form or as a coating on minerals, Class A and B characters return as useful guides to adsorption affinity, with Class A metals preferring O-containing surface functional groups and Class B metals preferring N- or S-containing groups.

If a soil is reacted with a series of aqueous solutions with increasing pH values while containing a metal cation at a fixed initial concentration, the amount of metal cation adsorbed by the soil will increase with pH to some maximum value n_M, unless complexing ligands in the soil solution compete overwhelmingly for the metal against surface functional groups. In the absence of soluble ligand competition, a graph of metal cation adsorbed against pH will have a

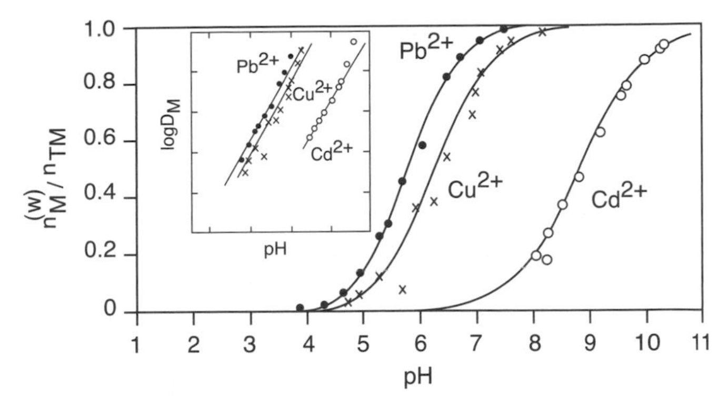

Figure 8.3. Adsorption edges for Pb^{2+}, Cu^{2+}, and Cd^{2+} interacting with poorly crystalline $Fe(OH)_3$. The inset shows graphs of the data according to Eq. 8.20 (Kurbatov plot). Data from Wang, Z.-J., and W. Stumm (1987) Heavy metal complexation by surfaces and humic acids: A brief discourse on assessment by acidimetric titration. *Netherlands J. Agric. Sci.* 35: 231–240.

characteristic sigmoid shape known as an *adsorption edge.* Adsorption edges for Pb^{2+}, Cu^{2+}, and Cd^{2+} on freshly precipitated $Fe(OH)_3$, as might be found in a flooded soil (Section 6.5), are shown in Figure 8.3. Often these curves are characterized numerically by pH_{50}, the pH value at which one half the value of n_M is achieved. It is observed typically that pH_{50} increases as the relative affinity of the metal cation for the soil decreases. For example, pH_{50} is often larger for Mn^{2+} than Cu^{2+}, and larger for Mg^{2+} than Ba^{2+}. In Figure 8.3 it is evidently larger for Cd^{2+} than for Cu^{2+}, and larger for Pb^{2+} than for Cu^{2+}. Nearly always, pH_{50} is well below the pH value at which significant hydrolysis of a metal cation occurs in aqueous solution.

A model equation to describe adsorption edges can be developed if a semilog graph of the *distribution ratio,*

$$D_i \equiv n_i/(n_{M_i} - n_i) \qquad \text{(adsorptive } i\text{)} \tag{8.19}$$

against pH is linear over a sufficiently broad range of the latter variable. Semilog graphs of D versus pH demonstrating a linear relationship, known as a *Kurbatov plot,*

$$\ln D_i = \alpha_i + \beta_i \mathrm{pH} \tag{8.20a}$$

are shown in the inset of Figure 8.3. A geometric interpretation of the empirical parameters α, β in Eq. 8.20a can be made as follows. The pH value at which half the moles of adsorptive i added are in an adsorbate form is defined as pH_{50}. Because $D_i = 1$ at this pH value, according to Eq. 8.19, it follows from Eq. 8.20a that

$$pH_{50} = \alpha_i/\beta_i \tag{8.21}$$

Therefore, Eq. 8.20a can be rewritten as

$$\ln D_i = \beta_i(pH - pH_{50}) \tag{8.20b}$$

or, after combining Eqs. 8.19 and 8.20b, as a model equation for n_i,

$$n_i = n_{Mi}\{1 + \exp[-\beta_i(pH - pH_{50})]\}^{-1} \tag{8.22}$$

The sigmoidal curve described by Eq. 8.22 can be interpreted with the help of a Schindler diagram (Section 7.5). Taking the data plotted in Figure 8.3 as an example, one can refer to the Schindler diagram for this system shown in Figure 7.8. The rather narrow range of pH over which $Fe(OH)_3$ can function as a cation exchanger is apparent, as is the conclusion that specific adsorption mechanisms must be operating in the reactions of Pb^{2+} and Cu^{2+}, because their adsorption edges plateau at pH $\lesssim$ p.z.n.c. while the adsorbent surface is still positively charged. The adsorption edge for Cd^{2+}, on the other hand, occurs mainly at pH > p.z.n.c. and, therefore, is consistent with a cation exchange mechanism.

A comparison of the apparent pH_{50} values for the adsorption edges in Figure 8.3 with the sequence of –log *K values for the three adsorptive metal cations shows that the two parameters are correlated positively (i.e., a high pH value for hydrolysis implies low adsorption affinity). This kind of correlation has been apparent in many studies of metal cation adsorption by oxyhydroxide minerals. In conceptual terms, it amounts to a general rule, that *metal cations that hydrolyze at low pH also will adsorb strongly* (i.e., will adsorb at pH values well below their –log *K value). From a coordination chemistry perspective, the complexation of a metal cation with OH^- in aqueous solution is analogous to the inner-sphere complexation of the metal cation to an ionized surface hydroxyl group, with the role of the proton in OH^- now being played by the metal in the adsorbent structure to which the surface hydroxyl is bonded: $\equiv Fe\text{–}O^- \Leftrightarrow HO^-$.

8.4 Anion Adsorption

The common soil anions Cl^- and NO_3^- adsorb mainly as diffuse-ion swarm species or as outer-sphere surface complexes. Evidence for this generalization comes from their readily exchangeable character and the frequent observation of *negative* surface excess when they are adsorbed by soils with low p.z.n.c., as encountered in the examples discussed in Section 8.1. Negative adsorption can occur only for adsorbate species in the diffuse-ion swarm. On the molecular scale, this phenomenon can be interpreted through the definition

$$n_i \equiv \frac{1}{m_s}\int_{Su} [c_i(x) - c_{0i}]\, dV \tag{8.23}$$

where $c_i(x)$ is the concentration (moles per unit volume) of anion i at point x in the aqueous solution portion of a suspension containing m_s kilograms of soil, c_{0i} is the concentration of ion i in the supernatant solution, and the integrals extend over the entire suspension volume. Equation 8.23 actually applies to any ion in the diffuse swarm. If negative adsorption occurs, $c_i(x) < c_{0i}$ and $n_i < 0$. This condition is produced by electrostatic repulsion of the ion i away from a surface of like charge sign (e.g., an anion in a soil containing significant humus or minerals with negative structural charge). An estimate of the average size of the interfacial region over which this repulsion is effective can be made by defining the *exclusion volume*:

$$V_{ex} \equiv \int \left[1 - \frac{c_i(x)}{c_{0i}}\right] dV/m_s = -n_i/c_{0i} \qquad (8.24)$$

In the first adsorption example discussed in Section 8.1, $V_{ex} = 0.0062$ mol kg^{-1}/20 mol m^{-3} = 3.1×10^{-4} m^3 kg^{-1} = 0.31 L kg^{-1}. (Note that $c_{0i} = 0.02$ molal $\approx$ 20 mol m^{-3}.) In suspensions of montmorillonite, this figure could be an order of magnitude larger for the same chloride concentration. In general, V_{ex} is the average volume of the region in the soil solution (per kilogram dry soil) in which $c_i(x)$ is smaller than its "bulk" value, c_{0i}. The observation of negative n_i and an appreciable V_{ex} is compelling evidence for significant diffuse-ion swarm species of an anion i.

Oxyanions, most notably arsenate, borate, phosphate, selenite, and carboxylate, are usually observed to adsorb as inner-sphere surface complexes. Several kinds of experimental evidence support this conclusion. Perhaps the most direct is the often-observed difficulty in desorbing anions like phosphate by leaching with anions like chloride. Another comparative type of evidence is the persistence of, for example, borate adsorption at pH > p.z.n.c., whereas chloride adsorption diminishes rapidly to zero at these pH values. Finally, spectroscopic methods have led to structural conceptualizations of adsorbed phosphate, selenite, borate, silicate, and molybdate ions like that shown for biphosphate shown in Figure 7.3. Although none of these pieces of evidence may be definitive when taken alone, combined they make a very strong case for ligand exchange (Eq. 3.12) as a principal mode of oxyanion (excepting nitrate, perchlorate, selenate, and possibly sulfate) adsorption by soil minerals. In general, for an anion $A^{\ell-}$ reacting with a Lewis acid site, the reaction scheme is

$$\equiv SOH(s) + H^+(aq) = \equiv SOH_2^+(s) \qquad (8.25a)$$

$$\equiv SOH_2^+(s) + A^{\ell-}(aq) = \equiv SA^{(\ell-1)-}(s) + H_2O(\ell) \qquad (8.25b)$$

If the Lewis acid site is present already, or if the concentration of A is very large, the protonation step in Eq. 8.25a is not required. Ligand exchange is favored by pH < p.z.n.p.c.

The graphs in Figure 8.4 illustrate the typical effect of pH on positive anion adsorption by soil particles. Fluoride and borate are representative anions of monoprotic acids, whereas phosphate represents anions of a polyprotic acid,

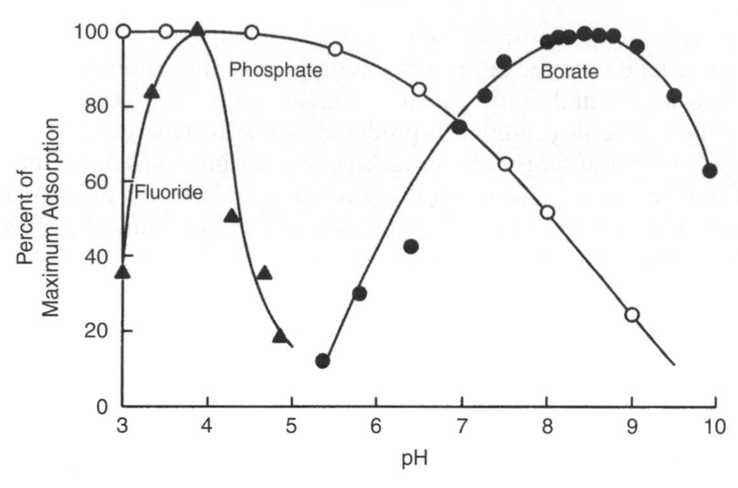

Figure 8.4. Sketches of adsorption envelopes for fluoride, phosphate, and borate anions. Note the distinct resonance feature in the envelopes for the monoprotic anions fluoride and borate.

a group that also includes arsenate, arsenite, carbonate, and molybdate. The monotonic graph of phosphate adsorption versus pH is termed an *adsorption envelope,* the inverse of the adsorption edge defined in Section 8.3. It can be described mathematically by eqs. 8.20 and 8.22 with $\beta_i < 0$ and with the absolute value of β_i then being used in Eq. 8.21. A monotonic decrease in the relative amount of anion adsorbed with increasing pH is observed for both strongly adsorbing anions and readily exchangeable anions, the difference between them appearing in the much higher pH_{50} value for anions that adsorb specifically. If an adsorptive anion does not protonate strongly (e.g., Cl^-, NO_3^-, SO_4^{2-}, and SeO_4^{2-}), the decrease in σ_H that always occurs with increasing pH produces a repulsion of the anion from soil particle surfaces that becomes dominant at pH > p.z.n.c. Therefore, a positive surface excess will decrease uniformly with pH, and pH_{50} will lie well below p.z.n.c. If an adsorptive anion of a polyprotic acid does protonate strongly (e.g., PO_4^{3-} and AsO_4^{3-}), it will adsorb according to the ligand exchange reactions in Eq. 8.25 and the decrease in σ_H will have less impact (mainly through reversing the reaction in Eq. 8.25a).

What about strongly adsorbing anions of monoprotic acids? Figure 8.5 shows the effect of pH on a calcareous Entisol reacting with borate in a NaCl background electrolyte solution. The relevant p.z.n.c. value is 9.5, and log K for $B(OH)_4^-$ protonation is 9.23. The corresponding Schindler diagram, shown in Figure 7.9, has a top band with a uniformly positive adsorbent surface charge indicated, a central band with a vertical line at pH 9.2, and a bottom band with a horizontal line extending over the very narrow range of pH between 9.2 and 9.5. Quite clearly, specific adsorption mechanisms are implicated in the reaction of borate with the soil. The relatively sharp peak in the adsorption

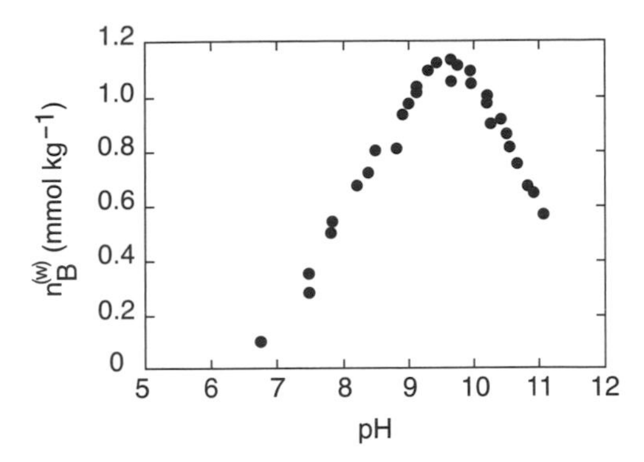

Figure 8.5. Adsorption envelope for borate reacting with an Entisol. Data from Goldberg, S., and R.A. Glaubig (1986) Boron adsorption on California soils. *Soil Sci. Soc. Amer. J.* **50**: 1173–1176.

envelope at a pH value approximately equal to log K for borate protonation, however, bears scrutiny. At pH values less than 9.2, borate anions do exist to some degree and are attracted to the positively charged soil adsorbent in increasing numbers as pH increases from 7 to 9. At pH values more than 9.2, the adsorptive is predominantly anionic, but now the adsorbent is also becoming increasingly negatively charged, leading to a sharp fall-off in the adsorption envelope at pH $>$ p.z.n.c. Thus, the *resonance feature* in Figure 8.5 can be interpreted as the net effect of an interplay between adsorptive charge and adsorbent charge. Because specific adsorption mechanisms play a major role in the reaction between the adsorptive and adsorbent, however, the resonance feature will be broadened relative to that resulting from charge relationships alone (cf. the resonance feature for fluoride in Fig. 8.3). Surveys of available data indicate that the strength of adsorption of an anion to an oxide mineral is indeed correlated positively with its log K value for protonation. From a local coordination perspective, *anions that protonate strongly will adsorb strongly*, with the complexation of a proton by the anion in aqueous solution being analogous to that between the anion and the positive surface site $\equiv S^+$ in Eq. 8.25b. In short, if an anion has a high affinity for the proton, it is expected to have a high affinity for a Lewis acid site.

8.5 Surface Redox Processes

Soil adsorption processes affect oxidation–reduction reactions in two important ways. One of them relates to surface-controlled dissolution reactions in which an adsorptive forms a surface complex with a metal cation exposed at the periphery of a mineral (Section 5.1), after which an electron transfer occurs between the metal and adsorbate prior to the detachment of the complex and its subsequent equilibration with the soil solution. Surface-controlled mineral dissolution promoted by an adsorptive ligand is illustrated in the lower half of Figure 5.1, but neither the metal in the adsorbent nor the adsorptive is redox active in this example (dissolution of gibbsite promoted by fluoride adsorption). The dissolution of redox-active Fe- and Mn-bearing minerals is described by several half-reactions listed in the middle of Table 6.1, but their coupling with the oxidation of a reductant species (e.g., Eq. 6.3 for the reductive dissolution of goethite) does not necessarily invoke adsorption as an intermediate step. (See also Problem 10 in Chapter 6.) However, *abiotic* reductive dissolution reactions usually involve the formation of surface complexes that serve as mediators of electron transfer.

Surface-controlled reductive dissolution reactions are distinguished by the formation of a surface complex between an adsorbent oxidant and an adsorptive reductant that facilitates a redox reaction, which then results in the dissolution of the adsorbent. Examples of these reactions include the reductive dissolution of Fe(III)- and Mn(IV)-bearing minerals by biomolecules (e.g., ascorbate or citrate) and the reductants in soil humus; by Fe^{2+} and Mn^{2+}; and by a variety of redox-active inorganic species, such as NH_4^+, $H_3AsO_3^0$, and $HSeO_3^-$. The potential for a surface-controlled reductive dissolution reaction can be examined by first evaluating whether adsorption of the reductant is favorable, using either a Schindler diagram for the adsorbent–adsorptive pair or detailed information about specific adsorption of the reductant, then evaluating whether a redox reaction between the pair has a favorable thermodynamic equilibrium constant. For example, similar to the bivalent metal cations with adsorption edges that appear in Figure 8.3, Mn^{2+} is adsorbed by goethite at alkaline pH ($pH_{50} \approx 8.7$, $-\log {}^*K = 10.6$). Whether the adsorbed Mn^{2+} can then reduce Fe(III) in the adsorbent can be evaluated by considering the reduction half-reactions (Table 6.1):

$$FeOOH(s) + 3H^+ + e^- = Fe^{2+} + 2H_2O(\ell) \qquad (8.26a)$$

$$MnOOH(s) + 3H^+ + e^- = Mn^{2+} + 2H_2O(\ell) \qquad (8.26b)$$

under the not unreasonable assumption that the oxidation of adsorbed Mn^{2+} results in rapid hydrolysis of the consequent adsorbed Mn^{3+} ($-\log {}^*K \approx -0.3$ for Mn^{3+}) to form a surface precipitate resembling the mineral manganite. The overall redox reaction, obtained by combining Eqs. 8.26a and 8.26b after reversing Eq. 8.26b,

$$FeOOH(s) + Mn^{2+} = MnOOH(s) + Fe^{2+} \qquad (8.26c)$$

has $\log K = -12$, according to the data given in Table 6.1. This value of log K predicts a highly *unfavorable* reaction (i.e., $[Fe^{2+}]/[Mn^{2+}] \approx 10^{-12}$ at equilibrium, implying a rather low yield of ferrous iron from reactant Mn^{2+}). The underlying reason for this result can be appreciated by constructing a redox ladder with "rungs" for the two redox couples: $FeOOH/Fe^{2+}$ and

$MnOOH/Mn^{2+}$ (Fig. 6.4). Because pE for the reductive dissolution of FeOOH lies well below that for MnOOH, electron transfer is favored from the former couple to the latter couple. Therefore, it is actually the reductive dissolution of manganite by Fe^{2+} that is the favorable reaction. Because p.z.n.c. $\approx$ 6.4 for manganite and –log *K = 9.4 for Fe^{2+}, a Schindler diagram predicts that adsorption of the reductant at alkaline pH should be facile, and manganite should be unstable in the presence of soluble ferrous iron.

Surface-controlled oxidative dissolution reactions are defined similarly to the reduction reactions, but with the direction of electron transfer reversed. Examples include the incongruent dissolution of Fe(s) (*zero-valent iron*, Section 6.4) by a broad variety of pollutant species (see Problems 8 and 9 in Chapter 6), of green rust (Section 2.4 and Problem 5 in Chapter 6) by a variety of oxyanions, and of Fe(II)-bearing primary and secondary minerals by a variety of pesticides and other pollutant compounds. For example, nitrate reduction (Section 6.2),

$$\frac{1}{8}NO_3^- + \frac{5}{4}H^+ + e^- = \frac{1}{8}NH_4^+ + \frac{3}{8}H_2O(\ell) \qquad (8.27a)$$

can be coupled with the oxidative dissolution of chloride-bearing green rust to form magnetite (Section 2.4),

$$\frac{3}{5}Fe_4(OH)_8Cl(s) = \frac{4}{5}Fe_3O_4(s) + e^- + \frac{8}{5}H^+ + \frac{8}{5}H_2O\,(\ell) + \frac{3}{5}Cl^- \qquad (8.27b)$$

The overall redox reaction,

$$\frac{3}{5}Fe_4(OH)_8Cl(s) + \frac{1}{8}NO_3^- = \frac{4}{5}Fe_3O_4(s) + \frac{1}{8}NH_4^+ + \frac{7}{20}H^+ + \frac{79}{40}H_2O\,(\ell) + \frac{3}{5}Cl^- \qquad (8.27c)$$

has log K = 14.90 + (3/5) [42.7 – (8/3) 18.16] = **11.46**, according to the data given in Table 6.1 and Problem 5 in Chapter 6, taking into account charge and mass balance as discussed in Special Topic 4 in Chapter 6. Thus, the oxidative dissolution of green rust by nitrate ions is highly favorable. The adsorption of nitrate by anion exchange with chloride is the likely first step of this process. Measurements of the yield of NH_4^+ and the consumption of Fe(II) for the reaction in Eq. 8.27c give a NH_4^+-to-Fe(II) molar ratio in agreement with the reaction stoichiometry, NH_4^+/Fe(II) = 1/8 ÷ 9/5 = 5/72, because 1 mol green rust contains 3 mol Fe(II).

Surface oxidation–reduction reactions are abiotic electron transfer processes in which the oxidant and reductant interact as adsorbate species (Fig. 8.6). In this case, *the adsorbent does not participate in the redox reaction.* Surface redox reactions are ubiquitous and important agents of transformation in soils and sediments. Their usual mechanism is a sequence initiated by inner-sphere surface complexation of either an oxidant or the reductant by an

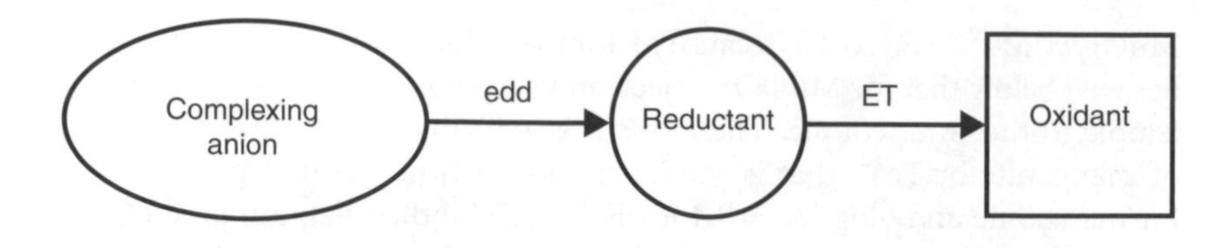

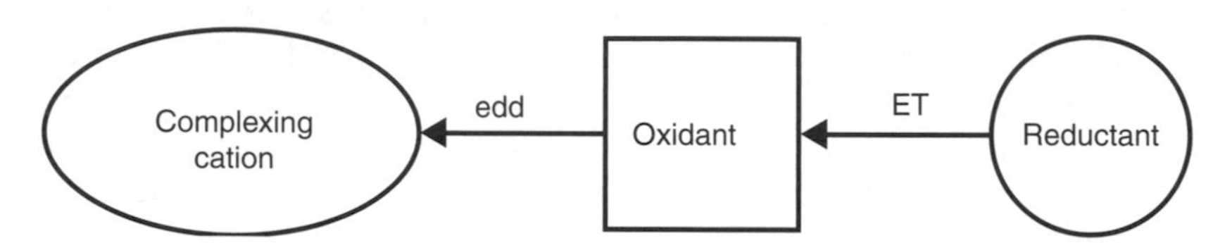

Figure 8.6. Conceptual scheme for surface oxidation–reduction reactions. The upper schematic illustrates electron density donation (edd) by a surface anionic site (complexing anion) to a cationic reductant adsorbed on the site. This donation of electron density enhances the ability of the reductant to transfer electrons (ET) to an oxidant that binds to the reductant to form a ternary complex on the surface. The lower schematic illustrates electron density donation by an adsorbed anionic oxidant to a cationic surface site (complexing cation), which then enhances the ability of the oxidant to accept electrons (ET) from a reductant that binds to it to form a ternary surface complex.

adsorbent. Then a complex forms between the adsorbed species and another reactant as a precursor to an electron-transfer step, after which this ternary surface complex (Section 7.1) becomes destabilized by the production of reduced and oxidized species. If the complex formed between oxidant and reductant is outer-sphere, the electron transfer step is termed a *Marcus process,* whereas if it is inner-sphere, the electron-transfer step is termed a *Taube process.* Electron transfer is likely to be the rate-limiting step in surface oxidation–reduction reactions governed by the Marcus process because of the intervening water molecules in an outer-sphere complex.

An example of a surface redox reaction can be developed by further consideration of Mn^{2+} adsorbed on goethite, because the adsorbent is stable against any reductive dissolution promoted by the adsorbate. A ternary surface complex involving a pair of goethite surface OH, Mn^{2+}, and O_2 will transform into oxidized Mn [e.g., Mn(III)] and reduced O_2 (to form H_2O) as products after electron transfer has occurred:

$$(\equiv FeO)Mn^0 + O_2 \rightarrow (\equiv FeO)_2Mn^0 \ldots O_2 \rightarrow \text{products} \qquad (8.28a)$$

where the dotted line represents O_2 bound to adsorbed Mn. This reaction is analogous to what predominates when Mn(II) is oxidized in aqueous solution:

$$Mn(OH)_2^0 + O_2 \rightarrow Mn(OH)_2^0 \quad \cdots \quad O_2 \rightarrow \text{products} \qquad (8.28b)$$

In both reactions, which are thought to involve Taube processes, the presence of two O-containing ligands in the initial complex with Mn^{2+} is required by the observed pH dependence of the overall redox reaction. *These ligands donate electron density to Mn^{2+} and thereby facilitate electron transfer to O_2.* This effect can be seen by comparing the second-order rate coefficients [*second* order because there are *two* reactants in Eq. 8.28 (Section 4.2)] for the oxidation of the three species Mn^{2+}, $Mn(OH)_2^0$, and $(\equiv FeO)Mn^0$. At 25 °C they are, respectively, $<10^{-10}$, 20.9, and 0.56 L mol^{-1} s^{-1}, which indicates clearly the great benefit of complexation. Note that this benefit is nearly two orders of magnitude larger for the soluble complex than for the corresponding surface complex. This kind of comparison, seen typically in surface redox reactions, evidently reflects the greater electron density donating power of soluble ligands.

In more quantitative terms, the effect of complexation can be expressed by the K value for the reduction half-reaction of the redox couple [e.g., $Mn(OH)_2^+$ /$Mn(OH)_2^0$], as discussed for complexes of iron in Section 6.4. Greater electron density donation would result in a smaller K value and, therefore, a greater stability of the *oxidant* in a redox couple, according to the reasoning given in Section 6.4. This relationship, in turn, implies that the second-order rate coefficient for oxidation of the reductant member of the couple will correlate *negatively* with K: the smaller K, the more stable the oxidant and the larger the rate of oxidation of the reductant. This kind of correlation is indeed found. For the example of Mn(II) oxidation, it implies that K for the soluble complex is smaller than K for the surface complex, because the rate coefficient for oxidation is larger for the former species. Smallest of all will be the rate coefficient for Mn^{2+} oxidations because water molecules in a solvation complex are rather weak electron density donors when compared with complexing ligands such as OH^- or $\equiv FeO^-$.

Of course, the adsorbed metal cation need not be Mn^{2+} (it could be Fe^{2+}, for example) and the adsorbed oxidant need not be O_2 [it could be Cr(VI) or U(VI), or even an organic compound]. The basic chemical concept is that a reductant species is adsorbed and forms a ternary surface complex with an oxidant species that then becomes unstable against a subsequent electron transfer. The role of the adsorbent in all this is *solely catalytic*: that of donating electron density to the reductant to facilitate its oxidation.

For that matter, the species bound directly to the adsorbent need not be a reductant. An oxidant could be adsorbed, then form a ternary surface complex with a reductant (Fig. 8.6). For example, $HCrO_4^-$ could be adsorbed and subsequently form a complex with an organic reductant (e.g., phenol) that then becomes unstable against electron transfer, with the result that Cr(III) is formed and the organic compound (e.g., a phenol or oxalate) is oxidized. In this case, a ligand exchange mechanism and formation of an inner-sphere surface complex by the oxidant (Eq. 8.25) will *withdraw electron density from it and thereby facilitate electron transfer to it from the reductant* it later complexes. This effect is analogous to that of protonation of an oxidant in aqueous solution, which is well-known to enhance electron transfer reactions (more so, typically, than does surface complexation, in consonance with the similar trend for adsorbed reductants). Once again, the adsorbent plays only a catalytic role: that of withdrawing electron density from the oxidant to facilitate its reduction.

For Further Reading

Bidoglio, G., and W. Stumm (eds.). (1994) *Chemistry of aquatic systems.* Kluwer Academic Publishers, Boston. Chapters by L. Charlet and by A. Stone, K. L. Godtfredsen, and B. Deng in this edited volume provide excellent overviews of adsorption phenomena and surface oxidation–reduction reactions respectively.

Essington, M. E. (2004) *Soil and water chemistry.* CRC Press, Boca Raton, FL. Chapter 7 in this comprehensive textbook gives a discussion of adsorption phenomena in soils quite parallel to but more detailed than that in the current chapter.

Huang, P. M., N. Senesi, and J. Buffle (eds.). *Structure and surface reactions of soil particles* John Wiley, Chichester, UK. The twelve chapters of this edited monograph provide an advanced survey of soil adsorption reactions, including spectroscopic methods and chemical modeling.

Sparks, D. L., and T. J. Grundl (eds.). *Mineral–water interfacial reactions.* American Chemical Society, Washington, DC. This symposium publication offers an eclectic, advanced discussion of specialized approaches to natural particle surface chemistry that extend the concepts discussed in the current chapter.

Sposito, G. (2004) *The surface chemistry of natural particles.* Oxford University Press, New York. Chapter 3 of this advanced textbook discusses the kinetics of specific adsorption, reductive dissolution, and surface oxidation–reduction reactions.

Problems

The more difficult problems are indicated by an asterisk.

1. Dry soil (350 mg) is mixed with 20 mL of a solution containing 4.00 mol m^{-3} KNO_3 at pH 4.2. After equilibration for 24 hours, a supernatant solution is collected and found to contain 3.96 mol m^{-3} KNO_3. Calculate

n_K and n_{NO_3} for the soil, in millimoles per kilogram. What is the p.z.n.c. of the soil?

2. The data in the table presented here refer to Cu(II) adsorption by an Aridisol. Plot an adsorption isotherm with the data and classify it according to the criteria discussed in Section 8.2.

n_{Cu} (mmol kg^{-1})	c_{Cu} (mmol m^{-3})	n_{Cu} (mmol kg^{-1})	c_{Cu} (mmol m^{-3})
6.87	1.87	25.39	28.11
10.64	3.06	34.14	77.68
18.05	9.19	37.34	155.1

3. Analyze the data in the following table [chlortetracycline (CT, an antibiotic) adsorption by an Alfisol] to classify the adsorption isotherm.

n_{CT} (μmol kg^{-1})	c_{CT} (μmol m^{-3})	n_{CT} (μmol kg^{-1})	c_{CT} (μmol m^{-3})
59	10	521	32
119	13	1025	76
231	18		

4. Calculate K_d as a function of n_{Cu} for the data in Problem 2, then select an isotherm equation to fit the data. Calculate the isotherm parameters and estimate the 95% confidence intervals for them.

5. Analyze the data in the table [Cd(II) adsorption by an Alfisol] to show that the van Bemmelen–Freundlich isotherm equation is appropriate to describe them. Calculate the parameters A and β and their 95% confidence intervals.

n_{Cd} (mmol kg^{-1})	c_{Cd} (mmol m^{-3})	n_{Cd} (mmol kg^{-1})	c_{Cd} (mmol m^{-3})
0.11	0.89	0.61	4.45
0.30	1.78	0.79	12.5
0.53	3.56	1.14	17.8

*6 a. Express K_d formally as a function of n to first order in a Taylor series, then derive Eqs. 8.14 and 8.15 using Eqs. 8.12 and 8.13.

b. Show that Eq. 8.16 leads to an infinite value of K_d as the surface excess tends to zero.

7. Plot an adsorption edge for Mg(II) on an Oxisol based on the data in the table presented here. Calculate pH_{50} given a maximum adsorption of 8 mmol kg^{-1} at pH 6. Evaluate the applicability of Eq. 8.20.

n_{Mg} (mmol kg^{-1})	pH	n_{Mg}(mmol kg^{-1})	pH
0.72	2.48	2.45	3.36
1.08	2.73	3.64	3.80
1.80	3.05	4.21	4.10
2.14	3.20	6.35	5.00

8. Estimate pH_{50} values for the adsorption edges in Figure 8.3 and perform a linear regression analysis of the relationship between pH_{50} and –log *K for the three metal cations. What value of pH_{50}, including an error of your estimate, is predicted for Mn^{2+}?

*9. Analyze the data in the table presented here (negative adsorption of Cl^- by a temperate-zone soil) to determine a power–law relationship between V_{ex} and the concentration of chloride in the supernatant solution. Mathematical modeling of negative adsorption in the diffuse-ion swarm based on Gouy–Chapman theory and Eq. 8.24 leads to the equation

$$V_{ex} = 2a_s/(\beta c)^{1/2}$$

where $\beta = 1.084 \times 10^{16}$ m mol^{-1} (at 25 °C) is a constant model parameter and a_s is specific surface area. Does the exponent in the power–law relationship you found agree with the model equation? If it does, apply the equation to estimate the specific surface area of the soil.

$V_{ex}(10^{-3}\ m^3\ kg^{-1})$	$c_{Cl}(mol\ m^{-3})$	$V_{ex}(10^{-3}\ m^3\ kg^{-1})$	$c_{Cl}(mol\ m^{-3})$
1.06	0.79	0.50	6.2
1.00	1.1	0.49	6.8
0.70	2.0	0.38	7.7
0.68	3.1	0.29	9.9
0.55	4.0	0.24	20.5

10. Plot an adsorption envelope for nitrate on an Oxisol using the data in the table presented here. Extrapolate the data to estimate n_{MNO_3}, then test the applicability of Eq. 8.20 (with $\beta_{NO_3} < 0$).

n_{NO_3} (mmol kg^{-1})	pH	n_{NO_3} (mmol kg^{-1})	pH
6.6 ± 0.3	2.5	3.5 ± 0.3	3.9
5.1 ± 1.0	2.8	3.0 ± 0.3	4.1
4.7 ± 0.7	3.2	1.5 ± 0.2	4.6
4.3 ± 0.8	3.6	1.2 ± 0.6	4.9
4.0 ± 0.1	3.7	0.6 ± 0.2	5.3

*11. Simon [Simon, N. S. (2005). Loosely-bound oxytetracycline in riverine sediments from two tributaries of the Chesapeake Bay. *Environ. Sci. Technol.* **39**:3480.] has extracted sorbed oxytetracycline, an antibiotic used in agriculture, from contaminated bay sediments using 1 mol dm^{-3} $MgCl_2$ solution adjusted to pH 8. The molecular structure of oxytetracycline is shown in Simon's Figure 2. Over what range of pH should the antibiotic be a cation? A neutral species? An anion? How would the mechanism of extraction using $MgCl_2$ likely differ in each range of pH? Why does Simon refer to the extracted antibiotic as "easily desorbed oxytetracycline?"

*12. It is a common observation that the adsorption edge for a bivalent metal cation (e.g., Zn^{2+}) on a soil mineral is shifted upward at low pH and downward at high pH after the mineral becomes coated by humus. Sketch a typical adsorption edge for a bivalent metal cation and a typical adsorption envelope for humus on a soil oxyhydroxide mineral, then develop a mechanistic explanation for this observation.

*13. When the antibiotic ciprofloxacin (Problem 12 in Chapter 3, Problem 13 in Chapter 7) is in the presence of MnO_2, the antibiotic disappears from solution and Mn^{2+} begins to appear in solution. The rate of antibiotic loss decreases as pH increases, but increases with the initial concentration of both the antibiotic and the Mn oxide. Discuss the hypothesis that the loss of the antibiotic results from a surface-controlled reductive dissolution reaction. What additional experiments would be useful for testing the hypothesis? (*Hint:* Begin by preparing a Schindler diagram for the antibiotic reacting with the Mn oxide.)

*14. Can the phenol hydroquinone (1,4-benzenediol; Eq. 6.15) be expected to provoke the surface-controlled reductive dissolution of birnessite? (*Hint:* Follow the approach outlined for manganite and Fe^{2+} in Section 8.5.)

*15. The second-order rate coefficient (k_L, in liters per mole per second) for Cr(VI) reduction by Fe(II) is found to be related to the pE_L value for a Fe(III)L/Fe(II)L couple (Eq. 6.20) by the equation

$$\log k_L = 8.13 - 0.60pE_L$$

a. Give a mechanistic interpretation of the decrease of log k_L with increasing pE_L.
b. The value of k_L for Cr(VI) reduction by adsorbed Fe(II) is about 8×10^3 L mol^{-1} s^{-1}. What is the pE value for the ≡Fe(III)/≡Fe(II) couple? How does it compare with that for $FeOH^{2+}/FeOH^{+}$? Give a mechanistic interpretation as part of your comparison.

9

Exchangeable Ions

9.1 Soil Exchange Capacities

The *ion exchange capacity* of a soil is the maximum number of moles of adsorbed ion charge that can be desorbed from unit mass of soil under given conditions of temperature, pressure, soil solution composition, and soil-solution mass ratio. In Section 3.3, a similar definition of the CEC of soil humus is stated and, in Chapter 8, the surface excess of an ion is related to the soil chemical factors that affect ion exchange capacities. In many applications, ion exchange capacity refers to the maximum positive surface excess of *readily exchangeable ions,* as defined in Section 7.2. These ions adsorb on soil particle surfaces solely via outer-sphere complexation and diffuse-ion swarm mechanisms (see Fig. 7.4).

Measurement of an ion exchange capacity typically involves replacement of the native population of readily exchangeable ions by an *index cation* or anion, then determination of its surface excess following the methodology discussed in Section 8.1. Detailed laboratory procedures for this measurement are described in *Methods of Soil Analysis* (see "For Further Reading" at the end of this chapter). For soils in which the readily exchangeable cations are monovalent or bivalent (e.g., Aridisols), the index cation can be Na^+ or Mg^{2+}, whereas for soils also bearing trivalent readily exchangeable cations (e.g., Spodosols), K^+ or Ba^{2+} is an index cation of choice (see also Section 3.3). Often NH_4^+ has been used as an index cation. Because this cation forms inner-sphere surface complexes with 2:1 layer-type clay minerals, like that shown for K^+ in Figure 7.4, and because it can even dissolve cations from primary soil minerals, the use of NH_4^+ to measure the soil CEC has potential for inaccuracy. The index anion of choice is typically ClO_4^-, Cl^-, or NO_3^-. Thus, for example, $MgCl_2$ could be selected as an index electrolyte for displacing readily exchangeable ions (see also Problem 11 in Chapter 8). A common modification of direct displacement of the native population of readily exchangeable ions is displacement after prior saturation of a soil adsorbent with an *indifferent electrolyte* (Section 7.4), such as $NaClO_4$ or LiCl.

A quantitative definition of ion exchange capacity can be developed in terms of the surface excess and charge balance concepts. Consider first a soil in which a net positive surface excess of anions is highly unlikely (e.g., the montmorillonitic Entisol discussed in Section 8.1). Suppose that the only adsorbed ions in this soil are Na^+, Ca^{2+}, and Cl^-. Then the CEC of the soil is defined by the charge balance condition

$$n_{NaT} + 2n_{CaT} - n_{ClT} - \text{CEC} \equiv 0 \qquad (9.1)$$

where n_{iT} (i = Na, Ca, or Cl) is the total moles of ion i per kilogram dry soil in a wet soil, as in Eq. 8.1. Equation 9.1 quantifies the role of soil particles bearing adsorbed cations as being on the same chemical footing as anions in the soil. The operational meaning of Eq. 9.1 is apparent, given a methodology for extracting the adsorbed ions, but its quantitative relation to the surface excess requires substitution of Eq. 8.1 for each participating ion:

$$\begin{aligned} \text{CEC} &= (n_{Na} + M_W m_{Na}) + 2(n_{Ca} + M_W m_{Ca}) - (n_{Cl} + M_W m_{Cl}) \\ &= n_{Na} + 2n_{Ca} - n_{Cl} + M_W(m_{Na} + 2m_{Ca} - m_{Cl}) \\ &= n_{Na} + 2n_{Ca} - n_{Cl} \equiv q_{Na} + q_{Ca} - q_{Cl} \qquad (9.2) \end{aligned}$$

where electroneutrality of the soil solution is invoked to eliminate the molalities and

$$q_i \equiv |Z_i| n_i \qquad (9.3)$$

is the *adsorbed ion charge* of species i. Now, the right side of Eq. 9.2 is equal to Δq in Eq. 7.2. Thus, *CEC is the net adsorbed ion charge evaluated under the condition that the net adsorbed anion charge is not a positive quantity.* Returning to the example of the Entisol in Section 8.1, we can calculate its CEC as

$$\text{CEC} = 6.07 + 2(58.23) - (-0.17) = \mathbf{122.7\ mmol_c\ kg^{-1}}$$

Note that the *negative* surface excess of Cl^- still contributes to the CEC. Formally, this is required by the condition of charge balance in Eq. 9.1, given the definition of the surface excess, but mechanistically it is a reflection of the fact that anion repulsion by a negatively charged particle surface is equivalent to cation attraction by the surface for species adsorbed in the diffuse-ion swarm.

The operational nature of CEC should not be forgotten. If a large concentration of the index cation is used in a solution at high pH (e.g., ≥ 8.2),

Table 9.1
Representative cation exchange capacities (in moles of charge per kilogram) of surface soils.[a]

Soil order	CEC	Soil order	CEC
Alfisols	0.15 ± 0.11	Mollisols	0.24 ± 0.12
Andisols	0.31 ± 0.18	Oxisols	0.08 ± 0.06
Aridisols	0.18 ± 0.11	Spodosols	0.27 ± 0.30
Entisols	0.20 ± 0.14	Ultisols	0.09 ± 0.06
Histosols	1.4 ± 0.3	Vertisols	0.50 ± 0.17
Inceptisols	0.21 ± 0.16		

[a]Based primarily on data compiled in Table 8.2 of Essington, M. E. (2004). *Soil and water chemistry.* CRC Press, Boca Raton, FL. CEC = cation exchange capacity.

the measured surface excess of the index cation should approximate closely the absolute value of the maximum negative intrinsic surface charge of a soil (Section 7.3). On the other hand, if the pH value or some other chemical property of the solution containing the index cation is arranged such that the maximum negative intrinsic surface charge is not neutralized by the adsorption of the index cation, then the measured surface excess of the latter will simply reflect the chemical conditions chosen. An example of this latter situation appears in Problem 1 of Chapter 7 for Na^+ adsorption by an Oxisol under varying pH and ionic strength. Both the maximum and the less than maximum CEC are useful in soil chemistry. The maximum negative intrinsic surface charge indicates the *potential* capacity of a soil for adsorbing cations, whereas a less than maximum negative intrinsic surface charge indicates the *actual* capacity of a soil for adsorbing cations under given conditions.

Table 9.1 lists representative CEC values for 11 soil orders, based primarily on measurements made using NH_4^+ as the index cation in a solution at pH 7. High variability of the CEC within each soil order is evident, but the very low values for Ultisols and Oxisols and the high values for Histosols and Vertisols are significant trends. Detailed studies of the CEC show that it is correlated positively with the content of humus, clay content, and soil pH, if an unbuffered solution containing the index cation is used in the measurement. The basis for the correlation with humus content—reflected dramatically in Table 9.1 by the CEC reported for Histosols—can be understood at once after comparison of the CEC values of humic substances (5–9 $mol_c\ kg^{-1}$, Section 3.3) with those for clay minerals like smectite and vermiculite (0.7–2.5 $mol_c\ kg^{-1}$, Section 2.3). The correlation with pH is understandable after reviewing the pH dependence of the net proton charge in Figure 3.3. Indeed, the pH dependence of a measured less than maximum intrinsic surface charge should mirror that of the adsorbed index ion charge (see Problems 1 and 3 in Chapter 7).

The composition of readily exchangeable ions in a soil can be determined by chemical analysis of the soil solution after reaction of the soil with index ions such as Li^+ and ClO_4^-. In alkaline soils, the readily exchangeable cations are Ca^{2+}, Mg^{2+}, Na^+, and K^+, decreasing in their contribution in the order shown. In acidic soils, the most important readily exchangeable metal cation is Al^{3+}, followed by Ca^{2+} and Mg^{2+}. Readily exchangeable Al(III), which likely includes Al^{3+}, $AlOH^+$, $Al(OH)_2^+$, and $AlSO_4^+$, can be measured by using K^+ as an index cation in an unbuffered KCl solution. The remaining exchangeable metal cations can then be determined by replacement with Ba^{2+}.

Comprehensive data compilations like those in Table 9.1 are not well established for the *anion exchange capacity* (AEC) of soils. The AEC tends to be important mainly for Spodosols, Ultisols, and Oxisols. Among these soil orders, AEC values in the range 1 to 50 $mmol_c\ kg^{-1}$ are representative. A quantitative definition of AEC is developed by generalizing Eq. 9.1. Consider the Oxisol discussed in Problem 1 of Chapter 7. Because both index ions have positive surface excess in this soil, the charge balance condition must be expressed in the form

$$n_{NaT} - n_{ClT} + AEC - CEC = 0 \tag{9.4}$$

Invoking the definition of the surface excess in Eq. 8.1, we can then derive the equation

$$q_{Na} - q_{Cl} = CEC - AEC \tag{9.5}$$

as a generalization of Eq. 9.2. Evidently, AEC in the soil simply equals q_{Cl} under given conditions of pH and ionic strength, with a maximal value near 10 $mmol_c\ kg^{-1}$ at pH 2.6 and $I = 30\ mol\ m^{-3}$. Similarly, CEC is the same as q_{Na}, reaching about 9 $mmol_c\ kg^{-1}$ at pH 5 and $I = 30\ mol\ m^{-3}$. More generally, *in the absence of negative adsorption,*

$$\Delta q_{ex} = CEC - AEC \tag{9.6}$$

where

$$\Delta q_{ex} \equiv \sigma_{OS} + \sigma_d \tag{9.7}$$

defines a special case of Eq. 7.2 appropriate to readily exchangeable ions. Equation 9.6 shows that *the net adsorbed ion charge of readily exchangeable ions equals the difference between CEC and AEC.* If it is known that anions are actually negatively adsorbed by a soil, then AEC is dropped from Eq. 9.6 and it reduces to Eq. 9.2 (under the conditions given for the example). If it is known that cations are negatively adsorbed by a soil, then CEC is dropped from Eq. 9.6. In either of these special cases, negative surface excess still contributes to the ion exchange capacity, as in the Entisol example.

9.2 Exchange Isotherms

An *exchange isotherm* is analogous to an adsorption isotherm (Section 8.2), except that the variables plotted are *charge fractions* instead of surface excesses and soil solution concentrations. The charge fraction of an adsorbed ion is defined by

$$E_i \equiv q_i/Q \tag{9.8}$$

where Q is the sum of adsorbed ion charges for each ion that undergoes exchange with ion i:

$$Q = \sum_k q_k \tag{9.9}$$

The charge fraction of an ion in aqueous solution is defined similarly as

$$\tilde{E}_i \equiv |Z_i| m_i \tilde{Q} \tag{9.10}$$

where m_i is the molality (or other concentration variable) of ion i and

$$\tilde{Q} = \sum_k |Z_k| m_k \tag{9.11}$$

An exchange isotherm, then, is a graph of E_i against $\tilde{E}_i$ under the same fixed conditions that apply to an adsorption isotherm. Evidently, the maximum range of a charge fraction is from zero to one.

Exchange isotherms for Ca → Mg exchange at pH 7 are illustrated in Figure 9.1 for two 2:1 clay minerals and two soils—a Vertisol and an Aridisol—with clay fractions that are dominated by the minerals indicated. One of the variables kept constant during the bivalent cation exchange reactions described by the isotherms was the charge fraction of adsorbed Na^+ (E_{Na}). Thus, the isotherms refer to both *binary* and *ternary* exchange systems. In natural soils, of course, ternary, quaternary, or even higher order exchange systems are the norm. A binary exchange reaction such as Ca → Mg is still useful for detailed laboratory study, however, but only under the critical assumption that naturally occurring, higher order *n*-ary exchange systems can be understood in terms of component binary exchange reactions. That this assumption may be true is indicated by the closeness of the exchange isotherms in Figure 9.1, which suggests that Ca → Mg exchange on the clay minerals and soils is largely independent of the presence of adsorbed Na^+ in the E_{Na} range investigated. Note that both Q and $\tilde{Q}$ are limited to contributions from Mg^{2+} and Ca^{2+} to allow direct comparison with binary exchange data ($E_{Na} = 0$).

The solid lines in Figure 9.1 are *thermodynamic nonpreference exchange isotherms*. For bivalent–bivalent exchange, and for any other exchange reaction involving ions having the same valence, the thermodynamic nonpreference isotherm is represented mathematically by the simple equation

$$E_i = \tilde{E}_i \tag{9.12}$$

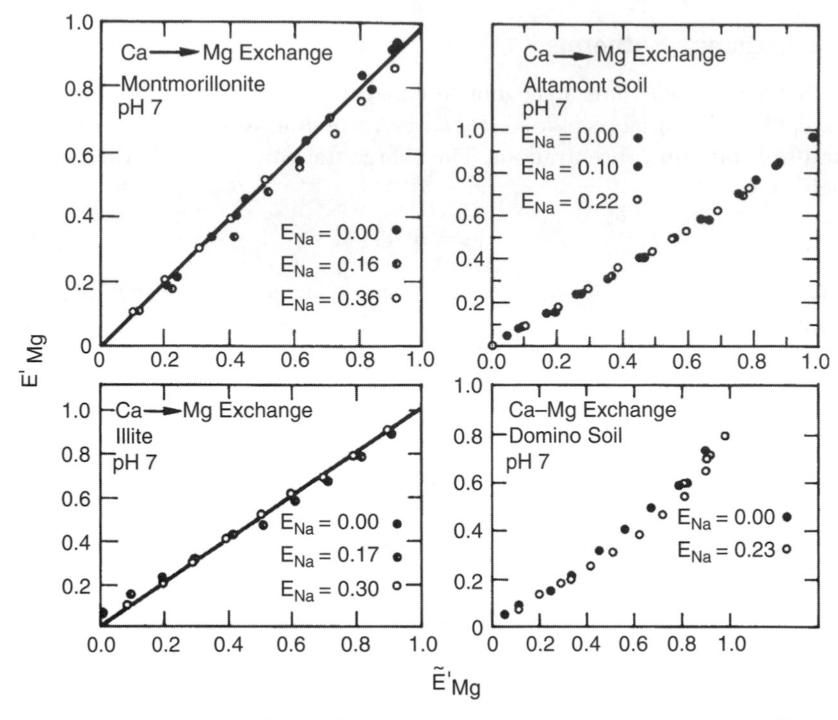

Figure 9.1. Exchange isotherms for Ca → Mg exchange at 25 °C on montmorillonite and a Vertisol (Altamont series), and on illite and an Aridisol (Domino series). Charge fractions were calculated considering only the surface excess and aqueous solution concentration of Ca and Mg, irrespective of whether Na was present. The thermodynamic nonpreference isotherm (Eq. 9.12) is indicated by a solid line. Data from Sposito, G., C. Jouany, K.M. Holtzclaw, and C.S. LeVesque (1983) Calcium-magnesium exchange on Wyoming bentonite in the presence of adsorbed sodium. *Soil Sci. Soc. Amer. J.* **47**:1081–1085; Fletcher P, Holtzclaw KM, Jouany C, Sposito G, and LeVesque CS (1984) Sodium-calcium-magnesium exchange reactions on a montmorillonitic soil: II. Ternary exchange reactions. *Soil Sci. Soc. Amer. J.* **48**:1022–1025; Sposito, G., C.S. LeVesque, and D. Hesterberg (1986) Calcium-magnesium exchange on illite in the presence of adsorbed sodium. *Soil Sci. Soc. Amer. J.* **50**:905–909.

which plots as a straight line that makes a 45° angle with both the x- and y-axes. On both montmorillonite and illite, the Ca → Mg exchange appears to show no preference for either cation at pH 7 and over the range of E_{Na} investigated. On the Vertisol (Atamont soil) this also may be true, although the data do appear to lie very slightly *below* a nonpreference line, indicating that Ca^{2+} may be slightly favored over Mg^{2+} in the exchange reaction. The situation for the Aridisol (Domino soil) is more clear: The data lie well below a nonpreference line, more so perhaps in the ternary system (E_{Na} = 0.23). These differences between the soils and the corresponding clay minerals may reflect the presence of humus, where carboxyl groups tend to have a greater affinity for Ca^{2+} than for Mg^{2+} because of the larger radius of the former cation (see Section 8.3).

For monovalent–bivalent exchange, the nonpreference isotherm is described by a more complicated equation:

$$E_{biv} = 1 - \left[\frac{A(1 - \tilde{E}_{biv})^2}{\tilde{E}_{biv} + A(1 - \tilde{E}_{biv})^2} \right]^{\frac{1}{2}} \quad (9.13)$$

where E_{biv} is the adsorbed charge fraction of the bivalent cation and $A \equiv \tilde{Q}\gamma^2_{mono}/2\gamma_{biv}$, with γ being a single-ion activity coefficient (Eq. 4.24). The curve resulting from Eq. 9.13 is illustrated in Figure 9.2 for $Na \rightarrow M^{2+}$ exchange (M = Ca or Mg) under the conditions $\tilde{Q} = 0.05\,mol_c\,L^{-1}$ and $\gamma^2_{Na}/\gamma_{biv} = 1.5$, so that $A = 0.0375$. If $\tilde{E}_{biv} = 0.2$, then, by Eq. 9.13, $E_{biv} = 1 - (0.024/0.224)^{\frac{1}{2}} = 0.67$, which agrees numerically with the solid curve in the figure. The derivation of Eq. 9.13 is outlined in Section 9.3. Suffice it to say here that the derivation is based on two assumptions: (1) the thermodynamic equilibrium constant for the exchange reaction has unit value and (2) the adsorbed cations behave as *an ideal solid solution* in the soil (Section 5.3).

According to Figure 9.2, which shows two more exchange isotherms for the Domino soil, the Ca and Mg isotherms lie *above* the nonpreference isotherm,

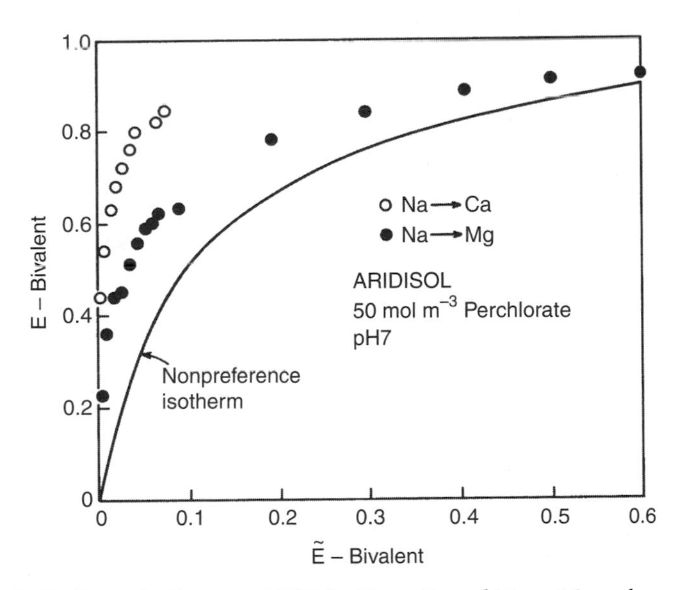

Figure 9.2. Exchange isotherms at 25 °C for Na → Ca and Na → Mg exchange on the Aridisol featured in Figure 9.1. The thermodynamic nonpreference isotherm is also indicated (Eq. 9.13).

indicating selectivity of the soil for bivalent cations relative to Na^+, which has the lesser valence. Evidently there is more of a selectivity difference between Ca^{2+} and Na^+ than Mg^{2+} and Na^+. This latter difference agrees with the Ca → Mg isotherm in the lower right quadrant of Figure 9.1. Note, however, that these conclusions depend on the chemical conditions under which the exchange reactions occur. The parameter A in Eq. 9.13 is related directly to the electrolyte concentration, such that E_{biv} increases as $\tilde{Q}$ decreases for a fixed $\tilde{E}_{biv}$. Therefore, a comparison like that in Figure 9.2 for $\tilde{Q} = 0.05\ mol_c\ L^{-1}$ may change completely as $\tilde{Q}$ changes to higher or lower values. As a general rule, when $\tilde{Q}$ increases, the nonpreference isotherm for monovalent–bivalent ion exchange becomes more like a straight line (i.e., more like the nonpreference isotherm for homovalent ion exchange). Thus, decreasing $\tilde{Q}$ or, more concretely, dilution of the aqueous solution containing the adsorptive ions will inevitably lead to a greater charge fraction of the adsorbed bivalent ion at a given value of $\tilde{E}_{biv}$, even though the adsorbent actually exhibits *no* preference for the bivalent cation over the monovalent cation. This trend is known as the *Scofield dilution rule.*

9.3 Ion Exchange Reactions

As stated in Section 9.1, the usual meaning of *ion exchange reaction* in soil chemistry is the replacement of one *readily exchangeable* ion by another. On the molecular level, this means that ion exchange is a surface phenomenon involving charged species in outer-sphere complexes or in the diffuse-ion swarm. In practice, this conceptualization is adhered to only approximately. Cation exchange reactions on soil humus, for example, include protons (Section 3.3) that may be adsorbed in inner-sphere surface complexes. Common extracting solutions for CEC measurements (e.g., $NH_4C_2H_3O_2$ or $BaCl_2$) may displace metal cations from inner-sphere surface complexes as well as readily exchangeable metal cations. Development of experimental methods that will quantitate only readily exchangeable ions or will partition adsorbed ions accurately into readily exchangeable and specifically adsorbed species remains the objective of ongoing research.

Ion exchange reactions on whole soils or soil separates (e.g., the clay fraction) cannot be expressed as chemical equations that show the detailed composition of the adsorbent. Soil adsorbents are very heterogeneous and, therefore, an approach similar to that used to describe cation exchange on soil humus (Section 3.3) must be adopted. The symbol X will denote the soil adsorbent in the same way that ≡SO was used to denote the humus adsorbent. This representation is meant to depict the ion exchange characteristics of soil only in some *average* sense, with chemical equations for ion exchange written by analogy with expressions like Eqs. 3.2 through 3.4. For example, the $Na^+ \rightarrow Ca^{2+}$ exchange reaction underlying some of the data in Figure 9.2 can be expressed as

$$Na_2X(s) + Ca^{2+} = CaX(s) + 2\,Na^+ \quad (9.14a)$$

$$2\,NaX(s) + Ca^{2+} = CaX_2(s) + 2\,Na^+ \quad (9.14b)$$

$$2\,NaX(s) + Ca^{2+} = 2\,Ca_{\frac{1}{2}}X(s) + 2\,Na^+ \quad (9.14c)$$

Equation 9.14 illustrates three alternative ways to represent the same cation exchange reaction. In Eq. 9.14a, X^{2-} denotes an amount of soil bearing 2 mol intrinsic negative surface charge (cf. Eq. 3.2), whereas in Eqs. 9.14b and 9.14c, X^- denotes an amount of soil bearing 1 mol intrinsic negative surface charge (cf. Eq. 3.3). As long as the amount of intrinsic surface charge is made clear, X can be used in either case as the symbol for the soil adsorbent, and in neither does the "valence" –1 or –2 have any molecular significance. Equation 9.14c differs from Eq. 9.14b by emphasizing that it is 1 mol Ca *charge* that reacts with 1 mol soil adsorbent charge. Thus Eq. 9.14b is expressed in terms of the moles of Ca and Na that react with 1 mol X, whereas Eq. 9.14c is expressed in terms of the moles of Ca and Na *charge* that react. The choice of which equation to use is a matter of personal preference, because both equations satisfy basic requirements of mass and charge balance (see Special Topic 1 in Chapter 1).

The kinetics of ion exchange can be described quantitatively in terms of the concepts developed in Special Topic 3 (Chapter 3) for adsorption–desorption reactions of cations. Essential to this approach is the decision regarding whether the rate-controlling step is diffusive transport or surface reaction. A variety of data suggests that *rates of ion exchange processes often are transport controlled, not reaction controlled.* This is patently true if the ions involved merely replace one another in the diffuse-ion swarm (Section 7.2). Thus, readily exchangeable ions most probably engage in reactions with rates that are transport controlled, whereas specifically adsorbed ions participate in reactions that are surface controlled (Table 9.2). Adsorption reactions that involve both solvated and unsolvated adsorbate species (e.g., the exchange of Na^+ in the diffuse-ion swarm for K^+ that forms an inner-sphere surface complex on a 2:1 clay mineral) could exhibit rates that are influenced by both diffusion and the kinetics of surface complexation reactions.

Even with diffusion taken as the rate-limiting process for ion exchange reactions, there remains a need to distinguish between film diffusion and intraparticle diffusion (the latter being ion diffusion into the pore space of an aggregate, with subsequent adsorption onto pore walls). This can be done experimentally by an *interruption test.* When an ion exchange reaction has been initiated, the adsorbent particles are separated physically from the aqueous solution phase, then are reimmersed in it after a short time interval. If film diffusion is the rate-limiting step, no significant effect of this interruption on the kinetics should be observed. If intraparticle diffusion is the rate-limiting step, the concentration gradient driving the ion exchange process should drop to zero during the interruption, and the rate of ion exchange should increase

Table 9.2
Comparing ion exchange with specific adsorption.

Property	Ion exchange	Specific adsorption
Surface species	Outer-sphere surface complexes and diffuse-ion swarm	Inner-sphere surface complexes
Adsorptive charge versus surface charge	Always opposite	Either opposite or the same
Kinetics	Transport control	Surface control
Cation affinity	Increases with Z and R[a]	Increases with log *K for hydrolysis[b]
Anion affinity	Increases with |Z| and R	Increases with log K for protonation[b]

[a]Abbreviations: R, ionic radius; Z, ionic valence.
[b]Equation 7.14.

after the exchanger particles are reimmersed in the aqueous solution phase, causing a gradient to be reestablished. In general, soil particles with large specific surface areas should favor film diffusion, whereas those with significant microporosity should favor intraparticle diffusion.

Equation S.3.8 in Special Topic 3 is a model rate law for transport-controlled adsorption of a cation i by a univalent charged site, $\equiv SO^-$. Adopting the convention in Eq. 9.14b for cation exchange (i.e., $\equiv SO^- \Leftrightarrow X^-$) and considering the example of $Li^+ \rightarrow Na^+$ exchange on a soil adsorbent, we can apply Eq. S.3.8 to both cations under the condition of equilibrium with respect to their adsorption reactions (rate = 0) to derive the ratio

$$\frac{[Na^+]}{[Li^+]} = \frac{k_{des}^{Na}/k_{ads}^{Na}[NaX]}{k_{des}^{Li}/k_{ads}^{Li}[LiX]} \tag{9.15a}$$

where the subscript *bulk* has been dropped to simplify notation. Equation 9.15a can be rearranged to form the expression

$$\frac{[NaX][Li^+]}{[LiX][Na^+]} = \frac{k_{ads}^{Na}/k_{des}^{Na}}{k_{ads}^{Li}/k_{des}^{Li}} \equiv K_V^{Li/Na} \tag{9.15b}$$

The parameter defined by ratios of rate coefficients is a conditional equilibrium constant (Section 4.2) for the cation exchange reaction:

$$LiX(s) + Na^+ = NaX(s) + Li^+ \tag{9.16}$$

Slightly generalized to replace concentrations with the *activities* of Li^+ and Na^+ in aqueous solution (Eq. 4.20), it is known as the *Vanselow selectivity coefficient*:

$$K_V^{Li/Na} = \frac{[NaX](Li^+)}{[LiX](Na^+)} \tag{9.15c}$$

(Note that $\gamma_{Na}^+ = \gamma_{Li}^+$ if Eq. 4.24 is used to calculate activity coefficients.)

Equation 9.15c is a model expression for an equilibrium constant to describe the reaction in Eq. 9.15b. In general, the ratio on the right side of Eq. 9.15c will not remain constant as the relative amounts of NaX(s) and LiX(s) change under varying aqueous solution composition; hence, it defines a *conditional* equilibrium constant, as discussed in Section 4.5. When the ratio on the right side of Eq. 9.15c does remain constant, the adsorbate is said to be an *ideal solid solution* comprising NaX(s) and LiX(s). This designation is consonant with the definition of an ideal solid solution given in Section 5.3, because [NaX]/[LiX] is equal to the stoichiometric ratio of NaX to LiX in the adsorbate and would be replaced by the activity ratio (NaX) to (LiX) if $K_V^{Li/Na}$ were a true thermodynamic equilibrium constant. Thus, *the Vanselow selectivity coefficient is an ion exchange equilibrium constant derived according to the ideal solid solution model of a soil adsorbate.*

Equation 9.15b also can be written in the form

$$K_V^{Li/Na} = \frac{E_{Na}\tilde{E}_{Li}}{E_{Li}\tilde{E}_{Na}} \tag{9.15d}$$

where the charge fractions E_i and $\tilde{E}_i$ (i = Na or Li) are defined in Eqs. 9.8 and 9.10 respectively. A rearrangement of Eq. 9.15d to solve it for E_{Na} yields

$$E_{Na} = \frac{K_V^{Li/Na}\,\tilde{E}_{Na}}{1 + (K_V^{Li/Na} - 1)\tilde{E}_{Na}} \tag{9.17}$$

after noting that $E_{Li} = 1 - E_{Na}$ and $\tilde{E}_{Li} = 1 - \tilde{E}_{Na}$. The Li → Na exchange reaction is said to be *selective* for Na^+ if $K_V^{Li/Na} > 1$ and selective for Li^+ if $K_V^{Li/Na} < 1$. The corresponding graphs of E_{Na} versus $\tilde{E}_{Na}$—exchange isotherms—will be convex toward the y-axis and convex toward the x-axis respectively, in keeping with these definitions of selectivity. If $K_V^{Li/Na} = 1$, there is said to be *no preference* for either cation, and the resulting exchange isotherm, equivalent to Eq. 9.12 with i = Na, is the *thermodynamic nonpreference exchange isotherm*. It plots as a straight line.

A Vanselow selectivity coefficient also can be developed for the Na → Ca exchange reaction in Eq. 9.14b:

$$K_V^{Na/Ca} = \frac{x_{Ca}(Na^+)^2}{x_{Na}^2(Ca^{2+})} \tag{9.18}$$

where $x_{Ca} \equiv [CaX_2]/([CaX_2] + [NaX])$ and $x_{Na} \equiv [NaX]/([CaX_2] + [NaX]) = 1 - x_{Ca}$ are termed the *mole fractions* of CaX_2 and NaX_2 in the adsorbate. (Note that the mole fraction ratio x_{Na} to x_{Li} appears in Eq. 9.15c.) Like the selectivity coefficient for Li → Na exchange, the value of $K_V^{Na/Ca}$ usually is not constant as the composition of the adsorbate changes, but Eq. 9.18 still can be used as a guide to selectivity. Upon introducing the definitions of charge fractions,

$$E_{Ca} = q_{Ca}/Q = 2[CaX_2]/(2[CaX_2] + [NaX]) = 1 - E_{Na}$$
$$\tilde{E}_{Ca} = 2[Ca^{2+}]/\tilde{Q} = 2[Ca^{2+}]/(2[Ca^{2+}] + [Na^+]) = 1 - \tilde{E}_{Na} \qquad (9.19)$$

and noting Eq. 4.20, one generalizes Eq. 9.13:

$$E_{Ca} = 1 - \left[\frac{(A/K_V{}^{Na/Ca})\,(1 - \tilde{E}_{Ca})^2}{\tilde{E}_{Ca} + (A/K_V^{Na/Ca})\,(1 - \tilde{E}_{Ca})^2}\right]^{1/2} \qquad (9.20)$$

When $K_V^{Na/Ca} = 1$, Eq. 9.20 reduces to the nonpreference isotherm in Eq. 9.13, with *biv* being Ca in this special case. If $K^{Na/Ca} > 1$, the exchange isotherm will lie above the curvilinear nonpreference isotherm, showing *selectivity* for Ca (or, more generally, *biv*), as in Figure 9.2. In this way, model expressions like Eqs. 9.18 and 9.20 permit the interpretation of exchange isotherms for univalent → bivalent ion exchange in soils of arbitrary texture and composition. The Vanselow model, however, is not necessarily a realistic description of the adsorbate in ion exchange reactions, but serves instead as both a useful approximation on which to base a semiquantitative understanding of these reactions as they occur in soils and a reference model for defining ion exchange selectivity.

9.4 Biotic Ligand Model

The *biotic ligand model* is a simplified chemical approach to characterizing the acute toxicity of borderline and Class B metals (Section 1.2) to organisms living in natural waters, sediments, or soils. (*Acute toxicity* refers to the effect on an organism caused by exposure to a single dose of a toxicant chemical over a period of time ranging from 24 to 96 hours.) The purpose of the model is to evaluate quantitatively the manner in which water and soil chemistry affect the short-term bioavailability of a toxic metal under conditions typical of acute toxicity tests, as implemented in environmental toxicology laboratories. Thus, the organisms considered with respect to toxic effects usually are those commonly used in such tests, and the primary goal of a biotic ligand model is to provide quantitative input into the development of water and soil quality criteria based on standard acute toxicity data. These data almost always are total concentrations of a toxicant that cause either death or significant impairment of the functioning of test organisms deemed by ecotoxicological practice to be reliable indicator species (see Problem 13 in Chapter 3). Although originally developed for application to aquatic organisms, the biotic ligand model has been extended to describe metal toxicity to microbes and plant roots in soils.

Biotic ligand models may differ regarding which mathematical formulation is used to describe dose–response relationships, or regarding how toxicant chemical species are identified and quantified, but they all invoke the same set of three hypotheses as their toxicological foundation. A comprehensive introduction to the biotic ligand model is given in a review article by Paquin et al. [Paquin, P. R., et al. (2002) The biotic ligand model: A historical overview. *Comp. Biochem. Physiol. Part C* 133:3.]

1. The free toxicant metal ion activity (section 4.3 to 4.5) in a natural water or the soil solution is determined by the chemical reactions depicted in the competition diagram shown in Figure 1.2. The timescales of these reactions are assumed to be incommensurate with the timescale of an acute toxicity test (i.e., other chemical forms of the toxicant metal than the free-ion species either have already equilibrated with it or can be assumed so kinetically inhibited as not to occur on the timescale of the toxicity test).

2. Toxic effects of a metal on a test organism are correlated positively with the concentration of the free metal ion species that is complexed by a ligand that is characteristic of the test organism, a so-called *biotic ligand.* Thus, a test organism (e.g., a species of microbes or roots in Figure 1.2, or of crustaceans, algae, or fish in a natural water) is assigned a metal-binding site analogous to those on soil particle surfaces. No assumption is made about the identity or molecular structure of the biotic ligand. However, it is assumed to react directly with free metal ion species in the aqueous solution phase contacting the test organism. In keeping with Hypothesis 1, the timescale for this reaction is very short when compared with that for toxic response.

3. Toxic response depends only on the fraction of binding sites on the biotic ligand that are occupied by the free-ion species of the toxicant metal. This fraction is determined by the competing reactions pictured in Figure 1.2 and by competition for the biotic ligand itself from other free-ion species (e.g., protons and Class A or B metals).

A biotic ligand model necessarily makes use of chemical speciation calculations as outlined in sections 4.3 and 4.4, but with pertinent precipitation–dissolution and adsorption–desorption reactions (chapters 5 and 8) included along with complexation reactions. All the caveats discussed in Section 4.4 apply to these calculations as well, so they must be borne in mind when the biotic ligand model is applied in a regulatory context. Adding to the approximate nature of the model is the characterization of adsorption reactions with particle surfaces, of which metal ion binding by the biotic ligand is one:

$$M^{m+} + \equiv BL^{-}(s) = \equiv BLM^{(m-1)}(s) \tag{9.21}$$

by analogy with the notation used to depict a reactive surface site in Special Topic 3 (Chapter 3), where $\equiv BL^{-}(s)$ is a biotic ligand site that adsorbs a metal cation M^{m+}. A conditional equilibrium constant for the reaction in Eq. 9.21 can be written by analogy with Eq. 4.7:

$$K_{cMBL} = \frac{[\equiv BLM^{(m-1)}]}{[M^{m+}][\equiv BL^{-}]} \tag{9.22}$$

By Hypotheses 1 and 2, K_{cMBL} and the conditional equilibrium constants for all of the reactions portrayed in Figure 1.2 along with those between the biotic ligand and the cations that compete with M^{m+} determine $[\equiv BLM^{(m-1)}]$, the concentration that correlates with toxic response.

Hypothesis 3 requires consideration of the speciation of the biotic ligand following the example in Eq. 4.13:

$$\begin{aligned} \equiv BL_T = [\equiv BL^{-}] + [\equiv BLM^{(m-1)}] + [BLH^{0}] \\ + [\equiv BLNa^{0}][\equiv BLCa^{+}] + \cdots \end{aligned} \tag{9.23}$$

with H^{+}, Na^{+}, and Ca^{2+} exemplifying competing cations. Of particular interest is the species distribution coefficient $\alpha_{\equiv BLM}$:

$$\alpha_{\equiv BLM} = \frac{[\equiv BLM^{(m-1)}]}{\equiv BL_T} = \frac{K_{cMBL}[M^{m+}]}{\{1 + K_{cMBL}[M^{m+}] + K_{cHBL}[H^{+}] + \cdots\}} \tag{9.24}$$

with a method of derivation that is described in Section 4.3. This distribution coefficient is introduced into a suitable dose–response expression to develop the predictive machinery of the model. It is evident from Eq. 9.24 that, in the absence of competing cations, the relation between $\alpha_{\equiv BLM}$ and $[M^{m+}]$ is analogous to the Langmuir isotherm equation (Eq. 8.7) and that, in the presence of competing cations, $\alpha_{\equiv BLM}$ is smaller than when they are absent. (Note that the impact of each competing cation is the product of an intensity factor and a capacity factor, as discussed in connection with Eq. 4.18.) Thus, by Hypothesis 3, the effect of cation competition is to diminish toxic response. Also according to Hypothesis 3, a unique value of $\alpha_{\equiv BLM}$ is associated with LC_{50} or any other concentration of the toxicant metal that causes a defined toxic effect (LC_{50} is the concentration that causes 50% mortality among the organisms used in an acute toxicity test; see Problem 13 in Chapter 3). This assertion implies that the conditional equilibrium constants in Eq. 9.24 do not vary with solution composition, an approximation similar to that made in the Vanselow model of ion exchange. Indeed, ratios of K_{cMBL} to the other conditional equilibrium constants in Eq. 9.24 describe competition in terms of the cation exchange reactions

$$\equiv BLH^0(s) + M^{m+} = \equiv BLM^+(s) + H^+ \quad (9.25a)$$

$$\equiv BLNa^0(s) + M^{m+} = \equiv BLM^+(s) + Na^+ \quad (9.25b)$$

$$\equiv BLCa^+(s) + M^{m+} = \equiv BLM^+(s) + Ca^{2+} \quad (9.25c)$$

and so on, in complete analogy with Eq. 9.15b. In this sense, cation exchange reactions (with $\equiv BL^- \Leftrightarrow X^-$) determine the extent to which the biotic ligand will adsorb M^{m+} and promote toxic effects on short timescales.

Validation of the biotic ligand model is performed both qualitatively and quantitatively. For example, at fixed pH and concentrations of competing metal cations, a LC_{50} value should increase with increasing dissolved humus concentration (*organic complexes* in Fig. 1.2) and, at fixed pH and humus concentration, it should also increase with increasing concentrations of competing cations. The first trend is a result of another ligand competing with the biotic ligand for M^{m+}, whereas the second trend is a result of metal competition from sites on the biotic ligand, both leading to a reduced value of $\alpha_{\equiv BLM}$ and, therefore, the need for a higher *total* concentration of M to cause a given percent mortality.

A more quantitative test of the model is provided by measurements of toxic effect in terms of the concentration of the free-cation species M^{m+}. According to Eq. 9.24,

$$[M^{m+}] = \frac{\alpha_{\equiv BLM}/K_{cMBL}}{1 - \alpha_{\equiv BLM}}[1 + K_{cHBL}[H^+] + K_{cNaBL}[Na^+] + \cdots] \quad (9.26)$$

Equation 9.26 implies that the concentration of M^{m+} causing a given toxic effect will increase linearly with the concentration of any competing cation, provided that Hypothesis 3 is accurate *and* that the conditional equilibrium constants do not vary significantly with solution composition. Hypothesis 3 can be examined indirectly, of course, by measuring mortality percentages or another toxic effect under varying total concentrations of metal M and testing goodness-of-fit for a proposed mathematical relationship between toxic effect and $\alpha_{\equiv BLM}$. A growing body of toxicological literature attests to the usefulness of $\alpha_{\equiv BLM}$ as a quantitative measure of acute toxicity, despite the simplifications inherent to the use of Eq. 9.24.

9.5 Cation Exchange on Humus

One of the most important reactions in Figure 1.2 that determines the free-ion concentration of an element in the soil solution is adsorption by particle surfaces, particularly where the surfaces comprise acidic organic functional groups in humus. These groups are not only more numerous (per unit mass) than those found on the surfaces of mineral particles (Section 3.3), but also are more complex in terms of molecular structure because of the supramolecular nature of humic substances, which constitute the major portion of humus (Section 3.2). This molecular-scale complexity poses a key challenge to quantitative modeling of cation exchange reactions that is parallel to that mentioned in connection with metal speciation calculations in Section 4.4: how to formulate metal cation interactions with humus carboxyl and phenolic OH groups to express the bound metal concentration in terms of conditional stability constants and free-ion concentrations.

Humic substances, whether dissolved or particulate, exhibit a variety of molecular-scale environments clustered together in a morass of organic tendrils and spheroids with labile conformations that depend on conditions of temperature, pressure, pH, and soil solution composition. Faced with this daunting heterogeneity, the modeling of metal complexation by these materials becomes an exercise constrained by parsimony, with the foremost challenge of striking a balance between the number of chemical parameters necessary to describe chemical speciation accurately and the varieties of functional group reactivity with metal cations that must be considered, even for just two classes of acidic group. One approach that shows promise for applications to soil solutions is the *NICA–Donnan model.* (NICA is the acronym for *nonideal competitive adsorption.*)

This model appears in Problem 6 of Chapter 3 for the specific application of describing σ_H and ANC of humic acid. The model expression for σ_H contains two terms, one for each class of acidic functional group. It is convenient, in making an acquaintance with the NICA–Donnan model, to restrict attention initially to just one such class. Then the equation for the moles of proton charge complexed by humus is

$$q_H(c_H) = b_H \frac{[(\tilde{K}_H c_H)^{\beta_H}]^p}{1 + [(\tilde{K}_H c_H)^{\beta_H}]^p} \tag{9.27}$$

where c_H is the concentration of protons in the aqueous solution phase near the adsorbent surface and $\tilde{K}_H$ is an *affinity parameter* analogous to K in the Langmuir equation (Eq. 8.7). According to the definitions given in Problem 7 of Chapter 3, the quantity q_H is the same as the total acidity of the class of functional groups to which it applies and σ_H is then equal to the difference between total acidity (q_H) and CEC (b_H). Applying these definitions to the NICA–Donnan expression for q_H, one finds an equation for σ_H that is identical in mathematical form to each of the two terms in the model equation discussed in Problem 6 of Chapter 3. Equation 9.27 is a version of the *Langmuir–Freundlich equation* (Eq. 8.16) discussed in Section 8.2. Its adjustable parameters are $\tilde{K}_H$, β_H, and p, the latter two of which arise from a factorization of the exponent β in Eq. 8.16. If $\beta_H p = 1$, then Eq. 9.27 reduces to the Langmuir equation.

Using the NICA–Donnan model, Eq. 9.27 is to be applied only to *complexed* protons. Protons adsorbed in the diffuse-ion swarm are accounted for by a version of surface charge balance as expressed in Eq. 7.5:

$$\sigma_p + V_{ex} \sum\nolimits_i Z_i (\bar{c}_i - c_{io}) = 0 \qquad (9.28)$$

where the sum is over all species in the diffuse-ion swarm (not only protons) with valence Z_i and concentration $\bar{c}_i$. This latter variable is identified as the average concentration of an ion i in the exclusion volume V_{ex} near the adsorbent surface (Eq. 8.24). Thus, V_{ex} represents the volume of aqueous solution (per unit mass of adsorbent) that encompasses the diffuse-ion swarm that balances the net total particle charge σ_p (Eq. 7.4). The second term on the right side of Eq. 9.28 is the net adsorbed charge contributed by the diffuse-ion swarm, given c_{io} as the bulk aqueous solution concentration of ion i, as in Eq. 8.24.

As indicated in Problem 9 of Chapter 8, V_{ex} is found to be inversely proportional to the inverse square root of c_o for a given ionic species. More generally, in log–log form,

$$\log\ V_{ex} = \alpha - \frac{1}{2} \log\ I \qquad (9.29)$$

where V_{ex} is in liters per kilogram, I is ionic strength in moles per liter (Eq. 4.22), and α is an adjustable parameter to be determined from measurements of the ionic strength dependence of the exclusion volume. The value of $\alpha \approx -0.53 \pm 0.03$—statistically the same as the coefficient of log I in Eq. 9.29—has been established in this way for a variety of humic substances.

Equations 9.27 (with one such equation for each class of acidic functional group), 9.28, and 9.29 constitute a mathematical model for proton adsorption by humic substances. The resulting optimized parameters, based on a large number of proton titration curves (Fig. 3.3) for humic and fulvic acids, are:

	b_1 (mol$_c$ kg^{-1})	log $\tilde{K}_1$	p_1	b_2 (mol$_c$ kg^{-1})	log $\tilde{K}_2$	p_2
Fulvic acid	5.88	2.34	0.38	1.86	8.60	0.53
Humic acid	3.15	2.93	0.50	2.55	8.00	0.26

in the notation of Problem 6 in Chapter 3, which drops the subscript H and combines the product $\beta_H p$ into a single exponent, p. (The results given here for humic acid also appear in Problem 6 of Chapter 3.) Note that the values of $(b_1 + b_2)$ fall well into the range of CEC typical for humic substances, that the same is true for b_2 and phenolic OH content, and that log $\tilde{K}_1$ and log $\tilde{K}_2$ do not differ substantially between the two types of humic substance, whereas b_1 does. The association of carboxyl groups with $\tilde{K}_1$ and phenolic OH with $\tilde{K}_2$ is facile.

Cation exchange is brought into the NICA–Donnan model by expanding Eq. 9.27 to include a metal cation M:

$$q_i(c_H, c_M) = b_i \frac{(\tilde{K}_i c_i)^{\beta_i}}{(\tilde{K}_H c_H)^{\beta_H} + (\tilde{K}_M c_M)^{\beta_M}} \frac{[(\tilde{K}_H c_H)^{\beta_H} + (\tilde{K}_M c_M)^{\beta_M}]^p}{1 + [(\tilde{K}_H c_H)^{\beta_H} + (\tilde{K}_M c_M)^{\beta_M}]^p} \tag{9.30}$$

where i = H or M and the additional parameters are interpreted analogously to those in Eq. 9.27. The full model equation is actually the sum of two terms like that in Eq. 9.30 (!), one for carboxyl groups and one for phenolic OH groups. This equation is complemented by eqs. 9.28 and 9.29 (with $\bar{c}_i$ in Eq. 9.28 equated to c_i in Eq. 9.30) in all applications. After multiplication by the solids concentration of humic substance (in kilograms per cubic decimeter), the right side of the equation becomes a molar concentration that can be substituted directly into a mass balance expression for the metal M as posed in a typical chemical speciation calculation (see, for example, Eq. 4.6).

The first factor on the right side of Eq. 9.30 is the maximum moles of proton or metal cation charge that can be *complexed* by the class of acidic functional group to which it applies. Experience with the model shows that this parameter, as would be expected, is proportional to the CEC of the acidic functional group (b_H), with the proportionality constant being the ratio of β_i to β_H. The exponents β_i (i = H, M), which take on values between 0 and 1, thus are relative *stoichiometric parameters* accounting for differences between H and M in respect to how many moles of ion charge are bound to one mole of acidic functional groups. (Thus, β_H was implicitly set equal to one in applying Eq. 9.27, leaving only the exponent, p.) The second factor on the right side then gives the fraction of complexes with the acidic functional group that contain the species i (i = H or M). (Note its similarity to Eq. 9.24.) The model affinity parameters for these two species are $\tilde{K}_H$ and $\tilde{K}_M$ respectively. They play the role of conditional stability constants, although no specific chemical reaction is associated with either of them in the model. A detailed discussion of these points and, more generally, the complete derivation of the NICA–Donnan model are given in a review article by Koopal et al. [Koopal, L. K., T. Saito, J. P. Pinheiro, and W. H. van Riemsdijk. (2005) Ion binding to natural organic matter: General considerations and the NICA–Donnan model. *Colloids Surf.* **265A**:40.], whereas a clear overview of the model is given by Merdy et al. [Merdy, P., S. Huclier, and L. K. Koopal. (2006) Modeling metal–particle interactions with an emphasis on natural organic matter. *Environ. Sci. Technol.* **40**:7459.].

The first two factors on the right side of Eq. 9.30 combine to describe a *capacity factor* for the complexation of species i (i = H or M). The third factor in the expression is an *intensity factor* that models the competition between protons and metal cations for complexation by a class of acidic functional groups. It contains a "smearing out" parameter, *p*, with values also between zero and one, that accounts for intrinsic variability in the affinity of the groups for either protons or metal cations caused by molecular-scale effects, such as local electrostatic fields created by the dissociation of groups near a group that has complexed a proton or metal cation, stereochemistry, or conformation. This kind of variability can in fact be represented mathematically by a distribution of affinity parameters with a median value that is $\tilde{K}_i$ (i = H or M) and with a breadth that is represented by the parameter p, with breadth increasing as p becomes smaller. This is similar to the interpretation of the exponent β in the Langmuir–Freundlich and van Bemmelen–Freundlich equations (Section 8.2).

Cation adsorption in the diffuse-ion swarm, including contributions from a background electrolyte, is considered only in Eq. 9.28. Ionic strength effects thus are assumed to be produced as a result of *screening* of the net proton charge by the background electrolyte cations (Section 7.2). Attracted by negative charge as quantified by σ_P (= σ_H), these cations diffuse in from the bulk electrolyte solution to approach dissociated acidic functional groups, with most of them swarming near the periphery of the humic substance within a distance of about 1 nm along an outward direction into the vicinal aqueous

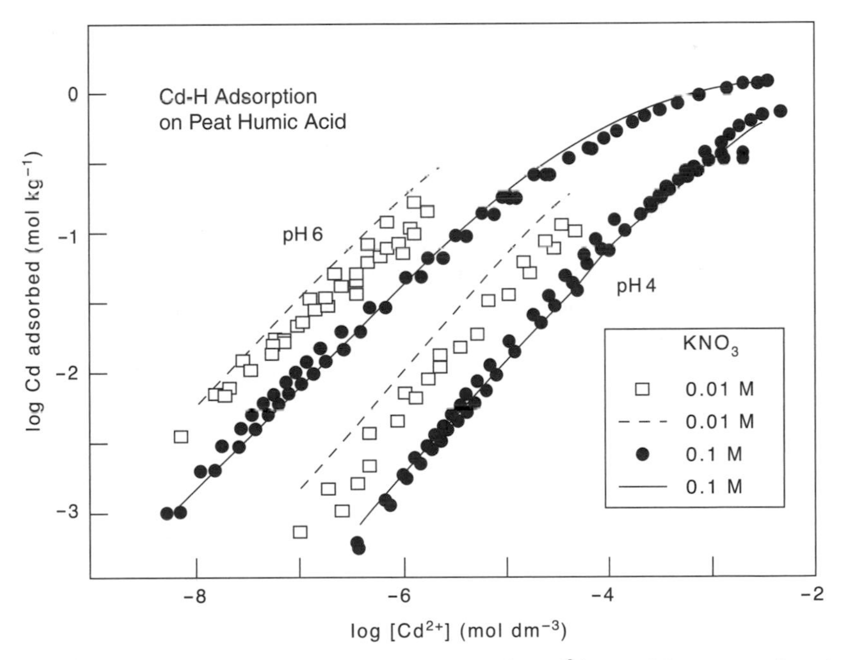

Figure 9.3. Log–log plot of adsorption isotherms for Cd^{2+} at 25 °C on a peat humic acid at two pH values and two ionic strengths (KNO_3 background electrolyte). The curves (dashed and solid lines) are corresponding plots of the adsorption isotherms predicted by the NICA–Donnan model. Data from Kinniburgh, D. G., et al. (1996) Metal binding by humic acid: Application of the NICA–Donnan model. *Environ. Sci. Technol.* 30:1687–1698.

solution. Lateral diffusive motions of the cations following this periphery are not restricted because the cations do not form complexes, but the electrostatic field created by the negative charge is diminished in strength by the diffuse swarm of cations that screens it. The picture here is roughly analogous to that of an electron cloud screening the nuclear charge in an atom.

Figure 9.3 shows an application of the extended version of Eq. 9.29 to describe the concurrent adsorption of protons and Cd^{2+} by a peat humic acid. The curves through the data points for two values of pH and ionic strength (KNO_3 background electrolyte) were calculated with $\alpha = -0.57$ in Eq. 9.29 and the parameter values:

Ion i	$\log \tilde{K}_{1i}$	β_{1i}	$\log \tilde{K}_{2i}$	β_{2i}
H	2.98	0.86	8.73	0.57
Cd	0.10	0.81	2.03	0.48
	$b_1 = 2.74$ mol$_c$ kg^{-1}	$p_1 = 0.54$	$b_2 = 3.54$ mol$_c$ kg^{-1}	$p_2 = 0.54$

Note that, as a result of cation exchange, q_{Cd} is decreased by increasing I and decreasing pH.

For Further Reading

Di Toro, D. M., H. E. Allen, H. L. Bergman, J. S. Meyer, P. R. Paquin, and R. C. Santore. (2001) Biotic ligand model of the acute toxicity of metals. *Environ Toxicol. Chem.* **20**:2383–2396. This review article gives a critical discussion of the biotic ligand model that takes the reader through each step typical of model development, with special emphasis given to calibrating chemical speciation calculations involving humus.

Essington, M. E. (2004) *Soil and water chemistry.* CRC Press, Boca Raton, FL. Chapter 8 in this comprehensive textbook gives a discussion of cation exchange in soils with many examples and careful development of the concept of the selectivity coefficient.

Milne, C. J., D. Kinniburgh, W. H. van Riemsdijk, and E. Tipping. (2003) Generic NICA–Donnan model parameters for metal–ion binding by humic substances. *Environ. Sci. Technol.* **37**:958–971. This article gives a critical compilation of NICA–Donnan model parameters for metal cations interacting with humic and fulvic acids.

Sparks, D. L. (ed.). (1996) *Methods of soil analysis: Part 3. Chemical methods.* Soil Science Society of America, Madison, WI. Chapters 40 and 41 of this standard reference describe tested laboratory methods for measuring ion exchange capacities and selectivity coefficients.

Sposito, G. (1994) *Chemical equilibria and kinetics in soils.* Oxford University Press, New York. Chapter 5 of this advanced textbook provides a comprehensive discussion of the kinetics and thermodynamics of ion exchange reactions.

Sposito, G. (2004) *The surface chemistry of natural particles.* Oxford University Press, New York. Chapter 4 of this advanced textbook contains a detailed description of the NICA–Donnan model as applied to cation adsorption by humic substances.

Problems

The more difficult problems are indicated by an asterisk.

1. After consulting *Methods of Soil Analysis* (listed in "For Further Reading"), discuss and compare the $BaCl_2$ (see also Section 3.3) and $NH_4C_2H_3O_2$ methods of measuring CEC as applied to a variable-charge soil (e.g., Spodosol or Oxisol).

2. Consult *Methods of Soil Analysis* to obtain details of the $CaCl_2/Mg(NO_3)_2$ and $BaCl_2$ methods of measuring CEC in soils. Compare the advantages and disadvantages of each method as applied to a soil with a mineralogy that reflects the early stage of Jackson–Sherman weathering (Table 1.7).

3. Calculate a conditional exchange equilibrium constant for the reaction $MgX_2(s) + Ca^{2+} = CaX_2(s) + Mg^{2+}$ based on the data in the table presented here. Plot $K_V^{Mg/Ca}$ against E_{Ca}. (Assume that single-ion activity coefficients can be calculated with the Davies equation.) Does the adsorbent exhibit preference?

c_{Mg} (mol m^{-3})	c_{Ca}	q_{Mg} (mol_c kg^{-1})	q_{Ca}	c_{Mg} (mol m^{-3})	c_{Ca}	q_{Mg} (mol_c kg^{-1})	q_{Ca}
20.4	2.4	0.21	0.078	10.1	12.3	0.093	0.21
17.8	4.8	0.17	0.11	4.9	17.0	0.053	0.26
14.8	7.2	0.14	0.15	2.4	19.6	0.027	0.31
12.4	9.7	0.12	0.17	1.2	20.9	0.016	0.30

*4. Show that the use of mole fractions in Eq. 9.18 is consistent with the assumption that the soil adsorbate is an ideal solid solution. (*Hint:* Reinterpret Eqs. 5.28c and 9.15c in terms of mole fractions and then reformulate the definition of an ideal solid solution.)

5. Plot an exchange isotherm like those in Figure 9.2 using the composition data on Na → Ca exchange in the table presented here. Include a nonpreference isotherm based on Eq. 9.13. (Take $\tilde{Q} = 0.05$ mol_c kg^{-1} and calculate single-ion activity coefficients with Eq. 4.24 for an ionic strength of 0.05 mol L^{-1}.)

m_{Na}	m_{Ca} (mol kg^{-1})	q_{Na}	q_{Ca} (mol$_c$ kg^{-1})	m_{Na}	m_{Ca} (mol kg^{-1})	q_{Na}	q_{Ca} (mol$_c$ kg^{-1})
0.0480	0.000136	0.100	0.037	0.0450	0.00164	0.047	0.103
0.0474	0.000320	0.074	0.056	0.0441	0.00212	0.039	0.108
0.0469	0.000717	0.060	0.081	0.0397	0.00475	0.016	0.121
0.0457	0.00118	0.049	0.095	0.0302	0.00976	0.009	0.134

6. Plot an exchange isotherm like those in Figure 9.2 using the data on Na → Mg exchange in the table presented here. Show that the isotherm is essentially a nonpreference isotherm, as described in Section 9.2. (Take $\tilde{Q} = 0.05$ mol$_c$ kg^{-1} and calculate single-ion activity coefficients with Eq. 4.24 for I = 0.05 mol L^{-1}.)

m_{Na}	m_{Mg} (mol kg^{-1})	q_{Na}	q_{Mg} (mol$_c$ kg^{-1})	m_{Na}	m_{Mg} (mol kg^{-1})	q_{Na}	q_{Mg} (mol$_c$ kg^{-1})
0.0495	0.00117	0.53	0.28	0.0340	0.0124	0.10	0.74
0.0474	0.00234	0.30	0.45	0.0291	0.0149	0.08	0.74
0.0440	0.00700	0.22	0.70	0.237	0.0174	0.06	0.78
0.0383	0.00940	0.23	0.86	0.0185	0.0197	0.06	0.95

*7. The data in the table presented on the next page show adsorbed cation charge resulting from the reaction between mixed perchlorate salt solutions ($\tilde{Q}$ = 0.05 mol$_c$ L^{-1}) and the silt plus clay fraction of a Vertisol containing 26 ± 7 g C kg^{-1} in addition to a high content of montmorillonite. Use appropriate statistical methods to determine whether a pH dependence exists for (a) CEC, (b) E_{Na}, or (c) preference for Ca. (*Hint:* Consider plotting each of the properties to be tested against E_{Mg} for each pH value, then follow with linear regression analyses.)

8. Derive Eq. 9.24b for a biotic ligand that binds a toxic metal M^{2+} and the competing cations H^+, Ca^{2+}, and Mg^{2+}, beginning your derivation with Eq. 9.23. Describe a method to test Hypothesis 3 of the biotic ligand model with the equation you derive.

9. A fundamental premise of the biotic ligand model is that acute toxic effect is correlated positively with the concentration of the uncomplexed (free) toxicant species in aqueous solution. Therefore, acute toxicity may be expressed in terms of the free species concentration that results in a loss of function or death (EC_{50} or LC_{50}) for half a population of test organisms

q_{Na} ($mol_c\ kg^{-1}$)	q_{Ca} ($mol_c\ kg^{-1}$)	q_{Mg} ($mol_c\ kg^{-1}$)
pH 4.7 ± 0.3		
0.16 ± 0.03	0.000	0.46 ± 0.02
0.16 ± 0.04	0.0897 ± 0.0007	0.391 ± 0.006
0.16 ± 0.03	0.16 ± 0.01	0.342 ± 0.009
0.17 + 0.03	0.206 ± 0.004	0.304 ± 0.007
0.15 ± 0.06	0.251 ± 0.007	0.255 ± 0.009
0.15 ± 0.04	0.298 ± 0.004	0.202 ± 0.005
0.17 ± 0.03	0.34 ± 0.01	0.148 ± 0.006
0.14 ± 0.02	0.386 ± 0.007	0.116 ± 0.004
0.13 ± 0.04	0.393 ± 0.008	0.074 ± 0.004
0.15 ± 0.03	0.43 ± 0.01	0.042 ± 0.002
0.13 ± 0.04	0.470 ± 0.002	0.000
pH 5.8 ± 0.1		
0.183 ± 0.006	0.000	0.481 ± 0.003
0.187 ± 0.007	0.100 ± 0.001	0.412 ± 0.003
0.165 ± 0.002	0.159 ± 0.001	0.352 ± 0.002
0.160 ± 0.005	0.2148 ± 0.0002	0.303 ± 0.0003
0.184 ± 0.003	0.266 ± 0.001	0.256 ± 0.001
0.176 ± 0.005	0.315 ± 0.001	0.210 ± 0.001
0.156 ± 0.006	0.352 ± 0.003	0.161 ± 0.003
0.153 ± 0.007	0.400 ± 0.001	0.120 ± 0.001
0.16 ± 0.01	0.439 ± 0.003	0.0789 ± 0.0003
0.16 ± 0.02	0.480 ± 0.001	0.044 ± 0.003
0.166 ± 0.004	0.522 ± 0.005	0.000
pH 6.9 ± 0.2		
0.144 ± 0.004	0.552 ± 0.003	0.000
0.146 ± 0.003	0.500 ± 0.005	0.0501 ± 0.0006
0.146 ± 0.006	0.459 ± 0.006	0.098 ± 0.001
0.150 ± 0.003	0.409 ± 0.003	0.144 ± 0.002
0.155 ± 0.003	0.358 ± 0.004	0.196 ± 0.002
0.154 ± 0.002	0.317 ± 0.002	0.243 ± 0.002
0.156 ± 0.004	0.268 ± 0.003	0.299 ± 0.002
0.156 ± 0.003	0.211 ± 0.002	0.346 ± 0.005
0.191 ± 0.004	0.154 ± 0.002	0.406 ± 0.005
0.159 ± 0.003	0.0956 ± 0.0007	0.482 ± 0.006
0.163 ± 0.002	0.000	0.596 ± 0.003

exposed to a toxicant for a short period of time (24–96 hours). Given the generic reaction between a bivalent metal cation and a biotic ligand,

$$\equiv BL^- + M^{2+} = \equiv BLM^+$$

and the model equilibrium constant in Eq. 9.22, show that EC_{50} for a toxic bivalent metal cation (e.g., Cd^{2+}) should increase linearly with the concentration of a nontoxic bivalent metal cation (e.g., Ca^{2+}) that can also bind to the biotic ligand.

10. The concentration of Cu^{2+} causing 50% immobilization of the freshwater test organism *Daphnia magna* (water flea), after 48 hours of exposure (EC_{50}) is observed to increase with the concentration of Ca^{2+} according to the linear regression equation

$$EC_{50} = 9.97 + 25.0\, c_{Ca}$$

where EC_{50} is in nanomoles per liter and c_{Ca} is in moles per cubic meter. Separate experiments show that EC_{50} is associated with $\alpha_{\equiv BLCu} = 0.47$. Calculate a value of the Vanselow selectivity coefficient for Cu → Ca exchange on the biotic ligand of *D. magna*. Explain why EC_{50} increases as the concentration of Ca^{2+} increases.

*11. The presence of a complexing ligand that can bind a toxic metal cation in competition with a biotic ligand should reduce the concentration of the free metal cation and thereby inhibit toxic effect. For example, the 48-hour LC_{50} for Ag toxicity to *D. magna* increases from 0.47 μg Ag L^{-1} to 1.2 μg Ag L^{-1} when the concentration of chloride ions is increased from 0.05 to 1.0 mM. (These LC_{50} values are expressed in terms of the *total* soluble Ag concentration that induces acute toxicity.) Given the value of the equilibrium constant for the formation of the soluble complex $AgCl^0$,

$$Ag^+ + Cl^- = AgCl^0 \quad K_s = 10^{3.31}$$

determine whether this increase in LC_{50} with chloride concentration is reasonable.

12. The concentration of Ni^{2+} causing a 50% reduction in normal barley root elongation (EC_{50}) is a linear function of the concentrations of protons, Ca^{2+}, and Mg^{2+} in the soil solution. Given the parameter values

$$\log K_{cNiBL} = 3.60 \pm 0.53,\ \log K_{cHBL} = 4.52 \pm 0.62,$$

$$\log K_{cCaBL} = 1.50,\ \log K_{cMgBL} = 3.81 \pm 0.60$$

estimate EC_{50} at pH 5.7 if $[Ca^{2+}] = 2 \times 10^{-3}$ mol L^{-1} and $[Mg^{2+}] = 2.5 \times 10^{-4}$ mol L^{-1}. Take $\alpha_{\equiv BLNi} = 0.05$.

13. The concentration of Cu^{2+} causing 50% immobilization of *D. magna* after 48 hours (Problem 11) is affected by H^+, Na^+, Ca^{2+}, and Mg^{2+} as competing cations. Given the parameter values

$$\log K_{cCuBL} = 8.02,\ \log K_{cHBL} = 5.40,\ \log K_{cNaBL} = 3.19,$$

$$\log K_{cCaBL} = 3.47,\ \log K_{cMgBL} = 3.58$$

calculate EC_{50} for pH ranging between 6.0 and 8.5 in a natural water having $[Na^+] = 0.002$ mol L^{-1}, $[Ca^{2+}] = 0.003$ mol L^{-1}, and $[Mg^{2+}]$

= 0.0006 mol L^{-1}. Does increasing pH increase or decrease the acute toxicity of Cu^{2+} to *D. magna?*

14. Benedetti et al. [Benedetti, M. F., W. H. van Riemsdijk, and L. K. Koopal. (1996) Humic substances considered as a heterogeneous Donnan gel phase. *Environ. Sci. Technol.* **30**:1805.] explain the origin of Eq. 9.28 in terms of the *Donnan model,* in which ions are distributed between a slurry containing charged particles and a supernatant solution (Section 8.1) according to the average properties of a diffuse-ion swarm. (Their Eq. 1 is the same as Eq. 9.28, with the correspondences $Q \Leftrightarrow \sigma_p$ and $\nu_D \Leftrightarrow V_{ex}$.) Experimental measurements of V_{ex} as a function of ionic strength are summarized in their Table 2 and their Figure 4 for humic and fulvic acids.

    a. Use linear regression analysis to estimate the parameter α and the coefficient of log I in Eq. 9.29 for the data on V_{ex} in Table 2 of the article by Benedetti et al. Be sure to include 95% confidence limits with your results.
    b. Typical values of the specific surface area for humic substances range from 500 to 800 $m^2\ g^{-1}$. Use the model equation for V_{ex} in Problem 9 of Chapter 8 to estimate the value of the parameter α in Eq. 9.28 based on this range of a_S values.

15. The table presented here lists value of log $\tilde{K}_M$ and β_M (Eq. 9.30) for bivalent metal (M) cation complexation by carboxyl groups in humic acids. Discuss the values of the product β_M log $\tilde{K}_M$ in terms of concepts presented in sections 8.3 and 9.3.

Cation	log $\tilde{K}_M$	β_M
Ba	−1.1	0.90
Ca	−1.37	0.78
Cd	−0.20	0.73
Co	−0.24	0.79
Cu	2.23	0.56
Hg	5.2	0.32
Mg	−0.6	0.77
Mn	−0.3	0.72
Ni	−0.26	0.64
Pb	1.25	0.60
Sr	−1.36	0.78
Zn	0.11	0.67

10

Colloidal Phenomena

10.1 Colloidal Suspensions

Colloids in soils are solid particles of low water solubility with a diameter that ranges between 0.01 and 10 μm (i.e., clay-size to fine-silt-size particles). The chemical composition of these particles may vary from that of a clay mineral or metal oxide to that of soil humus, or, more broadly, may be a heterogeneous combination of inorganic and organic materials. Regardless of composition, the characteristic property of colloids is that they do not dissolve readily in water to form solutions, but instead remain as identifiable solid particles in aqueous suspensions.

Colloidal suspensions are said to be *stable* (and the particles in them *dispersed*) if no measurable settling of the particles occurs over short time periods (e.g., 2–24 hours). Stable suspensions of soil colloids lead to erosion and illuviation because the particles entrained by flowing water or percolating soil solution remain mobile. Stable suspensions also have a secondary effect on the mobility of inorganic and organic adsorptives, especially radionuclides, phosphate, or pesticides, that can become strongly bound to soil colloids. Thus, colloidal stability is connected closely with particle and chemical transport.

Soil particles with a diameter that falls into the middle of the colloidal range, from approximately 100 nm to 1μm, are those observed to remain suspended in surface or subsurface waters for long periods of time. Colloids with a diameter that is less than this range coalesce and grow rather quickly to form larger particles, whereas colloids with a size that is larger than the midrange appear to settle rather quickly under the influence of gravity, at least in quiescent suspensions. Because of these observations, the study of soil colloidal phenomena has tended to focus on the behavior of midrange particles, including the influence of their surface chemistry, with the goal of pinpointing conditions that either ensure continued suspension or promote particle growth.

The process by which soil colloids in suspension coalesce to form bulky porous masses is termed *flocculation.* The particles formed during flocculation—also termed *coagulation*—and removed from suspension by settling are themselves candidates for further transformation into *aggregates,* the organized solid masses that figure in the structure, permeability, and fertility of soils. Flocculation processes are complicated phenomena because of the varieties of particle morphology and chemical reactions they encompass. From the perspective of kinetics, perhaps the most important generalization that can be made is the distinction between *transport-controlled* and *surface reaction-controlled* flocculation, parallel to the classification of adsorption processes described in Special Topic 3 (Chapter 3). Flocculation kinetics are said to exhibit transport control if the rate-limiting step is the movement of two or more particles toward one another prior to their close encounter and immediate coalescence to form a larger particle. Surface reaction control occurs if it is the particle coalescence process instead of particle movement toward collision that limits the rate of flocculation.

Three models of the transport-control mechanism for flocculation are in common use to interpret the kinetics of particle formation in colloidal suspensions (Fig. 10.1). The best known of these models is *Brownian motion* (*perikinetic flocculation*), which applies to quiescent suspensions of diffusing particles with a diameter that lies in the lower to middle portion of the colloidal range ($\lesssim 1\mu m$). Flocculation caused by stirring a colloidal suspension is described as *shear induced* (*orthokinetic flocculation*), whereas that caused by the settling of particles under gravitational or centrifugal force is described as *differential sedimentation.* In all three kinetics models, a second-order rate coefficient (Table 4.2) appears that is equal to the product of an effective cross-sectional area for two-particle collisions (a geometric factor) times an effective two-particle relative velocity (a kinematic factor). Large rate coefficients for flocculation thus are produced by optimal combinations of particle size (geometry) and opposing particle velocities (kinematics).

In a quiescent soil suspension, the motions of the particles are incessant and chaotic because of the thermal energy the particles possess. As shown by Albert Einstein in his doctoral dissertation, these Brownian motions in suspension are analogous to the diffusive motions of molecules in solution, with a diffusion coefficient expressed by the *Stokes–Einstein model:*

$$D = k_B T / 6\pi \eta R \qquad (10.1)$$

In Eq. 10.1, k_B is the Boltzmann constant (see the Appendix), η is the shear viscosity of water, and R is the radius of the colloidal particle (assumed effectively

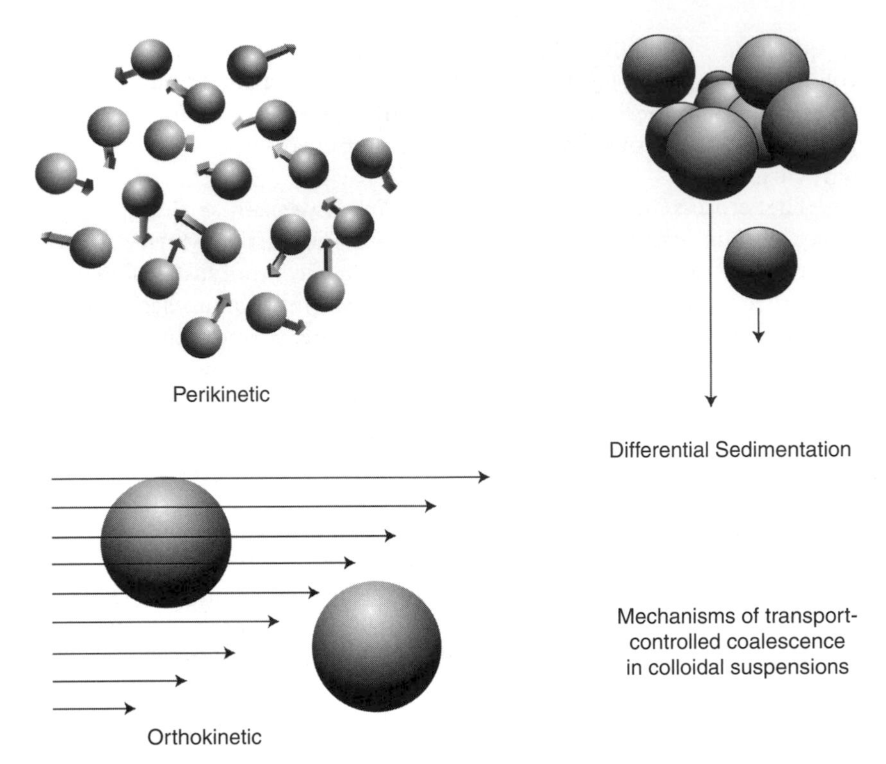

Figure 10.1. Three mechanisms of flocculation lead to rapid coalescence. Perikinetic flocculation usually is engaged in by colloids smaller than 1μm, whereas the other two mechanisms apply mainly to colloids larger than 1 μm.

spherical). The Stokes–Einstein relation indicates that a colloid will diffuse more rapidly if the temperature is high, if the fluid viscosity is low, or if the colloid is very small. Perikinetic flocculation postulates Brownian motion of colloidal particles that leads them to collide by chance, after which they coalesce instantaneously to form a *dimer.* The second-order rate coefficient describing this process is

$$k_p = 2\pi R_{11} \mathbf{D}_{11} \tag{10.2}$$

where R_{11} is the radius of the dimer and $\mathbf{D}_{11}$ is the diffusion coefficient of one of the colliding *monomers* relative to that of the other, as depicted from a reference point taken as their center of mass. In a first approximation, R_{11} is just twice the monomer radius, and $\mathbf{D}_{11}$ is just twice the monomer diffusion coefficient as modeled by Eq. 10.1. With these simplifications, the rate coefficient for dimer formation becomes

$$k_p^{SE} = 8\pi R \mathbf{D}_{SE} \equiv K_{SE} \tag{10.3}$$

where

$$K_{SE} = 4k_BT/3\eta \qquad (10.4)$$

is a constant parameter equal to 6.16×10^{-18} m^3 s^{-1} at 25 °C if water is the suspending fluid. Note that the use of the Stokes–Einstein relation leads to an exact cancellation of the monomer radius R from the perikinetic rate coefficient.

Orthokinetic flocculation postulates the capture of a monomer in the streamlines around another monomer while the former attempts to pass the latter (Fig. 10.1). This mechanism requires a fluid velocity gradient (or shear rate), G, that permits one monomer to overtake the other while they both are being convected by the fluid. The rate coefficient for orthokinetic aggregation is thus the product of the effective cross-sectional area of a dimer (proportional to the volume of a *capture sphere* enclosing the overtaken monomer and having a radius equal to that of the dimer formed) times the relative velocity of the overtaking monomer:

$$k_o = \frac{2}{3}R_{11}^3 G \qquad (10.5)$$

Usually, R_{11} is approximated once again by twice the monomer radius, such that

$$k_o = \frac{16}{3}GR^3 \qquad (10.6)$$

is the model rate coefficient for shear-induced flocculation. In this case, the geometric factor increases strongly with particle size. Typical values for G are in the range 1 to 10 s^{-1} for flowing natural waters. The importance of orthokinetic flocculation as the monomer radius increases can be seen by forming the dimensionless ratio of the right sides of Eqs. 10.3 and 10.6:

$$\frac{K_{SE}}{k_o} = \frac{\text{perikinetic rate}}{\text{orthokinetic rate}} = \frac{1.16}{GR^3} \qquad (10.7)$$

where R is in units of micrometers and G is in units of inverse seconds. Orthokinetic flocculation rates exceed those of perikinetic flocculation whenever $R > G^{-1/3}$ numerically (i.e., for monomers in the mid- to upper range of colloidal diameters, given the typical values of G).

Transport control of flocculation by differential sedimentation in a gravitational field is modeled by applying the well-known *Stokes law* for the terminal velocity of a particle settling in a viscous fluid to each particle in a pair of monomers with different radii, then multiplying the resulting difference in velocity of the two particles by the cross-sectional area of the dimer they form on collision:

$$k_{DS} = \frac{g}{9\eta}(\rho_s - \rho_f)\,\pi R_{12}^2 \left|R_{12}^2 - R_2^2\right| \qquad (10.8)$$

where g is the gravitational acceleration, and ρ_s and ρ_f are mass densities of the monomers and the fluid in which they are settling respectively. In this case, one monomer (1) overtakes the other (2) because it is larger and, therefore, has a larger terminal velocity. The dimer radius R_{12} may again be approximated by the sum of the monomer radii. Given the fourth-power dependence on monomer size and the typical magnitude of the constant prefactor in Eq. 10.8 (about $6 \times 10^6\ m^{-1}$), one deduces that the differential sedimentation mechanism only becomes important for particles larger than about $1\mu m$.

For colloidal particles smaller than $1\mu m$, the timescale for flocculation in a quiescent suspension can be estimated with the perikinetic rate coefficient in Eq. 10.3. According to Table 4.2, the half-life for flocculation is then the inverse of the product of the rate coefficient and an appropriate initial concentration of colloids. The rate at which the total number of particles per cubic meter of suspension, ρ, decreases because of a flocculation process can be described by the rate law:

$$\frac{d\rho}{dt} = -K_{SE}\,\rho^2 \tag{10.9}$$

a special case of Eq. 4.5 with $b = 2$. Equation 10.9, known as the *von Smoluchowski rate law,* contains the square of the number density ρ on the right side because *two* particles are involved in a collision, so the rate of flocculation depends on the number density of each. The corresponding half-life is

$$t_{1/2} = 1/K_{SE}\,\rho_o = 1.62 \times 10^{17}/\rho_o \tag{10.10}$$

where $t_{1/2}$ is in s and ρ_o is the initial number density. For example, if a suspension contains initially 10^{14} colloidal particles per cubic meter, it follows from Eq. 10.10 that $t_{1/2} \approx 1600$ s for the flocculation of these particles.

10.2 Soil Colloids

Colloids suspended in soil solutions will exhibit shapes and sizes that reflect both chemical composition and the effects of weathering processes. Kaolinite particles, for example, are seen in electron micrographs as roughly hexagonal plates comprising perhaps 50 unit layers (each layer is a wafer the thickness of a unit cell, about 0.7 nm), which are stacked irregularly and interconnected through hydrogen bonding between the OH groups of the octahedral sheet and the oxygens of the tetrahedral sheet (Section 2.3). In the soil environment, weathering produces rounding of the corners of kaolinite hexagons and coats them with iron oxyhydroxides and humus. Fracturing of the plates also is apparent along with a "stair-step" topography caused by the stacking of unit layers having different lateral dimensions. These heterogeneous features lead to flocculation products (*floccules*) that are not well organized. The fabric of the floccules consists of many stair-stepped clusters of stacked plates, interspersed with plates in edge–face contact (possibly because of differing surface charges on edges and faces) that are arranged in a porous three-dimensional network.

Similar observations have been made for 2:1 clay minerals. Illite, for example, is seen in electron micrographs as platelike particles stacked irregularly, although the bonding mechanism causing the stacking is an inner-sphere surface complex of K^+, not hydrogen bonding. These particles also exhibit a stair-step surface topography and frayed edges produced by weathering. Coatings of Al-hydroxy polymers and humus may have formed, with these features being made even more heterogeneous by a nonuniform distribution of isomorphic substitutions, with regions of layer charge approaching 2.0 grading to regions of layer charge near 0.5 (Section 2.3). These characteristics and a slight flexibility of the illite plates (probably caused by strains associated with isomorphic substitution) lead to floccules that are like those for kaolinite particles, but with greater porosity.

Smectite and vermiculite have a lesser tendency to form colloids comprising extensive stacks because their layer charge is smaller than that of illite and, therefore, is less conducive to inner-sphere surface complexation. They are also more flexible, probably because of stresses induced by their more extensive isomorphic substitutions in the octahedral sheet. Floccule structures built of these colloids exhibit irregularly shaped plates organized in a random framework of high porosity. Surface heterogeneities brought on by nonuniform layer charge and the sorption of Al-hydroxy polymers or humus add to the heterogeneity. Floccules of 2:1 clay minerals subjected to drying and rewetting cycles can form aggregates comprising regularly stacked layers. This parallel alignment of unit layers can be observed in very thick suspensions of Na-smectite and in any suspension of bivalent cation-saturated smectite. Stacked layers of Na-smectite are important in arid-zone soils because their ordered structure prevents the development of the large pores essential to soil permeability. They are created by the dewatering of suspensions originally containing dispersed unit layers (i.e., initially stable suspensions). Stacked layers of Ca-smectite (or any bivalent cation-saturated smectite) are organized by outer-sphere surface complexes of Ca^{2+} with pairs of opposing siloxane cavities. The octahedral salvation complex $Ca(H_2O)_6^{2+}$ is arranged in the interlayer region with its principal symmetry axis perpendicular to the siloxane surface. Four of the solvating water molecules lie in a central plane parallel to the opposing siloxane surfaces, whereas the remaining two water molecules reside in planes between the siloxane surfaces and the central plane to give an interlayer spacing of 1.91 nm (Fig. 10.2). An outer-sphere surface complex of this kind is a characteristic structure in suspensions of smectite bearing bivalent exchangeable cations.

Therefore, in relatively dilute, stable suspensions, Na-montmorillonite colloids will have a different structure from Ca-montmorillonite colloids. In stable suspensions of montmorillonite colloids bearing both Na^+ and Ca^{2+}, one would expect a continuous transition from stacked-layer particles to more or less single-layer particles as the charge fraction of exchangeable

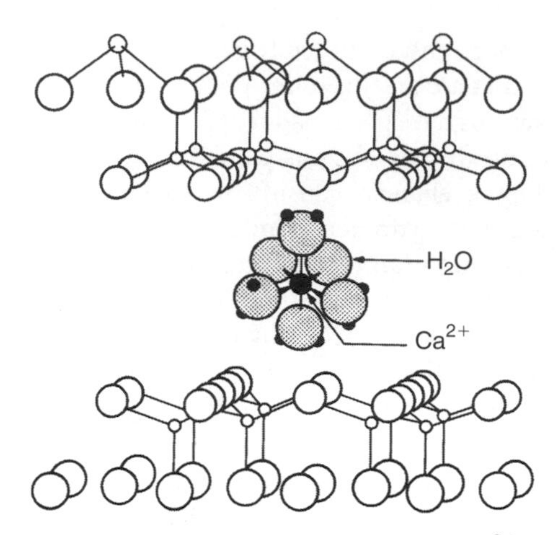

Figure 10.2. The outer-sphere surface complex between Ca^{2+} and opposing siloxane surfaces bounding the interlayer region in the three-layer hydrate of Ca-montmorillonite that occurs in aqueous suspensions.

Na^+ increases. Sharp increases in the number of single-layer particles are indeed observed when the charge fraction of Na^+ on the clay increases from 0.15 to 0.30, if the electrolyte concentration is low, indicating that the Ca-montmorillonite colloids are being broken up in favor of more or less single-layer particles. On the other hand, when $E_{Na} < 0.3$, these latter colloids are the favored entities, with any residual exchangeable Na^+ relegated to their external surfaces.

Heterogeneous soil colloids tend to harbor a variety of complex, irregular particles, including microbes, that are linked by tendrils of extracellular organic matter and cell wall compounds (Fig. 10.3). These colloids tend to coalesce rather slowly, unless the ionic strength is high, to form relatively compact floccules. Inorganic particles also can associate with smaller organic colloids by coalescence, with the latter particles possibly changing their conformation as a result of interactions with the charged surface of the inorganic partner. If the organic colloids are larger than the inorganic particles, or if they emanate long tendrils, the organic colloids may bind the smaller inorganic particles into a fibrous network of whorls with a complex overall morphology. These larger particles, in turn, may settle quickly.

10.3 Interparticle Forces

Regardless of how complex a soil colloid may be, it still is subject to the forces brought on by its fundamental properties of mass and charge. The property of mass gives rise to the *gravitational force* and the *van der Waals force.* The property of charge gives rise to the *electrostatic force.* The first two forces cause a colloidal suspension to be unstable, whereas the second can cause it to remain

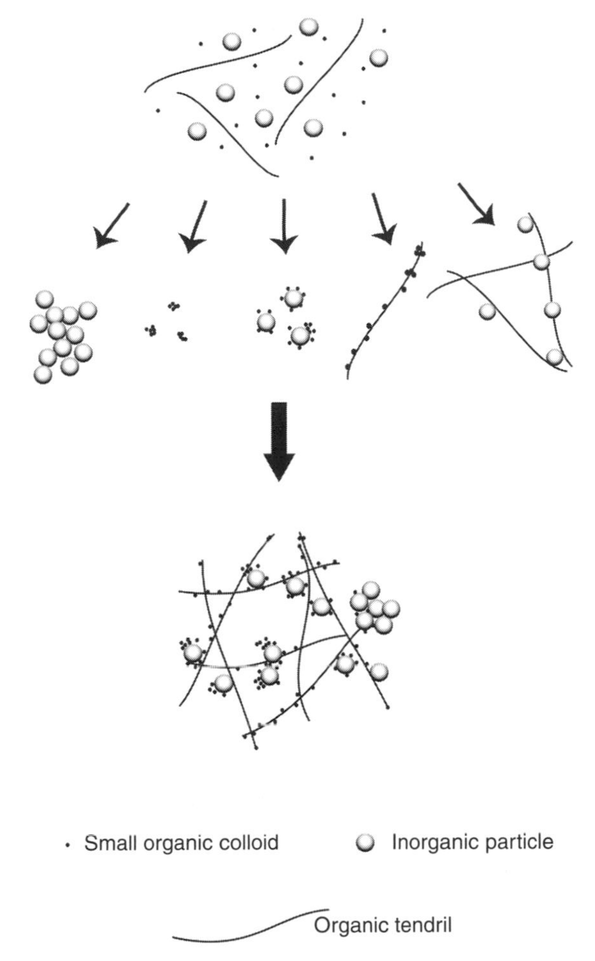

Figure 10.3. Formation and structure of heterogeneous natural colloids comprising small, roughly spherical particles, including microbes, and organic tendrils. Scheme after Buffle, J. *et al.* (1998) A generalized description of aquatic colloidal interactions. *Environ. Sci. Technol.* **32**:2887–2899.

stable. The gravitational force (corrected for the effect of buoyancy) initiates and sustains particle settling. This force is created simply by the gravity field of the earth. The other two forces are properly interparticle forces: They act between colloids either to attract them or to repel them.

The colloids in a stable soil suspension can be envisioned, at least in an ideal geometric sense, to be roughly spherical (humus and metal oxides) or to comprise one or more unit layers stacked together (clay minerals). If spherical colloids are large relative to the thickness of their diffuse swarm of adsorbed ions or, in the case of clay minerals, if layer stacking is not extensive or highly irregular, one can imagine the forces the particles exert on one another as coming from interacting planar surfaces. Interparticle forces can be discussed in detail on the basis of this geometric simplification.

The van der Waals interaction between soil colloids is exactly analogous to that between soil humus and organic polymers or clay minerals (sections 3.4 and 3.5). Over a time interval that is much longer than 10^{-16} s, the distribution of electronic charge in a nonpolar molecule is spherical. However, on a timescale $\leq 10^{-16}$ s (approximately the period of an ultraviolet light wave), the charge distribution of a nonpolar molecule will exhibit significant deviations from spherical symmetry, taking on a flickering, dipolar character. These deviations fluctuate rapidly enough to average to zero when observed over, say, 10^{-14} s (the period of an infrared light wave), but they persist long enough to attract or repel and, therefore, induce distortions in the charge distributions of neighboring molecules. If two nonpolar molecules are brought close together, each will induce in the other a fluctuating dipolar character and the *correlations* between these induced dipole charge distributions will not average to zero, even though the individual dipole distributions themselves will average to zero. The correlations between the two instantaneous dipole moments produces an attractive interaction with a potential energy that is proportional to the inverse sixth power of the distance of separation. The resulting attractive force is known as the *van der Waals dispersion force.* At small values of the separation distance, this interaction can be strong enough to cause particles to coalesce.

Suppose that a nonpolar molecule confronts the planar surface of a solid. The van der Waals dispersion energy for the attractive interaction between the single molecule and the planar solid surface can be shown to vary as the inverse third power of their distance apart. The inverse power is smaller than six because of the additive effect of van der Waals forces between the many atoms of the solid and the nonpolar molecule. Now suppose that, instead of a single nonpolar molecule, a solid surface comprising nonpolar molecules confronts the planar surface. A calculation of the van der Waals energy per unit area of surface then gives the equation

$$\text{van der Waals energy} = -\frac{\mathrm{A}}{12\pi d^2} \tag{10.11}$$

where d is the distance separating the planar solid surfaces and A is called the *Hamaker constant.* Equation 10.11 shows that the van der Waals dispersion energy (per unit area) falls off as the inverse square of the distance separating two opposing planar surfaces. Because it is additive for the many molecules in the two solids, this *attractive* interaction (hence the negative sign in Eq. 10.11) decreases with distance of separation much more slowly than the interaction between an isolated pair of molecules. The Hamaker constant, which gives a measure of the magnitude of the van der Waals energy at any separation distance, has a value near 2×10^{-20} J for soil colloids.

The van der Waals interaction is the cause of particle coalescence after a collision induced by Brownian motion, stirring, or settling, as described in

Section 10.1. If the particles each bear a net charge, however, their tendency to collide and stick together is strongly affected by an electrostatic force between them. A repulsive force arises if the sign of the charge on each particle is the same; otherwise, an attractive force arises that adds to the van der Waals attraction in promoting coalescence. The repulsive electrostatic interaction can be shown to be proportional to the product of three factors: the exclusion volume (Problem 9 in Chapter 8; see also sections 8.4 and 9.5), the bulk aqueous solution concentration of ions that *screen* the particle charge (Section 7.2), and an exponentially decreasing function of *d*. Expressed per unit area of particle surface and in the notation of the model equation for V_{ex} that was applied to monovalent ions in a diffuse swarm in Problem 9 of Chapter 8 and in Section 9.5,

$$\text{electrostatic repulsion energy} \propto \frac{c_o}{\kappa} \exp(-\kappa d) \tag{10.12}$$

where

$$\kappa \equiv (\beta c_o)^{1/2} = \left(2F^2 c_o / \varepsilon_o DRT\right)^{1/2} \tag{10.13}$$

is a *diffuse-swarm screening parameter* with an inverse that has the dimensions of length. In Eq. 10.13, F is the Faraday constant, ε_o is the permittivity of vacuum (see the Appendix), D is the dielectric constant of water (78.3 at 25 °C), R is the molar gas constant (as in Problem 9 of Chapter 1), and T is absolute temperature. The parameter κ determines both the spatial extent of V_{ex} [the average volume of aqueous solution that encompasses the (monovalent) ions in the diffuse swarm that screen the particle charge (Section 9.5)] and the spatial decay of the electrostatic repulsion energy as epitomized in the exponential factor appearing in Eq. 10.13. The value of κ^{-1} at 25 °C ranges from 1 to 30 nm as c_o ranges from 0.1 to 100 mol m^{-3}, which is typical of soil solutions. Thus, *nanometer length scales characterize the region around charged colloids in which the screening of particle charge by electrolyte ions is effective.*

As c_o is increased, it follows from eqs. 10.12 and 10.13 that the electrostatic repulsion between charged colloids will drop off ever more rapidly near the colloids, allowing them to approach more closely until they come under the dominating influence of the attractive van der Waals energy and coalesce. The smallest concentration of electrolyte, in moles per cubic meter, at which a soil colloidal suspension becomes unstable and begins to undergo perikinetic flocculation is called the *critical coagulation concentration (ccc)*. The value of the *ccc* will, in general, depend on the nature of the participating colloids, the composition of the aqueous solution in which they are suspended, and the time allowed for settling. A simple laboratory measurement of the *ccc* entails the preparation of dilute ($<$ 3 kg m^{-3} solids concentration) suspensions in a series of electrolyte solutions of increasing concentration. After about 1 hour of shaking and a subsequent standing period of 2 to 24 hours, the suspensions that have become unstable will show a clear boundary separating the settled solid mass from an aqueous solution phase, and the *ccc* can be bracketed between two values determined by the largest electrolyte concentration at which apparent flocculation did not occur and the smallest at which it did.

The study of soil colloidal stability has not yet produced exact mechanistic theories, but nonetheless, general relationships between stability, interparticle forces, and surface chemistry have been developed that are of predictive value. One of these relationships is the *Schulze–Hardy Rule.* This empirical generality concerning the *ccc*, first suggested by H. Schulze and generalized by W. B. Hardy more than a century ago, can be stated as follows:

> *The critical coagulation concentration for a colloid suspended in an aqueous electrolyte solution is determined by the ions with charge opposite in sign to that on the colloid (counterions) and is proportional to an inverse power of the valence of the ions.*

Published studies indicate that *ccc* values for monovalent counterions generally lie in the relatively narrow range of 5 to 100 mol m^{-3}, whereas those for bivalent counterions typically lie in the range of 0.1 to 2.0 mol m^{-3}, if the coalescing particles are inorganic, or largely so. This order-of-magnitude difference in *ccc* between monovalent and bivalent ions illustrates the Schulze–Hardy Rule qualitatively.

The relationship between *ccc* and counterion valence can be interpreted quantitatively on the basis of a simple consideration of length scales in the diffuse-ion swarm. Adsorption of counterions in the diffuse swarm acts to *screen* particle charge, in that the range of the electrostatic repulsion energy created by a charged particle is diminished by this adsorption. It is reasonable to assign to each counterion of charge Ze [e = protonic charge = F/N_A, where N_A is the Avogadro constant (see the Appendix)], a surface patch of equal and opposite charge that it screens. If this screening is to be effective, the coulomb potential energy restricting the counterion to the vicinity of the charged surface patch should be of the same order of magnitude in absolute value as the thermal kinetic energy of the counterion, $\frac{1}{2}k_BT$, where k_B is the Boltzmann constant (as in Eq. 10.1) and T is absolute temperature:

$$\frac{(Ze)^2}{4\pi\varepsilon_0 D L_S} \approx \frac{1}{2}k_B T \qquad (10.14)$$

The parameter L_s characterizes a *thermal screening length* with a magnitude that depends only on physical variables. Equation 10.14 can be rewritten in terms of the diffuse-swarm parameter β introduced in Problem 9 of Chapter 8 and defined in Eq. 10.13:

$$L_s \approx Z^2(\beta/4\pi N_A) \qquad (10.15)$$

At 25 °C, $\beta/4\pi N_A = 1.43$ nm, once again demonstrating that nanometer length scales are relevant to charge screening.

Effective charge screening by a counterion in the diffuse swarm must mean that the screening length for the electrostatic repulsion energy (Eq. 10.12) is comparable with L_s in Eq. 10.15. Otherwise, the influence of particle charge would "leak out" beyond the region of bound counterions. The screening length is, of course, κ^{-1}, where now we set $\kappa \equiv Z(\beta c_o)^{\frac{1}{2}}$, with the valence of the screening counterion added to the definition given in Eq. 10.13. Therefore, effective screening requires the constraint

$$\kappa L \approx 1 \tag{10.16}$$

from which an equation for the *ccc* can be derived by introducing Eq. 10.15 and the revised definition of κ:

$$ccc \approx \left(16\pi^2 N_A^2/\beta^3\right) Z^{-6} \tag{10.17}$$

The prefactor in Eq. 10.17 is equal to 45 mol m^{-3} at 25 °C. Given the typical range of *ccc* values for monovalent counterions (5–100 mol m^{-3}), the estimate of *ccc* provided by Eq. 10.17 for $Z = 1$ is reasonable. For bivalent counterions, the corresponding estimate of *ccc* is 0.7 mol m^{-3}, which also lies within the range of typical values: 0.1 to 2 mol m^{-3}. Equation 10.17 is a quantitative expression of the Schulze–Hardy Rule, which reveals it to be a manifestation of effective charge screening—a condition induced by decreasing the range of the electrostatic repulsion energy generated by a charged colloid until it is small enough to make the resultant coulomb attraction of a counterion toward the particle strong enough to quench the thermal kinetic energy that otherwise would send the counterion wandering off.

10.4 The Stability Ratio

Models of the second-order rate coefficient for transport-controlled flocculation, presented in Section 10.1 (eqs. 10.3, 10.6, and 10.8), show only a dependence on physical parameters, such as absolute temperature, fluid properties, and colloid size. These models evidently do not depend on chemical variables, such as background electrolyte concentration or pH, even though these latter variables must affect the flocculation of soil colloids, as the very definition of *ccc* plainly demonstrates. Under conditions in which *rapid* perikinetic flocculation is occurring, direct measurements of the rate of flocculation lead to k_p values in the range 1 to 4×10^{-18} $m^3\ s^{-1}$ at 25 °C for suspensions of synthetic mineral or organic colloids. This range of values is close to, but systematically smaller than $k_p^{SE} = 6.2 \times 10^{-18}$ $m^3\ s^{-1}$, as obtained from Eq. 10.4 applied at 25°C. Thus the Stokes–Einstein model of flocculation is generally consistent with observations of rapid perikinetic flocculation. The residual discrepancy between theory and experiment can be explained quantitatively in terms of the detailed mechanics of dimer formation: The dimer radius R_{11} is less than twice the monomer radius, and the relative diffusion coefficient D_{11} is less than twice the monomer diffusion coefficient, because of fluid mechanical effects that occur when two monomers are brought close together. These effects suffice to reduce k_p in Eq. 10.2 by the required factor of two to five relative to k_p^{SE} in Eq. 10.3.

Rapid perikinetic flocculation occurs at the *ccc*, and it is under this condition that Eq. 10.3 provides a useful model of the rate coefficient to be introduced into the von Smoluchowski rate law (Eq. 10.9). If the concentration of flocculating electrolyte is smaller than the *ccc*, the value of the second-order rate coefficient for flocculation also is found to be smaller (as would be expected from consideration of the method for measuring the *ccc* described in Section 10.3). Figure 10.4 shows the change in the ratio $k([KCl] = 80 \text{ mmol kg}^{-1})/k$ caused by an increase in electrolyte concentration for synthetic hematite colloids suspended in KCl solution at pH 6 and 25 °C. No change in the rate coefficient k for flocculation was observed at KCl concentrations above 80 mmol kg^{-1}; hence, all rate coefficients were normalized to this molal concentration. The ratio $k([KCl] = 80 \text{ mmol kg}^{-1})/k$ displays a gradual decline toward unit value as [KCl] is increased by more than an order of magnitude. To the extent that KCl behaves as an *indifferent electrolyte* (see Sections 7.4 and 9.1) in respect to hematite, this decline can be interpreted as an effect of electrolyte concentration on the diffuse-ion swarm, particularly for Cl^-, which should be the screening ion, given the positive surface charge on hematite expected at pH 6 (Section 7.4). Corresponding to the increase of concentration in Figure 10.3 there is a decrease by a factor of three in the screening length scale κ^{-1} (Eq. 10.13). This weakening of the

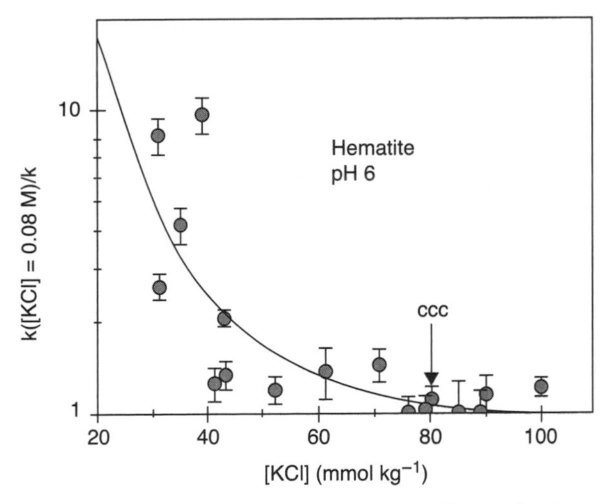

Figure 10.4. Dependence of the second-order rate coefficient for the coalescence of hematite colloids suspended in KCl solution at pH 6($\sigma_p > 0$) on KCl concentration. The value of the critical coagulation concentration (*ccc*) for the suspension is indicated by the arrow. Data from Chorover, J., J. Zhang, M.K. Amistadi, and J. Buffle (1997) Comparison of hematite coagulation by charge screening and phosphate adsorption. *Clays Clay Miner.* **45**: 690–708.

electrostatic repulsion energy (Eq. 10.12) near a hematite colloid is sufficient to provoke an increase in the rate coefficient for particle coalescence by an order of magnitude. Thus, Figure 10.4 is a graphic portrayal of the transition from surface reaction control to transport control in the flocculation of hematite colloids.

The quantity plotted against concentration in Figure 10.4 is termed the *stability ratio:*

$$W \equiv \frac{\text{initial rate of rapid flocculation}}{\text{initial rate of flocculation observed}} \tag{10.18}$$

where *rate* is interpreted as in Eq. 10.9 with ρ set equal to ρ_o. Given Eq. 10.10, W is also the ratio of a characteristic timescale for coalescence observed under prescribed conditions to that observed under conditions producing rapid flocculation. These latter conditions are associated with the value of a chemical variable, the electrolyte concentration, leading to an alternative definition of *ccc:*

$$\lim_{c \uparrow ccc} W = 1 \tag{10.19}$$

where c is the concentration of the counterion in an indifferent electrolyte.

Charge screening that induces rapid flocculation is the result of weak interactions between a diffuse swarm of adsorptive counterions and a charged particle surface (Section 10.3). Attractive van der Waals interactions, which are always present, then act to cause coalescence of the colloids. The mechanism by which rapid flocculation is brought on is not unique, however, because the repulsive coulomb interaction between two colloids having the same charge sign also can be weakened by *charge neutralization.* This mode of inducing rapid flocculation is the result of strong interactions between adsorptive counterions and a charged particle surface [i.e., those typically associated with specific adsorption processes (see Section 7.2 and Chapter 8)]. These include protonation and the inner-sphere surface complexation of metal cations or inorganic/organic anions.

Figure 10.5 illustrates the effect of pH on the second-order rate coefficient k for synthetic hematite colloids suspended in $NaNO_3$ solution at 25 °C. Rapid coalescence was observed at any pH value when [NaCl] = 100 mol m^{-3}, the value of k being 1.8×10^{-18} m^3 s^{-1}, which is close to that predicted by Eq. 10.4. At lower [NaCl], an effect of pH is apparent, with the value of k decreasing by up to three orders of magnitude as pH is varied below or above approximately 9.2. The graph in Figure 10.5 is essentially a plot of $-\log W$ versus pH, because $\log k = -\log W + \log k([NaNO_3] = 0.1M)$ according to the definition of W in Eq. 10.18. The value of pH required to produce a maximum value of k and, therefore, W = 1.0 is termed the *point of zero charge* (p.z.c.):

$$\lim_{pH \to p.z.c.} \log W = 0 \tag{10.20}$$

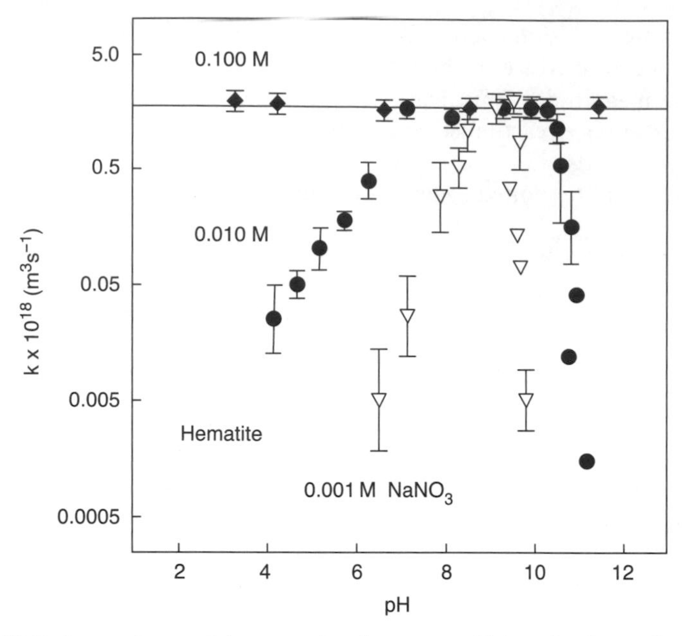

Figure 10.5. Dependence of the second-order rate coefficient for the coalescence of hematite colloids suspended in NaCl solution on pH at three electrolyte concentrations. The value of the *ccc* for the suspension is indicated by a horizontal solid line. Data from Schudel, M. *et al.* Absolute aggregation rate constants of hematite particles in aqueous suspensions. *J. Colloid Interface Sci.* **196**:241–253.

Direct measurement of p.z.c. (as p.z.s.e.) indicated that p.z.c. ≈ 9.2 (Section 7.4). To the extent that the condition $\sigma_p = 0$ is represented by Eq. 10.20, the terminology it introduces and that introduced in Section 7.4 are mutually consistent. Note that charge screening by the diffuse-ion swarm is *not* possible when pH = p.z.c. (see Eq. 7.5).

Unlike the behavior of W in response to increases in the concentration of an indifferent electrolyte (Fig. 10.4), W typically exhibits two branches as pH is increased from below to above the p.z.c. The "hairpin" – log W versus pH plot in Figure 10.5 broadens considerably as the concentration of the electrolyte is increased. This straightening-out effect can be understood as a synergism between pH and [$NaNO_3$] taken as controlling chemical variables. The "hairpin" is bent out into a straight horizontal line as [$NaNO_3$] approaches the *ccc* because charge screening is contributing more and more to the production of rapid flocculation at any pH value. The asymmetry of the hairpin about the p.z.c. signals the existence of physical factors controlling W instead of electrolyte concentration and pH (e.g., particle surface and morphological heterogeneities). The fact that there is a hairpin at all in Figure 10.5 derives from the change of σ_p from positive to negative as pH is increased and passes through the p.z.c. Coulomb repulsion will promote surface reaction control of flocculation irrespective of the sign of σ_p. Note that *anions become the counterions causing flocculation at pH < p.z.c., whereas cations become the counterions at pH > p.z.c.*

Figure 10.6 shows the effect of the strongly adsorbing anion, $H_2PO_4^-$, on the stability ratio of synthetic hematite colloids suspended in 1 mol m^{-3} KCl solution at pH 6 at 25 °C. The plot of log W versus log $[H_2PO_4^-]$ displays a characteristic "inverted hairpin" shape that signals particle charge reversal when log $[H_2PO_4^-] \approx -4.5$, under the experimental conditions selected ($\rho_o \approx 2.610^{16} m^{-3}$). This value of log $[H_2PO_4^-]$ is termed *the p.z.c. with respect to* $H_2PO_4^-$:

$$\lim_{\log[] \to \text{p.z.c.}} \log W = 0 \tag{10.21}$$

Note that *p.z.c. with respect to a strongly adsorbing ion is a negative quantity, whereas p.z.c. with respect to protons is a positive quantity.* Thus p.z.c. with respect to $H_2PO_4^-$ is −4.5.

Figure 10.7 illustrates the effect of the strongly adsorbing organic anions $CH_3(CH_2)_n COO^-$ (n = 2, 7, 9, 11) on the stability ratio for synthetic hematite colloids suspended in 50 mol m^{-3} NaCl at pH 5.2. The log–log plots indicate p.z.c. values in the range −5 to −3, and their inverted hairpin shape exhibits asymmetry about the p.z.c. similar to what is apparent in Figure 10.6. The p.z.c. value decreases as the number of C atoms in the anion increases, suggesting that hydrophobic interactions may play a role in promoting rapid flocculation.

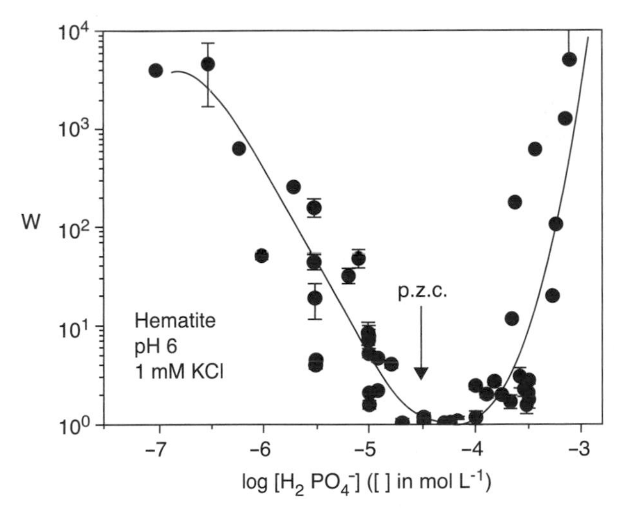

Figure 10.6. Dependence of the stability ratio for hematite colloids suspended in 1 mM KCL at pH 6 on the concentration of $H_2PO_4^-$ added to the suspension to induce flocculation. The value of the point of zero charge (p.z.c.) with respect to $H_2PO_4^-$ is indicated by the arrow. Data from Chorover, J. *op. cit.*

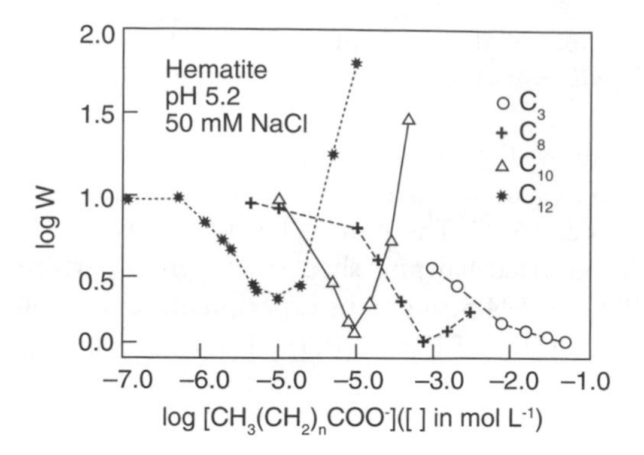

Figure 10.7. Dependence of the stability ratio for hematite colloids suspended in 50 mM NaCl at pH 5.2 on the concentration of aliphatic acid anions [$CH_3(CH_2)_nCOO^-$, $n = 2$, propionate; n = 7, caprylate; n = 9, caprate; n = 11, laurate] added to the suspension to induce flocculation. The minimum in each plot indicates the p.z.c. with respect to a given aliphatic anion. Note that rapid flocculation occurs at a NaCl concentration smaller than 100 mM, the nominal *ccc*. Data courtesy of Dr. J. J. Morgan

This result implies that humic substances also should be effective at promoting rapid flocculation of hematite colloids, as is indeed observed experimentally. Evidently a polymeric organic anion provokes rapid flocculation of positively charged colloids when strongly adsorbed at low concentrations.

Figures 10.5 to 10.7 show collectively that, in the presence of surface complex-forming ions, σ_p is the determining colloidal property for rapid flocculation. The existence of ions that can form surface complexes significantly can be detected by examining *ccc* as a function of the initial colloid concentration. If charge screening is the principal cause of flocculation, the *ccc* will be essentially independent of colloid concentration—at least over a severalfold change—whereas if surface complexation is the principal cause, the *ccc* will tend to increase with colloid concentration because the surface complexation capacity is also increased. If the surface-complexing ion is multivalent, like PO_4^{3-} or Al^{3+}, its strong adsorption can result in a reversal of the sign of σ_p. When this happens, ions that previously were of the same charge sign as the colloidal particles now become the flocculating ions. The mechanism of any flocculation induced by these ions can be either charge screening or strong adsorption.

When polymer ions (e.g., Al-hydroxy polymers or humus) form surface complexes with soil colloids, stability depends on surface charge density. If the extent of polymer adsorption is small, a soil colloidal suspension can become flocculated at a lower concentration of indifferent electrolyte (like NaCl) than in the absence of the polymer (Fig. 10.7)! In this situation, the addition of electrolyte brings the now less-repelling colloidal particles closer together until flocculation can occur at lower electrolyte concentrations than in the absence

Table 10.1
Chemical factors that affect the stability of soil colloidal suspensions.

Chemical factor	Affects	Promotes flocculation when	Promotes stability when
Electrolyte concentration	Charge screening	Increased	Decreased
pH value	σ_p	pH = p.z.c.	pH $\neq$ p.z.c.
Adsorption of small ions	σ_p	$\sigma_p = 0$	$\sigma_p \neq 0$
Adsorption of polymer ions	σ_p	$\sigma_p = 0$	$\sigma_p \neq 0$

of the polymer. Alternatively, the colloidal suspension may be stabilized by the repulsive electrostatic force between coatings of adsorbed polymers. Which phenomenon occurs depends on the pH value, the electrolyte concentration, and the configuration of adsorbed polymer ions.

The principal surface chemical factors that determine the stability of soil colloidal suspensions are summarized in Table 10.1, with conditions that lead to $W = 1$ listed in the third column. Particle surface chemistry universally affects colloid stability through changes in the strength of the repulsive electrostatic force. Rapid flocculation is the result of a reduction in this repulsive electrostatic force, whether through charge screening or surface complexation.

10.5 Fractal Floccules

Electron micrographs of floccules comprising specimen oxide minerals have been examined to determine the relationship between the number of *primary particles* N they contain and their spatial extent as expressed by some length scale, L. For example, L can be estimated by the geometric mean value of the longest linear dimension of a floccule and the dimension that is perpendicular to the axis of the former. Log–log plots of N versus L based on the examination of many floccules show that a linear relationship typically obtains, implying the power law:

$$N = AL^D \tag{10.22}$$

where A and D are positive parameters. This kind of power–law relationship between floccule primary particle number and length scale has also been observed in computer simulations of transport-controlled flocculation. In this case, the simulated value of D is 1.78 ± 0.04.

The power–law relationship in Eq. 10.22 has implications for measurements of floccule size and dimension during the flocculation process itself. If the principal contributor to floccule growth is encountered between particles of comparable size, the increase in N per encounter will be equal approximately to N itself, and if colloid diffusion is the cause of these encounters, the kinetics of flocculation will be described by a second-order rate law:

$$\frac{dN}{dt} \approx \frac{\text{increase in N per encounter}}{\text{timescale for encounter}} \approx NK_{SE}\rho = K_{SE}\rho_0 \qquad (10.23)$$

where K_{SE} is defined in Eq. 10.3 and $\rho_0 = N\rho$ is the initial number density, as in Eq. 10.10. Equation 10.23 implies

$$N(t) = N(0) + K_{SE}\rho_0 t \sim K_{SE}\rho_0 t \quad [t >> N(0)/K_{SE}\rho_0] \qquad (10.24)$$

Taken together, Eqs. 10.22 and 10.24 yield a relationship between floccule length scale and time:

$$L(t) \sim (K_{SE}\rho_0/A)^{1/D} \qquad t^{1/D}[t >> N(0)/K_{SE}\rho_0] \qquad (10.25)$$

Figure 10.8 is a log–log plot of the average floccule diameter (as measured by light-scattering techniques) versus time during the rapid flocculation of hematite colloids by Cl^- or $H_2PO_4^-$ (Figs. 10.4 and 10.6). The good linearity

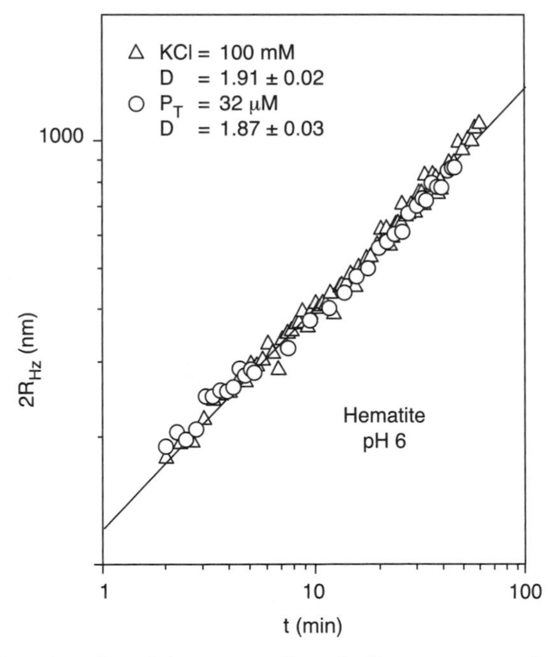

Figure 10.8. Log–log plot of the average floccule diameter versus time for hematite colloids suspended in either 100 mM KCL solution or 32μM KH_2PO_4 solution at pH 6. Floccule growth with time during rapid coagulation is the same in both suspensions, leading to the same mass fractal dimension (D) for the floccules. Data from Chorover, J. *op. cit.*

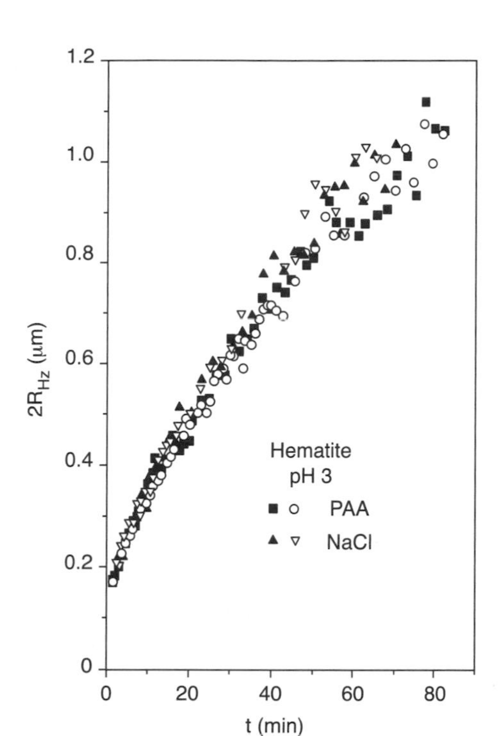

Figure 10.9. Plot of the average floccule diameter versus time for hematite colloids suspended in either NaCl solution at the *ccc* or polyacrylate (PAA) solution at the p.z.c. Floccule growth with time during rapid coagulation is essentially the same in both suspensions. Data from Ferretti, R., J. Zhang, and J. Buffle (1997) Kinetics of hematite aggregation by polyacrylic acid. *Colloids and Surfaces* **121A**: 203–215.

of the plot confirms Eq. 10.25, whereas the excellent superposability of the data shows the similar nature of the flocculation process at either the *ccc* or the p.z.c. with respect to $H_2PO_4^-$. The two resulting values of the exponent D agree within experimental precision. Comparable results have been reported for silica and goethite colloids flocculating in 1:1 electrolyte solutions. Floccule growth produced by rapid coalescence induced by the polymeric anion polyacrylate (PAA, a polymer of acrylate, $CH_2 = CHCOO^-$), is illustrated and compared with that caused by Cl^- for hematite colloids at pH 3 in Figure 10.9. The power–law shape of the time dependence of floccule diameter is apparent, as is the expected independence from the type of coagulating anion of the fractal dimension calculated with these data: $D = 1.88 \pm 0.02$.

The power–law relation in Eq. 10.22 can be interpreted physically as the characteristic of a *mass fractal.* The exponent D is then termed the *mass fractal dimension.* Some basic concepts about mass fractals are introduced in Special Topic 6 at the end of this chapter. Suffice it to say that Eq. 10.22 is a generalization of the geometric relation between the *number* of primary particles in a d-dimensional ($d = 1, 2$, or 3) floccule and its d-dimensional *size.* For example, imagine a one-dimensional floccule portrayed as a straight chain of circular primary particles, each of diameter L_0. The number of particles in a chain of length L is then

$$N(L) = L/L_0 \equiv AL \tag{10.26}$$

where $A \equiv 1/L_0$ in this case. Equation 10.26 has the appearance of Eq. 10.22, but with $D = 1$. If the floccule is two-dimensional, it can be represented by a parquet of circular primary particles packed together so that they touch. The number of particles in the cluster will then be

$$N(L) = g(L/L_0)^2 \equiv AL^2 \tag{10.27}$$

where now $A \equiv g/L_0^2$ and g is a geometric factor with a value that depends on exactly how the circular primary particles have been packed. In this case, Eq. 10.22 is recovered if $D = 2$. Evidently, Eq. 10.22 with $1 < D < 2$ represents a floccule with fractal dimension D with a structure that is intermediate between that of a chain and that of a parquet. If the floccule is a highly convoluted chain of particles, one that winds about in space but does not fill it, then it is reasonable to suppose that the size–dimension relation in Eq. 10.22 could describe it with a noninteger value of D. Its ability to fill the plane of view, irrespective of its shape, is quantified by how closely D approaches the value 2.0.

Experimental measurements of the fractal dimension of floccules formed by the rapid coalescence of specimen mineral colloids (oxides and clay minerals) typically fall in the range 1.2. to 1.9. These experimental values of D are comparable with 1.7 to 1.8, the range of fractal dimension calculated for floccules formed in a computer simulation in which colloids are permitted to diffuse randomly with a Stokes–Einstein diffusion coefficient (Eq. 10.1) until they collide, after which they coalesce instantly. The results also are comparable with the range of D values inferred for atmospheric aerosols: 1.7 to 1.8. Thus, the available data and calculations indicate that rapid coalescence leads to floccules with a size that is a power–law function of time and with a fractal dimension that lies in a narrow interval around 1.75.

Floccules formed under reaction control also are found to be mass fractals. Figure 10.10 illustrates this fact for the hematite suspensions with a stability ratio that is plotted against [KCl] in Figure 10.4. The values of D are seen to decrease from about 2.1 to near 1.7 as W declines by an order of magnitude. This decrease in the fractal dimension implies the concurrent development of floccules with a space-filling nature that decreases. Denser floccules formed more slowly permit time for colloids to seek out pathways of coalescence leading to more compact structures. A variety of studies has found that the fractal dimension of floccules formed under reaction control lies in the range 1.9 to 2.1. These values are in agreement with computer simulations of coalescence in which floccules with $D \approx 2.0$ to 2.1 are formed after assigning a very small probability to coalescence after the collision of two particles.

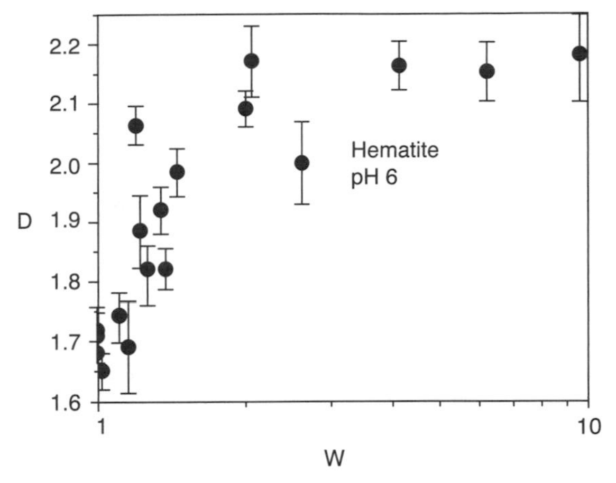

Figure 10.10. Dependence of the mass fractal dimension (D) on the stability ratio (W) for the hematite suspensions with flocculation kinetics behavior shown in Figure 10.4. Data from Chorover, J. *op. cit.*

Measurements of the fractal dimension of floccules formed in the presence of low concentrations of PAA, less than those required for rapid coalescence (Fig. 10.9), lead to $D \approx 1.9$ to 2.1 as well, with the correspondingly denser floccule structure then confirmed in electron micrographs. Thus, for PAA, a transition of D from values near 2.1 to values near 1.8 occurs as the concentration of the polymeric anion increases, in parallel with the trend in Figure 10.10.

As floccules go through drying and rewetting cycles to form aggregate structures (Section 10.2), it is possible that they may retain their mass fractal nature. An abundant body of literature now shows that this, indeed, is the case. Perhaps the most direct experimental demonstration of the fractal nature of soil aggregates is that based on the dependence of their bulk density on aggregate size. Bulk density (ρ_b) is defined by the equation

$$\rho_b \equiv \frac{\text{mass of solids}}{\text{volume}} \tag{10.28a}$$

where the numerator is the (dry) mass of the solid framework of an aggregate and the denominator is its total volume, including pore space. If the aggregate is a mass fractal, Eq. 10.22 applies and Eq. 10.28a becomes

$$\rho_b = \frac{BL^D}{\text{volume}} \quad (0 < D < 3) \tag{10.28b}$$

where B is a positive parameter that includes the density of the primary particles out of which the aggregate is built. Because the total volume of an aggregate characterized by the length L is proportional to L^3, Eq. 10.28b reduces

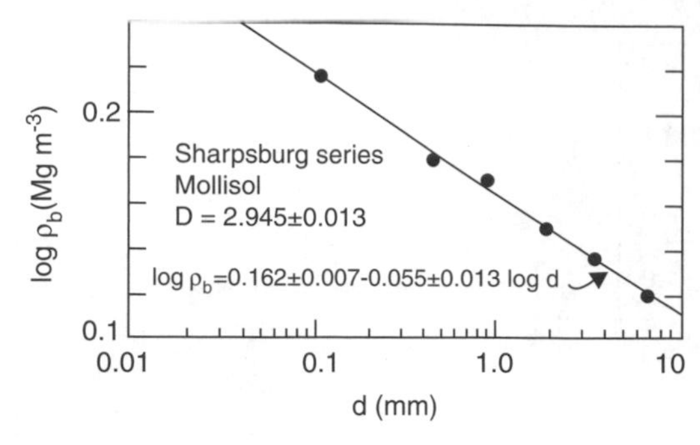

Figure 10.11. Dependence of the dry bulk density of natural aggregates in the Sharpsburg soil (fine, montmorillonitic, mesic Typic Argiudoll) on the average aggregate diameter, showing behavior expected for a mass fractal with fractal dimension D = 2.95. Data from Rieu, M., and G. Sposito (1991) Fractal fragmentation, soil porosity, and soil water properties: II. Applications. *Soil Sci. Soc. Am. J.* **55**:1239–1244.

to the power–law proportionality

$$\rho_b \propto L^{D-3} \qquad (10.28c)$$

Equation 10.28c shows that the bulk density of a mass fractal aggregate decreases with increasing size, because $D < 3$. [If $D = 3$, then the aggregate is not a fractal object, Eq. 10.28b becomes analogous to Eq. 10.27, and ρ_b is independent of aggregate size.] Equation 10.28c is illustrated in Figure 10.11 for a Mollisol with a clay fraction that is high in montmorillonite. Over the range of aggregate diameters between 0.05 and 7 mm, the log–log plot of ρ_b versus d (aggregate diameter) is linear with slope -0.055 ± 0.013 (i.e., $D = 2.95$). This larger value of the fractal dimension, typical of clayey aggregates, indicates their more space-filling structure.

For Further Reading

Baveye, P., J.- Y. Parlange, and B. A. Stewart (eds.). (1998) *Fractals in soil science.* CRC Press, Boca Raton, FL. The first and sixth articles in this compendium volume present comprehensive introductions to the applications of fractal concepts to soil aggregates.

Buffle, J., K. J. Wilkinson, S. Stoll, M. Filella, and J. Zhang. (1998) A generalized description of aquatic colloidal interactions. *Environ. Sci. Technol.* 32:2887–2899. A fine review of the structure and formation of floccules by colloids in natural waters.

Hunter, R. J. (2001) *Foundations of colloid science.* Oxford University Press, New York. A comprehensive standard textbook on colloid chemistry.

Sposito, G. (1994). *Chemical equilibria and kinetics in soils.* Oxford University Press, New York. Chapter 6 of this advanced textbook provides a discussion of the kinetics of flocculation, including fractal aspects.

Sposito, G. (2004) *The surface chemistry of natural particles.* Oxford University Press, New York. Chapter 5 of this advanced textbook contains a detailed description of flocculation and the light-scattering techniques used to quantify both floccule formation and structure.

Wilkinson, K. J., and J. Lead (eds.). (2007) *Environmental colloids and particles: Behaviour, separation, and characterisation.* Wiley, New York. The 13 chapters of this comprehensive edited monograph provide broad surveys of the current status of understanding colloidal structure and formation in natural waters as well as emerging experimental methodologies for exploring them.

Problems

The more difficult problems are indicated by an asterisk.

1. Calculate the diffusion coefficient of a soil colloid with radius 1μm that moves through water at 25°C. According to an analysis by Albert Einstein, the time required for a colloid to diffuse a distance Δx is $2(\Delta x)^2/3D$. Estimate the time required for the soil colloid to diffuse 10 μm and compare the result with the time required by an ion (see Special Topic 3 in Chapter 3 for a typical value of an ion diffusion coefficient).

2. A suspension consists of disk-shaped particles 1μm × 1μm × 10 nm, each with a mass density of 2.5×10^3 kg m^{-3}. Calculate the half-life for perikinetic flocculation at 25 °C in a quiescent suspension with an initial solids concentration of 1 kg m^{-3}.

3. The data in the table presented here give the number density in a kaolinite suspension during perikinetic flocculation. Calculate the half-life for flocculation.

$\rho \times 10^{-14}$ (m^{-3})	Time (s)	$\rho \times 10^{-14}$ (m^{-3})	Time (s)
5.00	0	2.52	335
3.90	105	2.00	420
3.18	180	1.92	510
2.92	255	1.75	600

4. Equation 10.9 applies to a suspension that initially contains colloids with the same radius R (i.e., a *monodisperse* suspension). If, instead, colloids of radii R_1 and R_2 are present initially (i.e., a *polydisperse* suspension), the

equation changes to have the form

$$\frac{d\rho}{dt} = -8\pi D_1 \bar{R} \rho^2$$

where

$$\bar{R} \equiv (R_1 + R_2)^2 / 4R_2$$

and D_1 is the Stokes–Einstein diffusion coefficient of a colloid with radius R_1. Calculate the rate coefficient at 25 °C for the flocculation of a mixture of 1 μm and 10 μm colloids, then compare the result with the rate coefficient for a monodisperse suspension of 1 μm colloids. Note the enhancement of the flocculation rate in the presence of the 10 μm colloids.

*5. Suspensions of Ca-montmorillonite in chloride solutions show an anion exclusion volume (see Section 8.4 and Problem 9 in Chapter 8) that decreases to a limiting value near $3 \times 10^{-4} m^3 kg^{-1}$ as the chloride concentration increases. Show that this limiting value implies complete exclusion of chloride from a region between stacked layers with the opposing siloxane surfaces separated by 1 nm. (See Section 2.3 for an estimate of the specific surface area of the clay mineral.)

6. Shown in the table presented here are values of W measured at pH 10.5 for a hematite suspension in the presence of varying concentrations of either NaCl or $CaCl_2$. Determine the *ccc* value in each electrolyte solution. Why are they different? How well do they conform to the Schulze–Hardy Rule (Eq. 10.17)?

W	[NaCl] ($mol\ m^{-3}$)	W	[$CaCl_2$] ($mol\ m^{-3}$)
15.0	13	213	0.1
1.9	20	73	0.2
1.1	48	5.8	0.4
		1.3	0.8

*7. A sample of slightly acidic soil (pH 6.4) with the exchangeable cation composition $E_{Na} = 0.20$ and $E_K + E_{Ca} + E_{Mg} = 0.75$ was found to disperse completely in water, whereas another sample of the soil taken from elsewhere in the profile (pH 5.0, $E_{Na} = 0.24$, $E_K + K_{Ca} + E_{Mg} = 0.61$) did not disperse. Use the concepts discussed in Section 10.4 to provide a chemical explanation for these observations.

*8. The *turbidity* of a colloidal suspension is its spatial *decay parameter* for the transmission of a light beam through it:

$$I_t = I_o \exp(-\tau \ell)$$

where I_o is the intensity of a beam incident on the suspension and I_t is the intensity of the beam after traveling a distance ℓ through the suspension while being scattered by floccules. Thus, τ is *mathematically* analogous to κ in Eq. 10.12, in that the inverses of both parameters are characteristic length scales over which a field phenomenon (electrostatic or electromagnetic) is attenuated. Models of the turbidity relate it to the number of floccules in a suspension, allowing its rate of change with time to serve as a quantitative measure of the rate of flocculation $d\rho/dt$, as in Eq. 10.9. Usually this rate of change is determined as the slope of the *initial* portion of a plot of τ versus time as a suspension flocculates. Use the slope data in the table presented here to estimate the *ccc* for a suspension of hematite colloids in NaCl solution at pH 4.7. Identify the flocculating ion. The value of $\Delta\tau/\Delta t$ corresponding to K_{SE} (*rapid* flocculation, as in Eq. 10.9) is 0.016 s^{-1} for unit length traversed by a light beam in the suspension. (*Hint:* Calculate W, then prepare a log–log plot of W against [NaCl] analogous to that in Fig. 10.6.)

$\Delta\tau/\Delta t$ (10^{-3} s^{-1})	[NaCl] (mol m^{-3})
0.818	48
1.00	60
2.34	72
3.34	84

9. Shown in the following table are values of the stability ratio for the hematite suspension in Problem 8, except that the NaCl concentration is fixed at 5 mM, and 0.1 mg L^{-1} fulvic acid was added prior to turbidity measurements. Determine the p.z.c. value for hematite under these conditions. In the absence of fulvic acid, p.z.c. $\approx$ 8.5. Does the p.z.c. value you determined differ from 8.5? Why or why not?

W	pH	W	pH
19.3	4.05	2.33	5.84
2.71	6.15	802	9.95
6.77	4.80	106	9.22
5.89	5.01	5.13	7.90

*10. A suspension of birnessite (see Section 2.4) colloids at pH 1.40 showed no observable particle migration as the result of an applied electric field. A similar suspension of this Mn oxide mineral in the absence of an applied electric field flocculated at pH 1.55. Explain why the two pH values can be interpreted as giving essentially the same estimate of p.z.c.

11. Shown in the table presented here are values of the stability ratio for the hematite suspension in Problem 9, except that the pH value is fixed at 6.9 and humic acid has been added. Determine a p.z.c. value for hematite flocculation by humic acid.

W	humic acid (μg L^{-1})	W	humic acid (μg L^{-1})
9.24	24.40	53.9	12.20
2.61	48.80	190	146.4
2.30	97.60	7.08	122.0
1181	244.0		

*12. Use concepts discussed in Sections 3.5 and 10.4 to explain the chemical basis for the statement: "Organic matter prevents the dispersion of dry soil aggregates; once the soil particles in the aggregates are forced apart (by shaking in suspension), however, the organic matter helps to stabilize the separated particles in suspension."

*13. The table presented here lists values of W at pH 4 for suspensions of kaolinite particles containing adsorbed fulvic acid. Calculate the *ccc* for each electrolyte solution. Why are the values of the *ccc* different? In your response, consider the Schulze–Hardy Rule and the relevant p.z.n.c. values for kaolinite and humus.

W	[$Ca(NO_3)_2$] (mol m^{-3})	W	[$Cu(NO_3)_2$] (mol m^{-3})	W	[$Pb(NO_3)_2$] (mol m^{-3})
91	0.35	111	0.10	91	0.10
56	0.50	32	0.20	13	0.20
15	0.80	5.5	0.35	2.7	0.35
1.9	1.0	2.0	0.50	2.0	0.50
1.9	1.5	1.9	0.80	1.9	0.80

14. The table presented on the next page shows the average number of silica particles in a floccule of radius R as measured during flocculation. Calculate the fractal dimension of the floccules and determine whether the flocculation process is transport controlled or surface reaction controlled.

R (nm)	N
1.0	30
2.0	120
3.8	500
8.0	2200
10.5	4400

15. The data in the following table show the dependence of aggregate size on bulk density for three soils of differing texture. Use linear regression analysis to estimate the mass fractal dimensions of the aggregates in the soils, including 95% confidence intervals on D.

	Bulk density		
Mean size (mm)	Fine sandy loam ($Mg\ m^{-3}$)	Silt loam ($Mg\ m^{-3}$)	Clay ($Mg\ m^{-3}$)
4.200	1.49	1.42	1.49
1.595	1.58	1.58	1.68
1.025	1.75	1.68	1.70
0.715	1.82	1.61	1.73
0.505	1.94	1.72	1.75
0.335	2.17	1.75	1.80
0.200	2.11	1.82	1.75
0.125	2.15	2.10	1.80

Special Topic 6: Mass Fractals

The term *fractal*, coined by Benoit Mandelbrot 40 years ago from the Latin adjective *fractus* (meaning *broken*), refers to the limiting properties of mathematical objects that exhibit the attributes of *similar structure* over a range of length scales; *intricate structure* which is scale independent; and *irregular structure* which cannot be captured entirely within the purview of classical geometric concepts, use of a spatial dimension that is not an integer.

A flocculation process involves the coalescence of primary particles into floccules. That this process can lead to a mass fractal can be illustrated by constructing clusters from a primary particle comprising five disks, each of diameter d_0 (Fig. 10.12). The primary particle has a diameter equal to $3d_0$. If five of these units are combined to form a cluster with the same symmetry

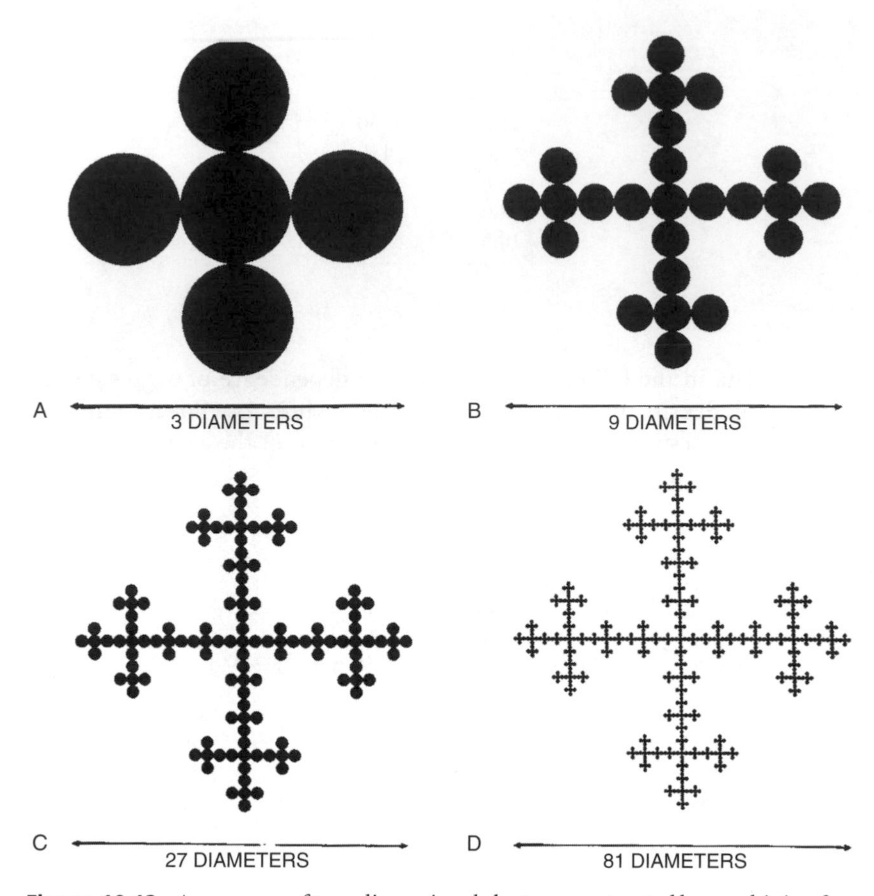

Figure 10.12. A sequence of two-dimensional clusters constructed by combining five-disk clusters (A) in such a way that their inherent symmetry is preserved at each level of combination (B–D). In the limit of infinite cluster size, a mass fractal is formed with fractal dimension $D = \ln 5/\ln 3 = 1.465$.

as a single unit (i.e., each unit in the cluster is arranged like a disk in the unit), the diameter grows to $9d_0$ (Fig. 10.12B). If five of these clusters are then combined in a way that preserves the inherent symmetry (Fig. 10.12C), the diameter increases to $27d_0$. The clusters formed in this process exhibit similar structure, complexity, and irregularity (Fig. 10.12D). Therefore they qualify as fractal objects.

The size of each cluster in the sequence as expressed through the number N of primary particles it contains is 5, 25, 125, and 625 for the four examples shown in Figure 10.12. Thus, $N = 5^n$, where $n = 1, 2, \ldots$, denotes the stage of cluster growth. The diameter $L = 3nd_0$, where $n = 1, 2, \ldots$, once again. The relationship between these two properties—cluster size and cluster dimension—can be expressed mathematically as in Eq. 10.22:

$$N(L) = AL^D \tag{S6.1}$$

where A and D are positive parameters. In the current example, Eq. S6.1 has the form

$$5^n = A(3^n d_0)^D (n = 1, 2, \ldots) \tag{S6.2}$$

Thus, $A = d_0^{-D}$ and

$$D = \ln 5/\ln 3 \approx 1.465 \tag{S6.3}$$

after substitution for A in Eq. S6.2 and solving the resulting expression for the exponent D. More generally, if r denotes the scale factor by which the diameter increases at each successive stage, then

$$D \equiv \lim_{n\uparrow\infty} \frac{1}{n} [\ln N(n)/\ln r] \tag{S6.4}$$

defines the *mass fractal dimension* of a cluster with a size N(n) that is at any stage. For the cluster in Figure 10.12, r = 3 and $N(n) = 5^n$, leading to the fractal dimension given by Eq. S6.3.

The mass fractal dimension of a floccule is a numerical measure of its space-filling nature. The clusters in Figure 10.11 occupy space in a plane (geometric dimension = 2). If they were formed by disks arranged in a single row, the mass fractal dimension that characterizes them should equal 1.0 because they would be effectively one-dimensional objects. On the other hand, if they were compact structures comprising closely packed disks, their mass fractal dimension would equal 2.0, indicating a complete paving of the plane and their space-filling nature. The mass fractal dimension of the clusters actually is near 1.5, meaning that the clusters have a porous structure that is not entirely space filling.

This porous structure of a fractal can be quantified by estimating its number density at any stage of growth. The bulk density (number per unit area) of any cluster in Figure 10.12 can be calculated with the equation

$$\begin{aligned}\rho_n &= N(n)/(\pi L(n)^2/4)\\ &= (4A/\pi)[L(n)]^{D-2} (D < 2)\end{aligned} \tag{S6.5}$$

where Eq. S6.1 has been used, the area occupied by a cluster being $\pi L^2/4$, and $L(n) = 3^n d_0$ is the diameter of the nth cluster. Because $D < 2$, it follows from Eq. S6.5 that ρ_n decreases as n increases. In general, for a mass fractal in E-dimensional space, the number density is given by a power–law expression like Eq. 10.28c:

$$\rho(L) = bA\, L^{D-E} (D < E) \tag{S6.6}$$

where b is an appropriate geometric factor (e.g., b $= 4/\pi$ in Eq. S6.5). Equation S6.6 shows that *bulk density always decreases as size increases.* This can be used as an experimental criterion for evaluating whether a floccule or an aggregate is a mass fractal object, as illustrated in Figure 10.11.

11

Soil Acidity

11.1 Proton Cycling

A soil is acidic if the pH value of the soil solution is less than 7.0. This condition is found in many soils, perhaps half of the arable land worldwide, particularly that under intensive leaching by freshwater, which always contains free protons at concentrations above 1 mmol m^{-3}. Soils of the humid tropics offer examples of acid soils, as do soils of forested regions in the temperate zones of the earth. Soils in peat-producing wetlands and those influenced strongly by oxidation reactions (e.g., rice-producing uplands) could be added as specific examples in which the biota plays a direct role in acidification.

The phenomena that produce a given proton concentration in the soil solution to render it acidic are complex and interrelated. (The quite separate issue of measuring this proton concentration is discussed in Special Topic 7 at the end of this chapter.) Those pertaining to sources and sinks for protons can be considered schematically as a special case of Figure 1.2, with "free cation or anion" in the center of the figure interpreted as H^+. In addition to the biogeochemical determinants of soil acidity, the field-scale transport processes *wetfall* (rain, snow, throughfall), *dryfall* (deposited solid particles), and *interflow* (lateral movement of soil water beneath the land surface on hill slopes) carry protons into a soil solution from external sources. Their existence and that of proton-exporting processes (e.g., volatilization, erosion) underscore the fact that the soil solution is an *open* natural water system subject to anthropogenic and natural inputs and outputs that may by themselves dominate the development of soil acidity. Industrial effluents (e.g., sulfur and nitrogen oxide gases or mining waste waters) that produce acidic deposition or infiltration and nitrogenous fertilizers with transformation and transport that produce acidic soil conditions are examples of anthropogenic inputs. Despite all this complexity, proton cycling in acidic soils at field scales has been quantified sufficiently well to allow some general conclusions to be drawn. Acidic deposition, production of $CO_2(g)$ and humus, and proton biocycling all serve to increase soil solution acidity, whereas proton adsorption and mineral weathering decrease it. Thus, over millennia, after readily "weatherable" minerals become depleted, freshwater leaching (Fig. 2.6) can produce highly acidic soils.

Carbonic acid ($H_2CO_3^*$) is a ubiquitous source of protons to soil solutions, but one with a concentration that varies spatially and temporally because of respiration processes. The formation of carbonic acid and its reactions in the soil solution are discussed in Section 2.5, Problem 15 in Chapter 1, and Problems 5 to 10 in Chapter 4. The key mathematical relationship with respect to soil pH is presented in Problem 8 of Chapter 4:

$$P_{CO_2}/\left(H^+\right)\left(HCO_3^-\right) = 10^{7.8} \qquad (T = 298.15\ K) \qquad (11.1)$$

Equation 11.1 shows that the partial pressure of CO_2 (in atmospheres) and the bicarbonate ion activity fully determine the pH value of the soil solution. Numerical calculation is facilitated by writing the equation in logarithmic form:

$$pH \equiv -\log\left(H^+\right) = 7.8 + \log\left(HCO_3^-\right) - \log\ P_{CO_2} \qquad (11.2)$$

The pH value of an acidic solution comprising only $H_2CO_3^*$, for which (H^+) closely approximates (HCO_3^-), can be calculated with Eq. 11.2 after P_{CO_2} is specified. For atmospheric air, $P_{CO_2} \approx 10^{-3.52}$ atm and pH = 5.7; for soil air in B horizons or in the rhizosphere, $P_{CO_2} \approx 10^{-2}$ atm and pH = 4.9; and for a flooded soil (Section 6.5), $P_{CO_2} \approx 0.12$ atm and pH = 4.4. Thus, pH values varying within 0.7 log units of 5.0 can be expected in soil solutions if carbonic acid dissociation is the dominant chemical reaction governing soil acidity.

A second major contributor to soil acidity is humus, whose proton exchange reactions were introduced in Section 3.3 and were described quantitatively in Section 9.5 using the NICA–Donnan model. Humic substances, a major fraction of humus, offer a large repository of acidic protons in the form of carboxyl groups with pH_{dis} values that are near 3.0 (Section 3.2). For soils in which organic C is cycled intensively (e.g., Alfisols, Mollisols, and Spodosols), the protonation–proton dissociation reactions of humus exert a strong influence on acidity. The key capacity factors governing this influence are *total acidity* (TA), *cation exchange capacity* (CEC), and *acid-neutralizing capacity* (ANC), as defined in Section 3.3 and in Problem 7 of Chapter 3:

$$TA = CEC - ANC \qquad (11.3)$$

where

$$\mathrm{ANC} \equiv -\sigma_{\mathrm{H}} \quad (\sigma_{\mathrm{H}} \leq 0) \tag{11.4}$$

when all quantities are expressed per unit mass of humus. Thus, TA (also termed *exchangeable acidity*) is a quantitative measure of the capacity of soil humus to donate protons under given conditions of temperature, pressure, and soil solution composition. Cation exchange capacity being the maximum moles of proton charge dissociable from unit mass of soil humus (i.e., the carboxyl and phenolic OH protons that are displaceable according to a cation exchange reaction like that in Eq. 3.4), TA can then be pictured as quantifying the exchangeable protons that remain after a given negative value of net proton charge has been reached. This net proton charge is necessarily balanced by adsorbed ion charge (Eq. 7.8) and, therefore, $-\sigma_{\mathrm{H}}(\sigma_{\mathrm{H}} < 0)$ provides a quantitative measure of the capacity of soil humus to replace adsorbed ions with protons under given conditions of temperature, pressure, and soil solution composition.

If the NICA–Donnan model is applied to describe the relationship between TA and CEC for, say, humus carboxyl groups, then Eq. 9.26 takes the form

$$\mathrm{TA} = \mathrm{CEC}\frac{\left(\tilde{K}_{\mathrm{H}} c_{\mathrm{H}}\right)^{p_{\mathrm{H}}}}{1 + \left(\tilde{K}_{\mathrm{H}} c_{\mathrm{H}}\right)^{p_{\mathrm{H}}}} \tag{11.5}$$

where $\tilde{K}_{\mathrm{H}}$ and p_{H} are adjustable parameters discussed in Section 9.5, in which specific values are given for humic substances. Equation 11.5 implies that TA increases with the concentration of protons (c_{H}) in the soil solution (i.e., it increases with decreasing pH). According to Eqs. 11.3 to 11.5,

$$\begin{aligned} \mathrm{ANC} = \mathrm{CEC} - \mathrm{TA} &= \mathrm{CEC}\left[1 - \frac{\left(\tilde{K}_{\mathrm{H}} c_{\mathrm{H}}\right)^{p_{\mathrm{H}}}}{1 + \left(\tilde{K}_{\mathrm{H}} c_{\mathrm{H}}\right)^{p_{\mathrm{H}}}}\right] \\ &= \frac{\mathrm{CEC}}{1 + \left(\tilde{K}_{\mathrm{H}} c_{\mathrm{H}}\right)^{p_{\mathrm{H}}}} \end{aligned} \tag{11.6}$$

in agreement with the model expression for σ_{H} given in Problem 6 of Chapter 3. Thus ANC increases as c_{H} decreases (i.e., it increases as pH increases), ultimately approaching CEC. Note that Eq. 11.6 implies $\mathrm{pH}_{\mathrm{dis}} \approx \log \tilde{K}_{\mathrm{H}}$.

Buffer intensity (β_H) is the derivative of ANC with respect to pH (Section 3.3 and Problem 8 in Chapter 3). Operationally, β_{H} is the number of moles of proton charge per unit mass that are dissociated from (complexed by) soil humus when the pH value of the soil solution increases (decreases) by 1 log unit. The buffer intensities of organic-rich surface horizons in temperate-zone acidic soils have maximal values in the range 0.1 to 1.5 $\mathrm{mol_c\ kg^{-1}\ pH^{-1}}$ around pH 5, *when expressed per unit mass of soil humus.* Thus, for example, the addition of 20 mmol proton charge to a kilogram of soil with a humus content $f_h = 0.1\ kg_h\ kg^{-1}$, for which the buffer intensity is 0.2 $mol_c\ kg_h^{-1}\ pH^{-1}$, would decrease the pH by 0.02 $mol_c\ kg^{-1}/(0.1\ kg_h\ kg^{-1} \times 0.2\ mol_c\ kg_h^{-1} pH^{-1})$ = 1.0 unit. The relationship exemplified by this calculation is

$$\Delta pH = \Delta n_A / f_h \beta_H \tag{11.7}$$

where Δn_A is moles of proton charge added or removed per kilogram *soil.* Note that, because β_H is pH dependent, ΔpH will be pH dependent. For example, using the model expression in Eq. 11.6, one finds

$$\begin{aligned}\beta_H \equiv \frac{dANC}{dpH} &\approx \frac{d}{dpH}\left[\frac{CEC}{1 + (\tilde{K}_H 10^{-pH})^{p_H}}\right] \\ &= CEC\frac{(\ln\ 10)^{p_H}\left(\tilde{K}_H 10^{-pH}\right)^{p_H}}{\left[1 + \left(\tilde{K}_H 10^{-pH}\right)^{p_H}\right]^2} \\ &= 2.303\ p_H \frac{TA}{CEC} ANC\end{aligned} \tag{11.8}$$

where the last step comes from ln 10 = 2.303 and an appeal to Eqs. 11.5 and 11.6. Equation 11.5 implies that TA/CEC is a monotonically decreasing function of pH that becomes negligibly small when pH > log $\tilde{K}_H$ (see also Problem 7 in Chapter 3), whereas Eq. 11.6 implies that ANC increases with pH as the mirror image of total acidity. It follows from Eq. 11.8 that β_H should then display a maximum at a pH value approximately equal to log $\tilde{K}_H$ appropriate for soil humus.

Adding complexity to this description of soil acidity and buffering are the roles that Al-hydroxy polymers and the weathering of Al-bearing minerals play (Section 11.3). Suffice it to say here that hydrolytic species of Al(III)—in aqueous solution, adsorbed on soil particles (especially particulate humus), or in solid phases—may strongly influence soil solution pH in mineral horizons of acid soils. The buffer intensity they provide, however, is typically an order of magnitude smaller than the values for soil humus.

The biological processes important in the development of soil acidity are ion uptake or release and the catalysis of redox reactions. Plants often take up more cations from soil than anions, with the result that protons are excreted to maintain charge balance. For example, under the anoxic conditions that prevail, peat bogs generate acidity because the vegetation takes up N either as NH_4^+ or as fixed $N_2(g)$, thus inducing excess cation uptake and a resultant excretion of protons to the soil solution. More generally, the rhizosphere may become acidified relative to the soil in bulk because of proton excretion or organic acid excretion, particularly those organic acids that have pH_{dis} values less than the ambient rhizosphere pH (Table 3.1). Under controlled experimentation, rhizosphere pH values as much as 2 log units less than bulk soil values have been measured. The influence on soil acidity from redox catalysis is discussed in Section 11.4. It pertains essentially to the transformations of C, N, and S.

11.2 Acid-Neutralizing Capacity

Acid-neutralizing capacity as defined for soil humus in Eq. 11.4 refers to the negative intrinsic surface charge produced by the exchange of complexed protons for metal cations that, in principle, can themselves be displaced subsequently by protons brought into the soil solution through any of the acidity-producing processes discussed in Section 11.1. This latter possibility is epitomized mathematically in the charge balance constraint that appears in Eq. 7.8. For each mole of adsorbed metal cation charge removed from soil humus, thereby causing Δq to decrease, a mole of complexed protons must be added, causing σ_H to increase. This shift in adsorbed species captures protons from the soil solution while releasing adsorbed metal cations into the soil solution.

Proton exchange reactions are not limited to soil humus, of course, and accordingly it is possible to extend the concept of ANC to an entire soil adsorbent. This is done through the combination of operational definitions of TA and CEC with a simple rearrangement of Eq. 11.3:

$$\text{ANC} \equiv \text{CEC} - \text{TA} \tag{11.9}$$

where total acidity is measured as the moles of titratable protons per unit mass displaced from a soil adsorbent by an unbuffered KCl solution—hence its alternative names: *exchangeable acidity* or *KCl-replaceable acidity.* Experiments with a variety of mineral soils have shown that the principal contribution to total acidity is made by readily exchangeable forms of Al(III): Al^{3+}, $AlOH^{2+}$, $Al(OH)_2^+$, and $AlSO_4^+$. The protons released when these species are displaced by K^+ and then hydrolyze in the soil solution are the titratable protons measured experimentally. On the other hand, for soil humus, the total acidity comprises mostly protons displaced from strongly acidic organic functional groups or from adsorbed Al- and Fe-hydroxy species. The pH dependence of TA for subsurface horizons of some acidic soils in the eastern United States is shown in Figure 11.1. The TA values were measured using unbuffered KCl solution, whereas the CEC values were measured using $BaCl_2$ solution buffered at pH 8.2. The ratio of TA to CEC declines sharply to zero as pH increases above 5. This trend is typical of acidic mineral soils. By contrast, TA for soil humus (Eq. 11.5) disappears well below pH 4.5. In Figure 11.1, TA/CEC = 0.5 at pH $\approx$ 4.8, whereas Eq. 11.5 predicts TA/CEC = 0.5 at pH $\approx \log \widetilde{K}_H \approx$ 2.3 to 2.9 for the strongly acidic functional groups in soil humic substances (Section 9.5). Acid-neutralizing capacity exists in a soil solution for the same fundamental reason that it exists on a soil adsorbent (i.e., the

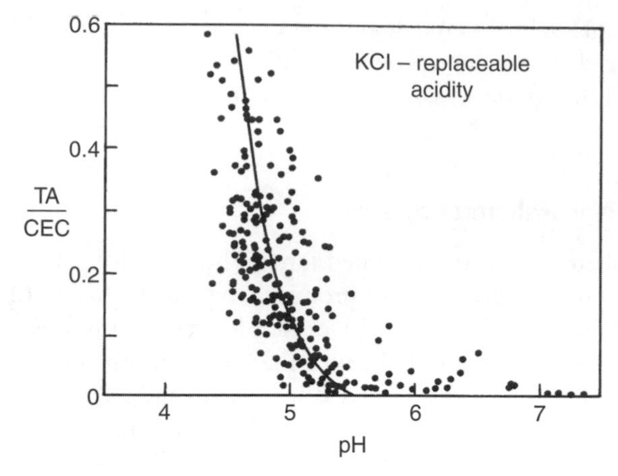

Figure 11.1. Total acidity (TA) as a fraction of cation exchange capacity (CEC) plotted against pH for acidic mineral soils of the eastern United States.

presence of functional groups that can complex protons under acidic conditions). For a soil solution in which carbonic acid is the only constituent that provides these groups (in the forms of bicarbonate and carbonate anions), ANC can be expressed by the equation

$$\mathrm{ANC} = \left[\mathrm{HCO_3^-}\right] + 2\left[\mathrm{CO_3^{2-}}\right] + \left[\mathrm{OH^-}\right] - \left[\mathrm{H^+}\right] \tag{11.10}$$

That Eq. 11.10 is analogous to Eq. 11.9 can be appreciated after noting the logical correspondences (see Problem 7 in Chapter 4):

$$\mathrm{CEC} \Leftrightarrow 2\mathrm{CO_{3T}} = 2\left[\mathrm{H_2CO_3^*}\right] + 2\left[\mathrm{HCO_3^-}\right] + 2\left[\mathrm{CO_3^{2-}}\right] \tag{11.11a}$$

$$\mathrm{TA} \Leftrightarrow 2\left[\mathrm{H_2CO_3^*}\right] + \left[\mathrm{HCO_3^-}\right] + [\mathrm{H^+}] - [\mathrm{OH^-}] \tag{11.11b}$$

and applying them to Eq. 11.9. According to Problem 9 in Chapter 4, the first two terms on the right side of Eq. 11.10 define the *carbonate alkalinity.* These two terms are analogous to the two terms representing acidic functional groups in soil humus that appear in the model equation for σ_H presented in Problem 6 of Chapter 3. Thus, carbonate alkalinity refers to the carbonate anion charge produced by the dissociation of complexed protons. The more general ANC given by Eq. 11.10 is termed the *alkalinity* of a soil solution. However, if $H_2CO_3^*$ and water are the only compounds present, ANC given by Eq. 11.10 equals *zero* because this equation is also the condition for charge balance in the solution. The same situation arises if the ANC of an aqueous *suspension* containing only particulate humus at concentration c_S is considered:

$$\mathrm{ANC} = -\sigma_H c_S - \sigma_d c_S + [\mathrm{OH^-}] - [\mathrm{H^+}] \quad (\sigma_H \leq 0) \tag{11.12}$$

where the first term on the right side is ANC contributed by the humus adsorbent (Eq. 3.7), the second term is titratable acidity contributed by protons adsorbed in the diffuse-ion swarm, and the last two terms are ANC contributed by the water in which the humus particles are suspended. Equation 11.12 is the condition for charge balance in the suspension; hence its ANC is equal to zero.

More generally, a soil solution will contain a variety of anions that protonate and a variety of metal cations that hydrolyze in the acidic pH range. For example, in the typical range $2 < pH < 6.5$,

$$\begin{aligned} ANC = {} & [HCO_3^-] + 2[CO_3^{2-}] + [HC_2O_4^-] + [H_2PO_4^-] + 2[HPO_4^-] \\ & + [OH^-] - [H^+] - 3[Al^{3+}] - 2[AlOH^{2+}] - [Al(OH)_2^+] \end{aligned} \quad (11.13)$$

which shows that protonating anions, including organic anions (Problem 1 in Chapter 4), increase ANC, whereas hydrolyzing metal cations decrease ANC of a soil solution. Charge balance in this example would be expressed typically by the equation

$$\begin{aligned} & [Na^+] + [K^+] + 2\,[Ca^{2+}] + 2\,[Mg^{2+}] + 2\,[Fe^{2+}] + 2\,[Mn^{2+}] \\ & + 3\,[Al^{3+}] + 2\,[AlOH^{2+}] + [Al(OH)_2^+] + [H^+] - [OH^-] \\ & - [Cl^-] - [NO_3^-] - 2\,[SO_4^{2-}] - [HCO_3^-] - 2\,[CO_3^{2-}] - [HC_2O_4^-] \\ & - 2\,[C_2O_4^{2-}] - [H_2PO_4^-] - 2\,[HPO_4^{2-}] = 0 \end{aligned} \quad (11.14)$$

The combination of Eqs. 11.13 and 11.14 leads to an alternative equation for ANC of a soil solution:

$$\begin{aligned} ANC = {} & [Na^+] + [K^+] + 2[Ca^{2+}] + 2[Mg^{2+}] + 2[Fe^{2+}] + 2[Mn^{2+}] \\ & - [Cl^-] - [NO_3^-] - 2[SO_4^{2-}] - 2[C_2O_4^{2-}] \end{aligned} \quad (11.15)$$

(Equations 11.13 to 11.15 neglect any nonhydrolytic metal complexes formed by the cations and anions considered; however, adding them is straightforward.) Equation 11.15 is illuminating in that it implies that removal of metal cation charge from a soil solution decreases its ANC, whereas removal of anion charge increases its ANC, *provided that* the metal cations do not hydrolyze and the anions do not protonate over the acidic pH range considered. These effects are analogous to those occurring on a humus adsorbent as epitomized in Eq. 7.8: Removing metal cation charge is equivalent to supplying proton charge, but removing anion charge is equivalent to removing proton charge from a soil solution. Conversely, adding anion charge through acidic deposition (NO_3^- and SO_4^{2-}) or biological production ($C_2O_4^{2-}$, oxalate) is equivalent to adding proton charge.

11.3 Aluminum Geochemistry

Low soil pH is accompanied by proton attack on Al-bearing minerals (Fig. 5.1) leading to the production of soluble Al(III) in the soil solution (Table 4.4). The free-ion species of this soluble Al will equilibrate with soluble complexes (e.g., Al-oxalate complexes, as in Eq. 1.4 and Table 4.4—see also problems 6 and 14 in Chapter 1); with the soil adsorbent; and, of course, with soil minerals. Aluminum solubility in acidic soils is influenced by a variety of minerals that are discussed individually in sections 2.3, 2.4, 5.2, and 5.4: gibbsite, kaolinite, allophane/imogolite, pedogenic chlorite or beidellite, and hydroxy-interlayer vermiculite or vermiculite. Dissolution reactions for these minerals are discussed in sections 2.3, 5.1, 5.2, and 5.4, as well as in problems 12, 14, and 15 of Chapter 2; and Problems 9, 10, 13, and 14 of Chapter 5. They are the essential input used to construct activity–ratio diagrams similar to that in Figure 5.5. The mineral with an activity–ratio line that lies highest in the diagram is assigned control of Al^{3+} activity, although metastability can intervene to require interpretation using the GLO Step Rule (Section 5.2).

Metastability, in fact, appears to be the rule with respect to Al solubility control in soils affected by acidic deposition or infiltration. The role it plays can be illustrated by consideration of the dissolution reactions of gibbsite, proto-imogolite allophane, and kaolinite:

$$\mathrm{Al(OH)_3\,(s) + 3H^+ = Al^{3+} + 3H_2O\,(\ell)} \tag{11.16a}$$

$$\frac{1}{4}\mathrm{Si_2Al_4O_{10}\cdot 5H_2O\,(s) + 3H^+ = Al^{3+}} + \frac{1}{2}\mathrm{Si(OH)_4^0} + \frac{7}{4}\mathrm{H_2O\,(\ell)} \tag{11.16b}$$

$$\frac{1}{2}\mathrm{Si_2Al_2O_5(OH)_4\,(s) + 3H^+ = Al^{3+} + Si(OH)_4^0} + \frac{1}{2}\mathrm{H_2O\,(\ell)} \tag{11.16c}$$

The dissolution equilibrium constants for these reactions each can vary over one or two log units, with the larger values associated with poorer crystallinity and, therefore, greater solubility of Al at a given pH value. Taking gibbsite and kaolinite as examples, one can derive activity–ratio equations for log $[(\text{solid})/(Al^{3+})]$ as described in Section 5.2:

$$\log\left[(\text{gibbsite})\,/\,(\mathrm{Al^{3+}})\right] = -\log {}^*K_{so} + 3\mathrm{pH} + 3\log(\mathrm{H_2O}) \tag{11.17a}$$

$$\log\left[(\text{kaolinite})\,/\,(\mathrm{Al^{3+}})\right] = -\log K_{so} + 3\mathrm{pH} + \log\left(\mathrm{Si(OH)_4^0}\right) + \frac{1}{2}\log(\mathrm{H_2O}) \tag{11.17b}$$

where the first term on the right side is determined by the degree of crystallinity of the mineral dissolving. For gibbsite, $\log^* K_{so} = 8.77$ if the mineral is reasonably well crystallized, 9.35 if it is in microcrystalline form, and 10.8 if it is amorphous. For kaolinite, log K_{so} = 3.72 if the mineral is well crystallized and 5.25 if it is not. This variability leads to the "windows" of mineral stability discussed in Section 5.2.

At pH 5 and above, Al tends to precipitate in acidic soils and, because unit water activities are expected, Eq. 11.17 specializes to the working equations

$$\log\left[(\text{gibbsite}) / (\text{Al}^{3+})\right] = 15.0 - \log\ {}^{*}K_{so} \quad (11.17c)$$

$$\log\left[(\text{kaolinite}) / (\text{Al}^{3+})\right] = 15.0 - \log\ K_{so} + \log\ \left(\text{Si(OH)}_4^0\right) \quad (11.17d)$$

After the values of the dissolution equilibrium constants are selected, Eqs. 11.17c and 11.17d can be plotted as in Figure 5.5. Figure 11.2 shows such an activity–ratio diagram with the gibbsite and kaolinite windows included. Allophane and the 2:1 clay minerals, also candidates for influencing Al solubility, would typically plot within the kaolinite window, but with less strong dependence on $\left(\text{Si(OH)}_4^0\right)$ than the smectite whose dissolution reaction appears in Eq. 5.18a because of lower Si-to-Al molar ratios, another characteristic of acidic soil environments. Allophane would show a weaker dependence on $\left(\text{Si(OH)}_4^0\right)$ than even kaolinite, but still would fall within the kaolinite window. A solubility window for solid-phase silica like that in Figure 5.5 also has been included in Figure 11.2, with amorphous silica depicted at its left boundary and quartz at its right boundary.

At pH 5, Al solubility control falls to well-crystallized kaolinite over the range of $\left(\text{Si(OH)}_4^0\right)$ shown (upper diagonal line). This result may be contrasted with that in Figure 5.5, which shows gibbsite taking over Al solubility

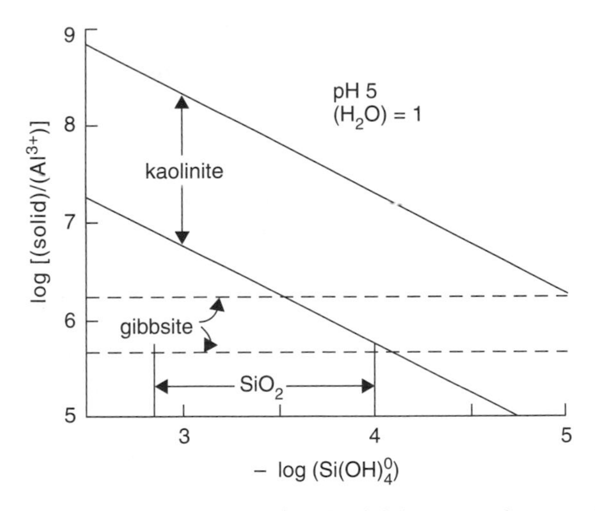

Figure 11.2. An activity–ratio diagram for Al solubility control at pH 5 by kaolinite or gibbsite, with solubility "windows" shown for each mineral and for silica solubility control by amorphous silica (left vertical line) to quartz (right vertical line).

control at $(\text{Si}(\text{OH})_4^0) \gtrsim 10^{-4.4}$. This difference arises solely because of the higher value of log $^*K_{so}$ for well-crystallized gibbsite (8.77 vs. 8.11) used in Figure 11.2, and it is significant: Silica leaching to a concentration less than 10 μmol L^{-1} is now required to stabilize well-crystallized gibbsite. On the other hand, if kaolinite is poorly crystallized, well-crystallized gibbsite takes over solubility control at $(\text{Si}(\text{OH})_4^0) \gtrsim 10^{-3.5}$ and even microcrystalline gibbsite can do this at $(\text{Si}(\text{OH})_4^0)$ below that sustained by quartz. Spodosols tend to support silica solubilities near that of quartz and, therefore, are expected to show Al solubility control by gibbsite, implying concentrations in the millimole per cubic meter range. Oxisols and Ultisols tend to show silica solubilities between those of quartz and amorphous silica, thus opening the door to Al solubility control by kaolinite, particularly if gibbsite precipitation is impeded by the presence of strongly complexing organic functional groups that render Al^{3+} unavailable for hydrolysis.

The near confluence of the lines in Figure 11.2 for quartz, poorly crystallized kaolinite, and microcrystalline gibbsite is noteworthy for theoretical reasons. If log $^*K_{so}$ for microcrystalline gibbsite were just 0.1 log unit smaller (i.e., slightly better crystallinity), the horizontal line depicting log [(gibbsite)/(Al^{3+})] for it would have intersected the vertical quartz solubility line at $(Al^{3+}) = 10^{-5.75}$, which is where the diagonal kaolinite line intersects it [introduce log $K_{so} = 5.25$ and log $(\text{Si}(\text{OH})_4^0) = -4$ into Eq. 11.17d]. Under this condition, eqs. 11.16a and 11.16b can be combined with the corresponding expression for quartz to yield the chemical equation

$$\text{SiO}_2\,(\text{s}) + \text{Al(OH)}_3\,(\text{s}) = \frac{1}{2}\text{Si}_2\text{Al}_2\text{O}_5\,(\text{OH})_4\,(\text{s}) + \frac{1}{2}\,\text{H}_2\text{O}\,(\ell) \qquad (11.18)$$

This reaction can be interpreted as the formation of an inorganic condensation polymer (Section 3.1), kaolinite, from its component oxide minerals. It is straightforward to show that log K = 0 for the reaction if log $^*K_{so} = 9.25$ for gibbsite dissolution. This value, in turn, would *require* $(H_2O) = 1$, as has been already assumed in Figure 11.2.

If one now imagines that the only aqueous solution species in the soil solution are those appearing in Eq. 11.16 [Al^{3+}, H^+, and $\text{Si}(\text{OH})_4^0$], then the four species appearing in Eq. 11.18 suffice as *components* for the system under consideration. That is, eqs. 11.16a and 11.16c, along with the dissolution reaction for quartz, are sufficient to describe the formation of all seven species from the four components. Thus, equilibrium is completely determined by the three relevant equilibrium constants and the fixed activities of the three minerals plus liquid water. If the activity of the liquid component were not 1.0, equilibrium as portrayed in Eq. 11.18 would not be possible. Indeed, log K = 0 for the reaction means that poorly crystallized kaolinite and water are not distinguishable thermodynamically from quartz and microcrystalline gibbsite, in the same sense that hydronium ion (H_3O^+) is not distinguishable thermodynamically from a water molecule and a proton (Table 6.2).

Besides the monomeric Al species indicated in Table 4.4, evidence exists for relatively stable polynuclear Al species, particularly in complexes with OH^- and organic anions. These polynuclear Al-hydroxy species, ranging from $Al_2\left(OH_2^{4+}\right)$ to $[AlO_4Al_{1.2}(OH)_{24}]^{7+}$, can engage in acid–base reactions as aqueous solutes or in adsorption reactions on both soil humus and soil minerals. The formation of hydroxy-interlayer vermiculite and pedogenic chlorite involves the adsorption of Al-hydroxy polymers as the first step (Section 2.3). These polymer coatings may affect Al solubility very similarly to the relationship between (Al^{3+}) and pH that obtains for gibbsite (Eq. 11.17a).

Given the kind of aqueous-phase speciation data listed in Table 4.4, the distribution of exchangeable cations can be calculated if a Vanselow selectivity coefficient has been measured (Section 9.3). Consider, for example, the analog of Eq. 9.14 for Ca–Al exchange:

$$3\ CaX_2(s) + 2\ Al^{3+} = 2\ AlX_3(s) + 3\ Ca^{2+} \tag{11.19}$$

The Vanselow selectivity coefficient (Eq. 9.18) is

$$K_V^{Ca/Al} = x_{Al}^2\ \left(Ca^{2+}\right)^3 / x_{Ca}^3\ \left(Al^{3+}\right)^2 \tag{11.20}$$

where the mole fraction x_{Ca} refers to CaX_2, and x_{Al} refers to AlX_3. If $K_V^{Ca/Al}$ has been measured and found not to depend significantly on x_{Ca}, then Eq. 11.20 can be used to calculate the mole fraction of exchangeable Al^{3+} that is in equilibrium with the soil solution. Suppose, for example, that, in the Spodosol to which Table 4.4 applies, $K_V^{Ca/Al} \approx 1.0$ and is independent of x_{Ca} (i.e., the Vanselow model applies). Then, with $(Al^{3+}) = 1.1 \times 10^{-6}$ and $(Ca^{2+}) = 2.64 \times 10^{-4}$, based on Table 4.4 (and Eq. 4.24 with $I = 2.03$ mol m^{-3}), one calculates from Eq. 11.20

$$1.0 = 1.18\,[x_{Al}^2/(1 - x_{Al})^3]$$

which implies that $x_{Al} \approx 0.41$. Thus, under the conditions given, neither exchangeable cation is predicted to dominate the soil adsorbent, although the value of $\tilde{E}_{Al}$ is only 0.005. [Note that $K_V^{Ca/Al} = 1.0$ corresponds to *no* thermodynamic preference in the cation exchange reaction (Section 9.3)!]

The competition between protons and Al^{3+} for carboxyl groups on soil humus can be described using the NICA–Donnan model as discussed in Section 9.5. For humic acid, model parameters for the proton ($\log \tilde{K}_i \equiv \log \tilde{K}_H = 2.93, p_1 \equiv \beta_H p = 0.5$) are in a table preceding Eq. 9.29. Those for Al^{3+} are $\log \tilde{K}_{Al} = 2.00$ and $\beta_{Al} = 0.25$. Given $\beta_H = 0.81$, derived from extensive data analysis yielding the model parameters tabulated in Problem 15 of Chapter 9, it follows that $p = 0.62$ in Eq. 9.29. According to Table 4.4, $c_H \approx 48$ mmol m^{-3} and $c_{Al} \approx 1.8$ mmol m^{-3}. With these data as input, the second factor on the right side of Eq. 9.29 is found to be 0.607, whereas the third factor equals 0.423. Their product is the charge fraction of complexed

Al^{3+} (q_{Al}/b_{Al}), about 0.26 under the conditions given (i.e., pH 4.3 and micromolar Al^{3+}). Thus about one fourth of the carboxyl groups are predicted to complex Al^{3+} at pH 4.3. [Note that $\beta_{Al}/\beta_H = 0.31$, as expected from the difference in valence of the two competing cations, such that $b_{Al} \approx 0.3b_H$, where $b_H \equiv b_1 = 3.15$ mol$_c$ kg^{-1} according to the table preceding Eq. 9.29.]

Mineral dissolution reactions like those in Eq. 11.16 consume protons and bring metal cations and neutral silica into the soil solution, thereby contributing to its ANC while depleting that of the soil solids. This depletion is reflected in a common field observation that the decrease in ANC of soil solids is approximately equal to the net proton flux consumed by mineral weathering. The same effect accompanies the cation exchange reaction in Eq. 11.19 (understood to proceed from left to right), because Al^{3+} removal from the soil solution increases its ANC and decreases the ANC of the soil adsorbent (Eq. 11.9; loss of Al^{3+} through adsorption implies an increase in total acidity.) In this "trade-off" sense, the complexation of Al^{3+} by soil humus, although it reduces the toxic effect of this metal cation, does not alter the total acidity of the humus, or the ANC of the soil solution, if it proceeds by a proton exchange reaction, as in the example based on Eq. 9.29.

11.4 Redox Effects

In Section 6.2, it is emphasized that most of the reduction half-reactions that occur in soils result in proton consumption (Eq. 6.7). Therefore, an important source of ANC in soil solutions are the redox reactions that feature a net proton consumption and effective microbial catalysis. More specifically, if a selected reduction half-reaction couples strongly to the oxidation of soil humus, as exemplified by the reverse of the penultimate reaction in Table 6.1, and if the stoichiometric coefficient of H^+ in the selected reduction half-reaction is larger than 1.0, the resulting redox reaction will deplete the soil solution of protons. For example, denitrification, depicted by the redox reaction

$$4\,NO_3^- + 5\,CH_2O + 4\,H^+ = 2\,N_2(g) + 5\,CO_2(g) + 7\,H_2O\,(l) \qquad (11.21)$$

consumes 1 mol H^+ mol^{-1} N produced as $N_2(g)$. Similarly, the reductive dissolution of $Fe(OH)_3(s)$ in a flooded soil consumes 2 mol H^+ mol^{-1} Fe^{2+} produced. Both of these redox reactions increase the ANC of the soil solution (Eq. 11.15). [The production of $CO_2(g)$ in Eq. 11.21 does not affect ANC because of the condition imposed by charge balance on a solution of $H_2CO_3^*$ (Eq. 11.10).] Note that denitrification causes the removal of a nonprotonating anion without the simultaneous loss of a nonhydrolyzing metal cation, whereas reductive dissolution causes the addition of a nonhydrolyzing metal cation without the simultaneous production of a nonprotonating anion (see Eq. 11.15).

Field studies have indicated the importance of the reactions in Table 6.1 to the generation of ANC in the soil environment. Indeed, reductive dissolution reactions involving Fe and Mn are examples of an increase in ANC caused by mineral weathering. Because the seven principal redox-active elements in soils (Section 6.3) also are essential elements for the nutrition of green plants, however, the ultimate impact of their redox reactions cannot be estimated without full consideration of their biogeochemical cycles. For example, the uptake of nitrate or sulfate by plant roots is balanced by the exudation of hydroxide ion or bicarbonate into the soil solution, thus increasing ANC (Eq. 11.13). The uptake of cations like Na^+ or Fe^{2+} by plant roots is balanced by their exudation of H^+, which decreases ANC. This latter process usually dominates with respect to the net moles of ion charge entering roots, with a resultant acidification of the soil solution and, ultimately, of the rhizosphere (Section 11.1). Assimilatory nitrate reduction by bacteria increases ANC, as does dissimilatory nitrate reduction (Section 6.2). A more subtle example is provided by the transformation of urea, $(NH_2)_2CO$, the nitrogenous fertilizer most widely applied in agriculture worldwide, especially in wetlands rice cultivation. Urea is converted rapidly to NH_4^+ and $CO_2(g)$ by hydrolysis and protonation:

$$(NH_2)_2CO + 2\,H^+ + H_2O\,(\ell) = 2\,NH_4^+ + CO_2\,(g) \qquad (11.22)$$

This conversion consumes 1 mol H^+ per mole of N produced as NH_4^+. Subsequent uptake of ammonium by plant roots provokes the exudation of charge-balancing protons that replace those consumed in urea hydrolysis. If instead the NH_4^+ produced by urea hydrolysis is oxidized to form NO_3^-, 2 mol H^+ are released into the soil solution per mole of N produced as nitrate ions:

$$NH_4^+ + 2\,CO_2\,(g) + H_2O\,(\ell) = NO_3^- + 2\,CH_2O + 2\,H^+ \qquad (11.23)$$

This means that the overall transformation of urea to nitrate would yield a net 1 mol H^+ to the soil solution. However, the subsequent uptake of the nitrate by plant roots results in the exudation of OH^- or HCO_3^- that neutralizes the effect of H^+ released by the oxidation of NH_4^+. Thus the ANC of the soil solution has not been changed by the reactions in eqs. 11.22 and 11.23, if the nitrate ions produced are entirely consumed by plants. Even if loss by denitrification (Eq. 11.21) is the fate of the nitrate produced, this result holds true, because denitrification consumes directly the excess 1 mol H^+ released by the combination of reactions in eqs. 11.22 and 11.23. Nitrate loss from the soil solution by leaching, however, can reduce ANC if charge-balancing metal cations are leached as well, thus leaving a net 1 mol H^+ added to the soil solution.

Besides producing $CO_2(g)$, which has no net effect on ANC, mineralization of soil humus can reduce the ANC of a soil solution through the production of NO_3^- and SO_4^{2-} (Eq. 11.15). This effect can be diminished by removal of aboveground biomass before mineralization (e.g., harvesting of agricultural crops). Reduced-N fertilizers, like NH_4NO_3 and $(NH_4)_2SO_4$, can decrease soil solution ANC via oxidation, as in Eq. 11.23. Long-term field studies on fertilized plots have shown that NH_4NO_3 and $(NH_4)_2SO_4$ applications in particular can decrease ANC greatly through nitrification. Ammonium sulfate and reduced-S species, often introduced into soils by dry deposition, can produce decreases in ANC by subsequent oxidation. These processes may account for half the input of protons into soil from atmospheric deposition sources. The examples here thus serve to illustrate the broad scope of redox effects on soil acidity, as well as the strong interrelatedness of the proton cycling components identified in Section 11.1.

11.5 Neutralizing Soil Acidity

The processes that increase the pH value of a soil solution are mineral weathering, anion uptake by the biota, protonation of anions or surface functional groups, adsorption of nonhydrolyzing metal cations, and reduction half-reactions. In acidic soils, these processes may not be adequate to maintain pH in an optimal range if acidic deposition intrudes or acidifying fertilizers are applied. When soil pH is such that the corresponding total acidity exceeds about 15% of the CEC (Fig. 11.1), a variety of serious problems for plant and microbial growth (e.g., Al, Fe, and Mn toxicity or Ca, Mg, and Mo deficiency) is expected. Under this deleterious condition, soil amendments to decrease total acidity must be considered.

The practice of neutralizing soil acidity is formalized in the concept of the *lime requirement.* This key parameter is defined formally as the moles of Ca^{2+} charge per kilogram soil required to decrease the total acidity to a value deemed acceptable for an intended use of the soil. Because of the relatively unique relationship between total acidity and pH typified by Figure 11.1, decreasing total acidity to zero and, therefore, increasing pH to 5.5 and above, offers a general, straightforward criterion for application of the lime requirement concept to a broad range of mineral soils. Typically the lime requirement is expressed in the convenient units of centimoles of charge per kilogram and is found to have a value somewhere between that of the total acidity and the CEC of a soil as measured with a buffered solution of $BaCl_2$ (Section 11.2). It is clear that the lime requirement will depend on soil parent material mineralogy, content of clay and humus, and the Jackson–Sherman weathering stage of a soil (Section 1.5). Special consideration also must be given to soils enduring chronic proton inputs from acidic deposition and acidifying fertilizers.

Methods for measuring the lime requirement are described in *Methods of Soil Analysis* (see "For Further Reading" at the end of this chapter). The procedures that have been used range from field applications of $CaCO_3$ (involving years to achieve a steady state), to laboratory incubations of soil samples with $CaCO_3$ (involving weeks to months), to soil titrations with $Ca(OH)_2$ over several days, to rapid soil equilibrations (< 1 hour) with buffer solutions with a composition that has been optimized for use with a given group of soils. Considerations such as the number of soil samples to be analyzed with the accuracy of the estimated lime requirement enter into the choice of method.

The fundamental chemical reaction underlying the concept of the lime requirement appears in Eq. 11.19. This reaction, reversed so that $CaX_2(s)$ and Al^{3+} are the products, can be coupled with the dissolution reactions of a Ca-bearing mineral added as an amendment, and an appropriate Al-bearing mineral that precipitates, to produce an overall reaction that removes Al^{3+} from the soil solution (see Problem 12 in Chapter 2 for a case involving beidellite as an adsorbent). For example, if $CaCO_3(s)$ is added and $Al(OH)_3(s)$ precipitates in response, one can combine eqs. 5.3, 11.16a, and 11.19 to obtain the overall reaction

$$\begin{aligned} &2\ AlX_3\ (s) + 3\ CaCO_3\ (s) + 3\ H_2O\ (\ell) \\ &\quad = 3\ CaX_2\ (s) + 2\ Al(OH)_3\ (s) + 3\ CO_2\ (g) \end{aligned} \tag{11.24}$$

Note that the equilibrium constant for this reaction provides a relationship between adsorbate composition and the partial pressure of carbon dioxide, if the two mineral phases and liquid water have unit activity. Evidently the reaction will not affect the ANC of the soil solution, if it goes to completion, but the ANC of the soil adsorbent will be increased. If the Al^{3+} that exchanges for Ca^{2+} does not precipitate, on the other hand, the ANC of the soil solution will decrease, as can be deduced from Eq. 11.13.

If the reaction in Eq. 11.24 occurs in a soil, the activities of Al^{3+} and Ca^{2+} in the soil solution are governed by Eq. 5.10 and an expression for the thermodynamic cation exchange constant (Section 9.3):

$$K_{ex}^{Ca/Al} = (AlX_3)^2\left(Ca^{2+}\right)^3 / (CaX_2)^3\left(Al^{3+}\right)^2 \tag{11.25}$$

These two equations can be combined to derive the relationship

$$\begin{aligned} pH + \frac{1}{2}\ \log\left(Ca^{2+}\right) &= \frac{1}{6}\ \log\left[{}^{*}K_{SO}{}^{2}K_{ex}^{Ca/Al}\right] \\ &\quad + \frac{1}{3}\ \log\left[(CaX_2)^{3/2} / (AlX_3)\right] \end{aligned} \tag{11.26}$$

The left side of Eq. 11.26 is called the *lime potential*, an activity variable equal to one half the common logarithm of the IAP of $Ca(OH)_2(s)$ (Section 5.1). As a rule of thumb, soils that have adequate ANC have lime potentials greater than 3.0. Equation 11.26 shows that the lime potential depends sensitively on the activities of $CaX_2(s)$ and $AlX_3(s)$ on the soil adsorbent. The application of Eq. 11.26 to soil acidity perforce requires a relationship between these two activities and the charge fractions of $CaX_2(s)$ and $AlX_3(s)$ (Section 9.2). For example, the linear regression equation

$$\begin{aligned} \log\left[\left(Ca^{2+}\right)^3 / \left(Al^{3+}\right)^2\right] &= -0.74 + 1.02 \pm 0.07 \\ &\quad \times \log\left[E_{CaX_2}^3 / E_{AlX_3}^2\right] \quad \left(R^2 = 0.86\right) \end{aligned} \tag{11.27}$$

describes the relationship between soil solution activities and charge fractions of adsorbed Ca and Al in the A horizons of Spodosols. This empirical equation is tantamount to setting $K_{ex}^{Ca/Al} \approx 0.2$ in Eq. 11.25 while replacing activities on the soil adsorbent by charge fractions. If $^*K_{so} \approx 10^{9.35}$ is assumed (Section 11.3), then Eq. 11.26 reduces to the predictive equation

$$\mathrm{pH} + \frac{1}{2}\log\left(\mathrm{Ca}^{2+}\right) = 3.0 + \frac{1}{3}\log\left[E_{\mathrm{CaX_2}}^{3}/E_{\mathrm{AlX_3}}^{2}\right] \qquad (11.28)$$

It follows from Eq. 11.28 that the lime potential will be more than 3.0 if the charge fraction of adsorbed Ca^{2+} is $\gtrsim 0.57$.

For Further Reading

Alpers, C. N., J. L. Jambor, and D. K. Norstrom (eds.). (2000) *Sulfate minerals.* Mineralogical Society of America, Washington, DC. Chapter 7 of this edited workshop volume gives a comprehensive review of Fe and Al mineral formation in acidic waters dominated by sulfate inputs.

Ehrenfeld, J. G., B. Ravit, and K. Elgersma. (2005) Feedback in the plant–soil system. *Annu. Rev. Environ. Resour.* **30**:75–115. This thought-provoking review describes, among other things, the influence of plants on soil acidity and redox conditions.

Essington, M. E. (2004). *Soil and water chemistry.* CRC Press, Boca Raton, FL. Chapter 10 of this textbook provides a thorough introduction to soil acidity, including that induced by sulfur oxidation in pyritic materials.

Rengel, Z. (ed.). (2003) *Handbook of soil acidity.* Marcel Dekker, New York. This edited monograph provides useful surveys of the causes, effects, and management of soil acidity at an advanced level.

Sparks, D. L. (ed.). (1996) *Methods of soil analysis: Part 3. Chemical methods.* Soil Science Society of America, Madison, WI. Chapter 17 of this standard reference describes methods to measure the lime requirement of soils.

Sposito, G. (ed.). (1996) *The environmental chemistry of aluminum.* CRC Press, Boca Raton, FL. The 10 chapters of this edited volume provide a detailed account of Al geochemistry in acidic soils and waters, including the effects of acidic deposition.

The following articles give valuable research-oriented discussions of chemical processes in acidic soils and waters.

Chadwick, O. A., and J. Chorover. (2001) The chemistry of pedogenic thresholds, *Geoderma* **100**:321–353.

Driscoll, C. T., et al. (2001) Acidic deposition in the northeastern United States: Sources, inputs, ecosystem effects, and management strategies. *Bioscience* **51**:180–198.

Problems

The more difficult problems are indicated by an asterisk.

1. Use the concept of charge balance and the data in Problem 7 of Chapter 4 to verify that $(H^+) = (HCO_3^-)$ in a pure solution of $H_2CO_3^*$. [*Hint:* Calculate (H^+) for the range of P_{CO_2} typical of soils using Eq. 11.2 and the assumption that $(H^+) = (HCO_3^-)$. Then show that, for the range of (H^+) calculated, $[OH^-]$ and $[CO_3^{2-}]$ are negligible.]

2. State whether the addition of a small amount of each of the following compounds to a soil solution will increase, decrease, or not change its ANC. Explain your conclusion in each case.

   a. CO_2 c. $Si(OH)_4$ e. Na_3PO_4
   b. $NaNO_3$ d. H_2SO_4 f. CH_3COOH

3. The buffer intensity of soil collected from an A horizon ($f_h = 0.10$) is 1.5 $mol_c\ kg_h^{-1}\ pH^{-1}$ at pH 4.0. Calculate the moles of proton charge per kilogram soil that must be removed to raise the pH value by 0.5 log units.

*4. In the table presented here are data on the buffer intensity at pH 4.5 for surface and subsurface horizons of Spodosols with varying humus content. Calculate β_H from the relationship between soil buffer intensity and humus content. What change in pH is expected for an input of 92 mmol of proton charge per kilogram of soil?

$f_h(kg_h kg^{-1})$	Buffer intensity ($mol_c\ kg^{-1} pH^{-1}$)	$f_h(kg_h kg^{-1})$	Buffer intensity ($mol_c\ kg^{-1} pH^{-1}$)
0.103	0.0088	0.054	0.0054
0.162	0.0166	0.102	0.0084
0.030	0.0068	0.038	0.0054
0.902	0.0738	0.401	0.0416
0.947	0.0824	0.870	0.1122

5. The ratio of ANC to CEC for O horizons of Swiss Inceptisols was observed to depend linearly on pH:

$$\frac{ANC}{CEC} = 0.57\ pH - 1.38 \quad (R^2 = 0.90)$$

   a. Calculate the soil buffer intensity range that corresponds to CEC in the range for Inceptisols (Table 9.1).
   b. Plot TA/CEC against pH in the range 3.0 to 4.0. At what pH value is TA/CEC = 0.5?

6. Given that NH_4^+ uptake by plants usually produces excess cation over anion uptake, and that NO_3^- uptake usually produces the opposite effect, what is the expected change in rhizosphere pH from the uptake of each N species?

7. Long-term field experiments indicate that acidic soils receiving nitrogen fertilizer as $(NH_4)_2SO_4$ decrease in pH, whereas those receiving $NaNO_3$ increase in pH. Give an explanation for these results in terms of the ANC of the soil solution and all the processes described in Section 11.1. (You may neglect deposition processes.)

*8. The following table shows the changes in exchangeable metal cation charge ($mmol_c\ kg^{-1}$) in the O horizons of several forest soils as their pH values were decreased gradually to pH 3.0 by addition of dilute HCl solution.

   a. Calculate the loss of soil adsorbent ANC resulting from the acid input.
   b. Estimate the buffer intensity of each soil at pH 3.

   (*Hint:* Assume that CEC as well as ANC changes after the pH value is decreased to 3.0)

Initial pH	Al^{3+}	Ca^{2+}	Fe^{2+}	H^+	K^+	Mg^{2+}	Mn^{2+}
3.49	0.0	−5.8	−0.8	+5.6	+2.6	−1.6	+0.1
3.60	−5.5	+4.2	−0.6	−4.6	+1.3	−1.6	−0.2
3.64	−2.1	−29.8	−0.2	+9.3	−1.4	−6.0	−0.8
3.97	+6.1	−13.6	+0.4	+16.2	−5.5	−3.4	−2.8
4.19	+13.9	−67.2	+1.5	+10.1	−3.7	−12.2	−5.8

*9. Prepare an activity–ratio diagram like that in Figure 11.2 for the soil solution described in Table 4.4. You may ignore the difference between activity and concentration. Plot a point on the diagram representing (Al^{3+}) and $(Si\ (OH)_4^0)$. What solid phase is predicted to control Al solubility?

10. The cation exchange reaction between Ca^{2+} and Al^{3+} in the O horizons of Spodosols was found to be described well by the linear regression equation

$$\log\left[\left(Ca^{2+}\right)^3 / \left(Al^{3+}\right)^2\right] = -2.03 + 0.95 \pm 0.06$$

$$\times \log\left[E_{CaX_2}^3 / E_{AlX_3}^2\right] \qquad (R^2 = 0.90)$$

where E is a charge fraction that serves as a model of the activity of an exchangeable cation species on a soil adsorbent.

   a. Estimate the value of $K_{ex}^{Ca/Al}$ for the cation exchange reaction. Which cation is preferred by the soil adsorbent?

b. Given that $\tilde{Q} = 3 \text{ mol}_c \text{ m}^{-3}$, prepare an exchange isotherm with the charge fractions of Ca^{2+} used as plotting variables.

11. Manganese toxicity in acidic soils is associated with $(Mn^{2+}) \approx 10^{-3.7}$ in the soil solution. Use relevant information in Table 6.1 to estimate the pH value below which Mn toxicity should occur as pE ranges between 8 and 12.

12. Ammonium sulfate is applied to an acidic soil at the rate of 236 mg $(NH_4)_2SO_4$ per kilogram of soil. Calculate the lime requirement for neutralizing the total acidity expected if complete nitrification occurred, without uptake of NH_4, and the protons produced were entirely adsorbed by the soil.

13. Calculate the lime requirement for the Inceptisols described in Problem 5 to increase their pH value from 3.5 to the point at which total acidity equals zero. Take CEC $= 160 \text{ mmol}_c \text{ kg}^{-1}$.

14. Apply Eq. 11.6 to calculate the lime requirement for soil humic acid carboxyl groups, as described by parameters given in Section 9.5, to increase pH from pH_{dis} to the point at which TA ≈ 0.

*15. Calculate the lime potential of the soil solution described in Table 4.4, then estimate the corresponding ANC/CEC of the soil adsorbent using Eq. 11.27. Neglect the difference between activity and concentration.

Special Topic 7: Measuring pH

As mentioned in Special Topic 5 (Chapter 6), Arnold Beckman developed an instrument for measuring pH based on a glass electrode for detecting protons in aqueous solution. This electrode comprises an outer membrane that adsorbs protons and an inner solution containing a high concentration of Cl^- in contact with a Ag/AgCl electrode. An electrode potential is created because of proton concentration differences across the glass membrane, proton diffusion processes within the membrane, and the oxidation half-reaction

$$Ag\,(s) + Cl^- = AgCl\,(s) + e^- \qquad \log K = 3.75 \qquad (S7.1)$$

at the inner electrode. The pE value for the Ag/AgCl couple is fixed if the activity of Cl^- is maintained constant (Section 6.1), hence the presence of Cl^- in the inner solution. An electric potential difference between the glass electrode and a reference electrode [usually the calomel electrode, at which the reduction half-reaction

$$\frac{1}{2}\, Hg_2Cl_2\,(s) + e^- = Hg\,(\ell) + Cl^- \qquad \log K = 4.13 \qquad (S7.2)$$

takes place] then can be created by a difference in concentration of protons between a soil solution and the inner solution of the glass electrode, provided that a KCl "salt bridge" has been placed between the two electrodes to prevent equilibration of the Cl^- concentration in the soil solution with that in the reference electrode. The salt bridge is a porous plug intervening between the soil solution and a saturated solution of KCl that bathes the calomel electrode. Its purpose is to prevent mixing processes that would allow soil solution Cl^- to equilibrate with the reference electrode and, therefore, contribute to the overall electric potential difference that one wishes to attribute solely to protons.

Electric potential differences (E) in the glass–calomel electrode system are modeled by an equation similar to Eq. 6.18:

$$E = A + E_J + \frac{RT}{F} \ln\ 10\ \text{pH} = A + E_J + 0.05916\ \text{pH} \qquad \text{(S7.3)}$$

where A is a constant parameter that depends on the redox reactions and ion concentrations in the electrodes and E_J is the *liquid junction potential* created by the variation in chemical composition that occurs across the salt bridge. The value of E_J thus depends on the details of the steady-state charge transfer through the liquid junction between a soil solution and the KCl salt bridge—a necessary evil if the electrode system is to respond only to changes in proton concentration. To measure proton activity, E is calibrated directly in terms of the *assigned* pH values of buffer solutions. Therefore, by convention, only *relative* values of pH can be measured:

$$\text{pH (soil solution)} = \text{pH (buffer)} + \frac{E\ \text{(buffer)} - E\ \text{(soil solution)}}{0.05916} \qquad \text{(S7.4)}$$

at 298 K, where the E values are in volts. [This convention differs very much from that used for electron activity, which assigns pE = 0 to the reduction of the proton (third reaction in Table 6.1) at pH = 1 and P_{H_2} = 1 atm, according to Eq. 6.10.]

A principal difficulty in applying eqs. S7.3 and S7.4 to soil solutions is uncertainty regarding the magnitude of E_J. Because soil solution compositions differ greatly from those of pH buffers, the ions diffusing across the salt bridge in the former differ from those in the latter, and the corresponding liquid junction potentials can be quite different as well. It is virtually impossible to know precisely how large this difference in E_J will be, because no method exists for measuring or calculating E_J accurately. If the difference in E_J implicit in Eq. S7.4 is indeed large and unknown, the pH value of the soil solution measured by a glass electrode would be of *no chemical significance.* This conclusion is even stronger if one attempts to apply Eq. S7.4 to soil pastes or suspensions, because then E_J is certainly very different in the soil system from what it is in a standard buffer solution.

To see this issue in more detail, consider the situation in which the difference in E between a standard buffer solution and a soil solution arises solely because of a difference in liquid junction potential, ΔE_J, thus leading to an *apparent* pH difference:

$$\Delta pH_J \equiv \Delta E_J/59.16 \qquad (S7.5)$$

where ΔE_J is in millivolts. Equation S7.5 is derived from combining Eqs. S7.3 and S7.4 under the assumed conditions. A difference in liquid junction potential equal to 10 mV (about 5% of the electrode potential corresponding to the half-reactions in eqs. S7.1 and S7.2) leads to $\Delta pH_J = 0.1$, a rather large systematic error. If, as a rule of thumb, pH measurements in soil solutions are judged to be no more *accurate* than 0.05 log units, this degree of accuracy would require liquid junction potential differences to be no larger than 3 mV. (Of course, whether a pH value can be read from a pH meter to three decimal places becomes irrelevant in this case!)

Compounding this problem is the difference in E_J between a soil solution and a soil suspension caused by the presence of charged colloidal particles in the latter. Suppose that, in fact, pH were the same in both systems. An application of Eq. S7.3 then yields the expression

$$E_{So} - E_{Su} = E_{JSo} - E_{JSu} \qquad (S7.6)$$

where So refers to the soil solution and Su refers to the soil suspension. Equation S7.6 describes the result of comparing the output of glass electrode systems dipping respectively into a soil slurry and its supernatant solution, with the slurry and solution in contact, such that equilibration of the protons in the two has taken place. Thus, under equilibrium conditions, the electric potential difference between the electrode pairs is determined solely by electric potential differences developed at two liquid junctions that involve KCl salt bridges. The two E_J values will differ because of the effect of soil colloids. The fact that this difference can develop is known as the *suspension effect.* It is described in a classic article by Babcock and Overstreet [Babcock, K. L., and R. Overstreet. (1953) On the use of calomel half cells to measure Donnan potentials. *Science* **117**:686.] and recently is discussed in great detail by Oman et al. (Oman, S. F., M. F. Camões, K. J. Powell, R. Rajagopalan, and P. Spitzer. (2007) Guidelines for potentiometric measurements in suspensions. *Pure Appl. Chem.* **79**:67.], who point out that, in addition to anomalous liquid junction potentials, nonunique electric potentials of the kind that often plague electrochemical pE measurements (Section 6.3) also can occur for glass electrodes inserted into soil suspensions.

12

Soil Salinity

12.1 Saline Soil Solutions

A soil is designated *saline* if the conductivity of its aqueous phase (EC_e) obtained by extraction from a saturated paste has a value more than 4 dS m^{-1}. (For a discussion of this measurement, see *Methods of Soil Analysis*, referenced at the end of this chapter. The SI units of conductivity are defined in the Appendix.) About a quarter of the agricultural soils worldwide are saline, but values of $EC_e > 1$ dS m^{-1} are encountered typically in *arid-zone soils,* with a climatic regime that produces evaporation rates that exceed precipitation rates on an annual basis. Ions released into the soil solution by mineral weathering, or introduced there by the intrusion of saline surface waters or groundwater, tend to accumulate in the secondary minerals formed as the soils dry. These secondary minerals include clay minerals (Section 2.3), carbonates and sulfates (Section 2.5), and chlorides. Because Na, K, Ca, and Mg are relatively easily brought into solution—either as exchangeable cations displaced from smectite and illite, or as structural cations dissolved from carbonates, sulfates, and chlorides—it is this set of metals that contributes most to soil salinity. The corresponding set of ligands that contributes then would be CO_3, SO_4, and Cl. Thus arid-zone soil solutions are essentially electrolyte solutions containing chloride, sulfate, and carbonate salts of Group IA and IIA metals.

According to Eq. 4.23, a conductivity of 4 dS m^{-1} corresponds to an ionic strength of 58 mol m^{-3} (i.e., $\log I = 1.159 + 1.009 \log 4 = 1.77$). This salinity is 10% of that in seawater, high enough in an agricultural context that only crops that are relatively salt tolerant can withstand it. Moderately salt-sensitive crops are affected when the conductivity of a soil extract approaches 2 dS m^{-1}, corresponding to an estimated ionic strength of 29 mol m^{-3}. Salt-sensitive crops are affected at 1 dS m^{-1} (I = 14 mol m^{-3}). Thus, with respect to salinity tolerance, a soil can be saline at any ionic strength greater than 15 mol m^{-3} if the plants growing in it are stressed. The visual evidence of this is a reduction in crop growth and yield caused by a diversion of energy from normal physiological processes to those involved in the acquisition of water under osmotic stress.

The chemical speciation of a saline soil solution can be calculated as described in sections 4.3 and 4.4. Total concentration data and the percentage speciation that are representative of the saturation extract of an irrigated Aridisol are listed in Table 12.1. Notable in the table are the dominance of soluble Ca over Mg; the relatively complicated speciation of Ca, Mg, HCO_3^-, SO_4, and PO_4; and the high free-ion percentages for Na, K, Cl, and NO_3. (Note that organic complexes are unimportant for the metal cations considered.) Neutral sulfate complexes reduce the contribution of SO_4 to the ionic strength—and, therefore, the conductivity—by more than one fourth. The computed ionic strength of the soil solution in Table 12.1 is 23 mol m^{-3}, which corresponds to an estimated conductivity of 1.6 dS m^{-1}, which is high enough to affect salt-sensitive crops.

The *alkalinity* of a saline soil solution can be defined by Eq. 11.3, with neglect of organic and Al(III) species, of course, but with inclusion of $B(OH)_4^-$, leading to the expression

$$\text{Alkalinity} = [HCO_3^-] + 2[CO_3^{2-}] + [H_2PO_4^-] + 2[HPO_4^-] + 3[PO_4^{3-}] + [B(OH)_4^-] + [OH^-] - [H^+] \quad (12.1)$$

Table 12.1
Composition and speciation of an Aridisol soil solution (pH 8.0).

Constituent	C_T (mol m^{-3})	Percentage speciation
Ca	5.9	Ca^{2+} (79%), $CaSO_4^0$ (17%), $CaHCO_3^+$ (2%)
Mg	1.3	Mg^{2+} (83%), SO_4^0 (14%), $MgHCO_3^+$ (1%)
Na	1.9	Na^+ (99%)
K	1.0	K^+ (99%), KSO_4^- (1%)
CO_3	3.0	HCO_3^- (92%), $H_2CO_3^0$ (2%), $CaHCO_3^+$ (4%), $CaCO_3^0$ (2%)
SO_4	4.4	SO_4^{2-} (72%), $CaSO_4^0$ (23%), $MgSO_4^0$ (4%)
Cl	5.0	Cl^- (99%)
NO_3	0.28	NO_3^- (99%)
PO_4	0.065	HPO_4^{2-} (46%), $CaHPO_4^0$ (30%), $MgHPO_4^0$ (9%), $H_2PO_4^-$ (5%), $CaPO_4^-$ (9%)
B	0.038	$B(OH)_3^0$ (3%), $B(OH)_4^-$ (6%)

for a soil solution like that described in Table 12.1. (As mentioned in Section 11.3, the concentrations of metal complexes of the anions on the right side of Eq. 12.1 can be included readily.) In the example of Table 12.1, the alkalinity equals 2.8 mol m^{-3}, with 99% derived from bicarbonate. Bicarbonate alkalinity ranging from 1 to 4 mol m^{-3} is common in saline soils.

The pH value of a saline soil solution with an alkalinity derived from bicarbonate is governed by Eq. 11.2:

$$\text{pH} = 7.8 + \log\left(\text{HCO}_3^-\right) - \log\ P_{\text{CO}_2} \tag{12.2}$$

If the Davies equation (Eq. 4.24) is used to calculate the activity coefficient of HCO_3^-, and if the Marion–Babcock equation (Eq. 4.23) is used to relate ionic strength to conductivity, then Eq. 12.2 takes the form

$$\text{pH} = 7.8 + \Delta(\kappa) + \log\left[\text{HCO}_3^-\right] - \log\ P_{\text{CO}_2} \tag{12.3}$$

where

$$\Delta(\kappa) \approx -0.512\left[\frac{0.12\sqrt{\kappa}}{1 + 0.12\sqrt{\kappa}} - 0.0043\kappa\right] \tag{12.4}$$

is a very small correction for ionic strength effects [$\Delta(\kappa) < -0.1$ for $\kappa < 4$ dS m^{-1}]. Equation 12.3 shows that soil solution pH is determined by the bicarbonate alkalinity and the $CO_2(g)$ pressure in atmospheres. Conversely, the $CO_2(g)$ pressure at equilibrium can be calculated with Eq. 12.3 from measured pH and alkalinity values. Given $P_{CO_2} = 10^{-3.5}$ atm, the atmospheric value, the range of $\left[HCO_3^-\right]$ quoted earlier leads to pH values in the range 8 to 9. For the soil solution of Table 12.1, pH = 8.0, $\left[HCO_3^-\right] = 0.00276$ mol dm^{-3}, and a $CO_2(g)$ pressure of $10^{-2.9}$ atm is calculated. Note that Eq. 12.3 predicts increasing soil solution pH with either increasing bicarbonate alkalinity or decreasing $CO_2(g)$ pressure. This relationship has often been observed in arid-zone soils.

12.2 Cation Exchange and Colloidal Phenomena

Exchange reactions among the cations Na^+, Ca^{2+}, and Mg^{2+} are of great importance in arid-zone soils. These reactions, in the convention of Eq. 9.14b, can be expressed as

$$2\,\text{NaX(s)} + \text{Ca}^{2+} = \text{CaX}_2\text{(s)} + 2\text{Na}^+ \tag{12.5a}$$

$$2\,\text{NaX(s)} + \text{Mg}^{2+} = \text{MgX}_2\text{(s)} + 2\text{Na}^+ \tag{12.5b}$$

$$\text{MgX}_2\text{(s)} + \text{Ca}^{2+} = \text{CaX}_2\text{(s)} + \text{Mg}^{2+} \tag{12.5c}$$

where X represents 1 mol intrinsic negative surface charge. (Note that any one of these reactions can be obtained by combining the other two.) Exchange isotherms based on the reactions in Eq. 12.5 are shown in Figure 9.2 for an Aridisol. They indicate, as observed typically for arid-zone soils, that Ca and Mg are preferred over Na, and that Ca is preferred slightly over Mg by the soil adsorbent. A conditional exchange equilibrium constant for the reaction in Eq. 12.5a is presented in Eq. 9.18, and analogous expressions apply to the reactions in Eqs. 12.5b and 12.5c. Although these Vanselow model selectivity coefficients do vary with adsorbate composition, the variability is small enough to neglect to a first approximation. For example, if $K_V^{Na/Ca} \approx 16$, then Eq. 9.20 predicts $E_{Ca} \approx 0.8$ when $\tilde{E}_{Ca} \approx 0.05$ [along with $\tilde{Q} = 0.05\ mol_c\ dm^{-3}$ $(A = 0.037)$], which agrees with the exchange isotherm for Ca in Figure 9.2.

By the same process used to derive Eq. 9.20 from Eq. 9.18, one can rewrite Eq. 9.18 in the form

$$K_V^{Na/Ca} = \Gamma \frac{[Na^+]^2 / [Ca^{2+}]}{4\,E_{Na}^2}\left(1 - E_{Na}^2\right) \qquad (12.6)$$

where $\Gamma = \gamma_{Na}^2/\gamma_{Ca}$ and γ is a single-ion activity coefficient (Eq. 4.24). Equation 12.6 contains two important chemical variables. The expression

$$SAR \equiv 10^{3/2}\,[Na^+] / [Ca^{2+}]^{1/2} \qquad (12.7)$$

defines the *sodium adsorption ratio* (SAR) and $100\ E_{Na} \equiv ESP$ defines the *exchangeable sodium percentage* (ESP). With these two definitions, Eq. 12.7 becomes the expression

$$K_V^{Na/Ca} = 2.5\,\Gamma\left(\frac{SAR}{ESP}\right)^2\left[1 - (ESP/100)^2\right] \qquad (12.8)$$

Given a value for $K_V^{Na/Ca}$, Eq. 12.8 provides a unique relationship between ESP and SAR. This relationship is essentially a linear one for SAR $< 20\ mol^{1/2}m^{-3/2}$ and $K_V^{Na/Ca} > 1$. The variable defined by Eq. 12.7 is equivalent to using SI units of moles per cubic meter instead of moles per liter:

$$SAR \equiv c_{Na}/\sqrt{c_{Ca}} \qquad (12.9)$$

where c is concentration in SI units (see the Appendix). Because direct measurements of the free-ion concentrations of Na^+ and Ca^{2+} are not common in field work, SAR in Eq. 12.9 usually is replaced by the variable SAR_p:

$$SAR_p \equiv Na_T/\sqrt{Ca_T} \qquad (12.10)$$

where the subscript p indicates the *practical* SAR, and *total* concentrations in SI units are indicated on the right side. This variable is typically smaller than SAR because of greater soluble complex formation by Ca^{2+} than Na^+. (For example, it is 10% smaller for the soil solution speciated in Table 12.1). Statistical analyses of the SAR_p–SAR relationship in soil saturation extracts indicate that SAR_p is about 12% smaller than SAR, on average. This small difference and the deviation of Eq. 12.8 from a 1:1 line are often neglected in applications to irrigation water quality evaluation, with the result that Eq. 12.8 simplifies to the expression

$$ESP \approx SAR_p \quad (SAR_p < 30) \qquad (12.11)$$

Equation 12.11 has long been used in field studies of ESP. Given the several approximations leading to it, along with its intended application to evaluate irrigation water quality, the expedient assumption that $K_V^{Na/Ca} \approx K_V^{Na/Mg}$ for $Na \rightarrow Mg$ exchange also is made, such that SAR_p can be generalized to

$$SAR_p \equiv Na_T/(Ca_T + Mg_T)^{1/2} \qquad (12.12)$$

and incorporated into Eq. 12.11. Although the chain of assumptions linking Eq. 12.11 to Eq. 12.6 via Eq. 12.12 is somewhat tenuous and long, it is well defined enough to make the conceptual basis of the ESP–SAR_p relation apparent.

The important differences between monovalent and bivalent cations in respect to the stability of colloidal suspensions are discussed in Chapter 10 (see Eq. 10.17 and Problem 6). Laboratory studies of the stability ratio for suspensions of soil colloids based on turbidity measurements (see Problems 8 and 9 in Chapter 10) suggest that stable suspensions are the rule if ESP $\gtrsim$ 15%. Studies of the permeability characteristics and aggregate structures in arid-zone soils have substantiated this effect of adsorbed Na^+, leading to the designation *sodic* for a soil in which the ESP value is larger than about 15. In these soils, if the ionic strength is low, colloids will tend to disperse in the soil solution, and a reduction in permeability will occur because of aggregate failure and swelling phenomena (Section 10.2). However, the dispersive effect of exchangeable Na^+ will be observed only if the electrolyte concentration is less than that required to maintain the integrity of soil aggregate structures. The upper limit of this concentration is the *ccc* (Section 10.3), but the electrolyte concentration at which a noticeable (say, 15%) decline in soil permeability occurs may be lower than the *ccc*. Regardless of the exact value of this threshold electrolyte concentration, the key point is that *soil salinity tends to counteract the effect of exchangeable sodium on soil aggregate structure.*

Equation 10.17, which quantifies the Schulze–Hardy Rule (Section 10.3), implies that Ca-saturated colloids flocculate at electrolyte concentrations about 60 times smaller than those required to flocculate Na-saturated colloids. Thus ESP (or SAR) must be related to colloidal stability and, therefore, aggregate failure. This kind of relationship has been found in a large number of experiments with soils. Examples are shown in figures 12.1 and 12.2. They are termed *Quirk–Schofield diagrams,* graphs of the electrolyte concentration (or conductivity) below which significant deterioration in aggregate soil structure should occur (as measured, for example, by a 15% loss of permeability), plotted against the SAR_p value above which the same deterioration in soil structure

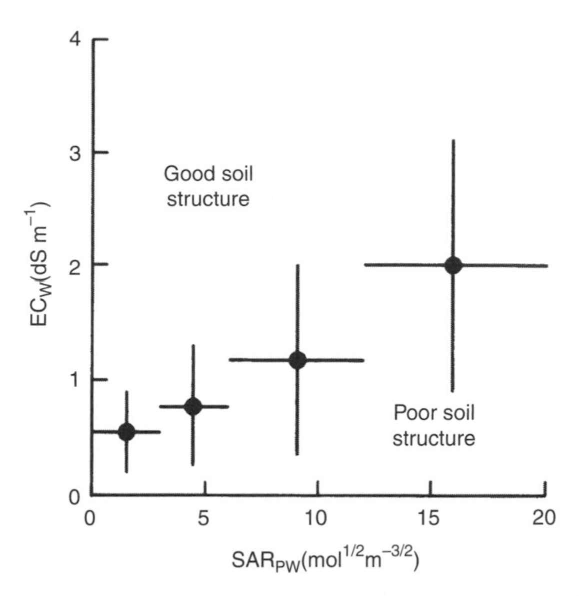

Figure 12.1. A Quirk–Schofield diagram for California soils based on the properties of water applied for irrigation (conductivity vs. practical sodium adsorption ratio). Note the large error bars to allow for effects of pH and soil variability. Data from Shainberg, I., and J. Letey (1984) Response of soils to sodic and saline conditions. *Hilgardia* 52(2):1–55.

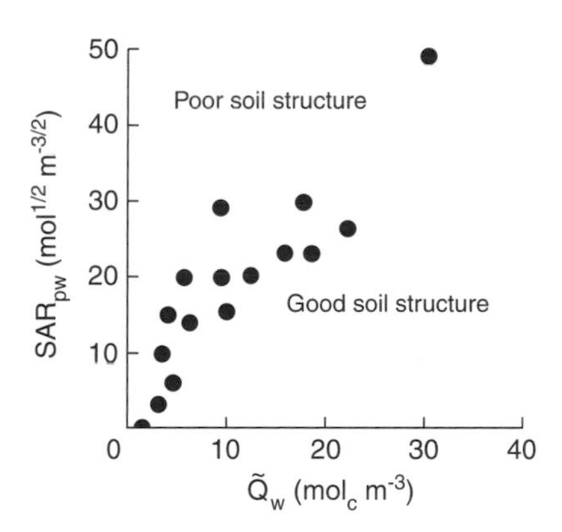

Figure 12.2. A Quirk–Schofield diagram for irrigated soils based on the properties of irrigation water (practical SAR vs. the threshold electrolyte charge concentration causing a 15% reduction in soil permeability). Note that the graph has coordinate axes reversed from those in Figure 12.1. Data from Quirk, J.P. (2001) The significance of the threshold and turbidity concentrations in relation to sodicity and microstructure. *Aust. J. Soil Res.* **39**:1185–1217.

should take place. Figures 12.1 and 12.2 are Quirk–Schofield diagrams based on experiments relating mainly to California soils. A "window," allowing for variability among soils, separates regions of expected good and poor soil structure as expressed by permeability characteristics. Note that any soil for which the SAR_p electrolyte concentration combination falls into the poor soil structure region could be termed *sodic*, insofar as soil permeability is concerned. An SAR_p value of 15 would lead to poor soil structure only if the electrolyte concentration dropped below about 10 mol m^{-3} (EC $\lesssim$ 1.5 dS m^{-1}). On the other hand, an apparently low SAR_p value of 3.0 would still lead to poor soil structure if the electrolyte concentration dropped below about 2 mol m^{-3} (EC $\lesssim$ 0.2 dS m^{-1}). A saline soil as defined conventionally should not have poor structure unless the SAR_p value rises well above 30.

As discussed in Section 10.4, the stability ratio for a colloidal suspension is also affected by pH and strongly adsorbing ions (see Table 10.1). Laboratory studies of colloidal suspensions from arid-zone soils indicate that pH effects are relatively minor at pH $>$ 6, but that polymeric ions (e.g., Al-hydroxy polymers or humus) exert strong effects, with dissolved humus enhancing stability considerably (see Problem 12 in Chapter 10).

12.3 Mineral Weathering

Soils in arid regions are often at the early stage of the Jackson–Sherman weathering sequence (Table 1.7) and, therefore, they contain silicate, carbonate, and sulfate minerals that are relatively susceptible to dissolution reactions in percolating water. The composition and structures of these minerals are described in sections 2.2 and 2.5. Their dissolution reactions are discussed in sections 5.1 and 5.5. (See also Problem 15 in Chapter 1; Problem 11 in Chapter 2; and problems 3, 4, 6, and 8 in Chapter 5.) Laboratory studies have shown that these reactions may add 3 to 5 $mol_c m^{-3}$ in charge concentration to percolating waters, with most of the addition coming from Ca, Mg, and HCO_3 under alkaline soil conditions (see Table 12.1). A dissolution reaction of the easily weatherable silicate mineral anorthite, which can produce this effect, is illustrated in Eq. 2.8 and discussed further in Problem 6 of Chapter 5.

Soil mineral weathering that increases the salinity of the soil solution and enriches it in Ca and Mg has important implications for the colloidal phenomena discussed in Section 12.2. If water entering a soil has a very low electrolyte concentration (e.g., rainwater or irrigation water diverted from pristine surface waters), a very small SAR_{pw} in the water would be sufficient to cause problems with soil structure and permeability (figs. 12.1 and 12.2). For example, SAR_{pw} values as low as 3.0 can be deleterious to soil structure if the applied water EC_w is around 0.5 dS m^{-1}. On the other hand, infiltrating water that causes soil minerals to dissolve and increase the conductivity of the soil solution to near 1.0 dS m^{-1} would make only SAR_p values $>$ 5.0 of concern. Moreover, if most of the increase in electrolyte concentration came from Ca^{2+}

and Mg^{2+}, then the SAR_p value would actually drop in the equilibrated soil solution (Eq. 12.12), further diminishing the chance of adverse soil structure effects. The conclusion to be drawn is that *soils containing easily weatherable minerals will be less sensitive to percolating low-salinity waters than those that are depleted of easily weatherable minerals.*

Increasing salinity tends to enhance the solubility of weatherable minerals. This effect can be predicted on the basis of the ionic strength dependence of single-ion activity coefficients (Eq. 4.24). Consider, for example, the mineral gypsum ($CaSO_4 \cdot 2H_2O$), with the solubility product constant defined in Eq. 5.8:

$$K_{so} = \left(Ca^{2+}\right)\left(SO_4^{2-}\right) = 2.4 \times 10^{-5} \tag{12.13}$$

The solubility of gypsum is related to the *concentration* of Ca^{2+} (see Problem 3 in Chapter 5) and, therefore, to the *conditional* solubility product, K_{soc}:

$$K_{soc} = \left[Ca^{2+}\right]\left[SO_4^{2-}\right] = K_{so}/\gamma_{Ca}\gamma_{SO_4} \tag{12.14}$$

where the γ's are single-ion activity coefficients. It follows from Eqs. 4.24 and

$$\log\ K_{soc} = \log\ K_{so} + 4.096\left[\frac{\sqrt{I}}{1+\sqrt{I}} - 0.3I\right] \tag{12.15}$$

where I is ionic strength in moles per cubic decimeter. This equation shows that the conditional solubility product will *increase* with ionic strength, as long as the term in the square root of I exceeds 0.3 I (i.e., as long as $I < 2\ mol\ dm^{-3}$). For example, experimental studies of the enhancement of gypsum solubility in NaCl solutions show that when the concentration of NaCl increases up to $0.5\ mol\ dm^{-3}$, the total concentration of Ca in equilibrium with gypsum more than doubles.

The weathering reactions of calcite ($CaCO_3$) are of great importance in arid-zone soils. As discussed in Section 5.1, the dissolution of this mineral far from equilibrium is surface controlled and, therefore, follows a zero-order kinetics expression (Eq. 5.2, with Ca^{2+} replacing A). The net rate of precipitation–dissolution near equilibrium can be expressed analogously to Eq. 5.16 (see Problem 4 in Chapter 5):

$$\frac{d\left[Ca^{2+}\right]}{dt} = k_p K_{soc}\left(1 - \Omega\right) \tag{12.16}$$

where k_p is a rate coefficient for precipitation that depends on pH and specific surface area and $K_{so} = 3.3 \times 10^{-9}$ is the solubility product constant for well-crystallized calcite undergoing the dissolution reaction

$$CaCO_3(s) = Ca^{2+} + CO_3^{2-} \tag{12.17}$$

$\left(Ca^{2+}\right)\left(CO_3^{2-}\right)$ is the IAP, and $\Omega = \left(Ca^{2+}\right)\left(CO_3^{2-}\right)/K_{so}$. Equation 12.16 describes the net rate of precipitation of calcite if $\Omega > 1$ (supersaturation).

For that case, if $\Omega < 10$ and pH > 8, $k_p K_{soc} \approx 2.5 \times 10^{-9}$ mol L^{-1} s^{-1}. For an initial Ca^{2+} concentration of, say, 0.001 mol dm^{-3}, Table 4.2 indicates a half-life on the order of hours $\left(t_{1/2} = \left[Ca^{2+}\right]_0 / 2k_p K_{so}\Omega\right)$.

Another useful rate expression analogous to that in Eq. 5.16 is obtained by transforming the IAP according to the calcite dissolution reaction in Eq. 5.32:

$$CaCO_3\,(s) + H^+ = Ca^{2+} + HCO_3^- \tag{12.18}$$

Given the equilibrium constant for the bicarbonate formation reaction

$$H^+ + CO_3^{2-} = HCO_3^- \quad K_2 = 10^{10.329} \tag{12.19}$$

where $K_2 = \left(HCO_3^-\right) / \left(H^+\right)\left(CO_3^{2-}\right)$, one can rewrite Eq. 5.16 in the alternative form

$$\frac{d\left[Ca^{2+}\right]}{dt} = k_p K_{soc}\left[1 - \left(Ca^{2+}\right)\left(HCO_3^-\right)/K_2 K_{so}\left(H^+\right)\right] \tag{12.20}$$

The relationship $(H^+) \equiv 10^{-pH}$ and the *Langelier Index*, pH–pH_s, where pH_s is defined by the equation

$$\left(Ca^{2+}\right)\left(HCO_3^-\right)/K_2 K_{so} \equiv 10^{-pH_s} \tag{12.21}$$

may be used to transform Eq. 12.20 into the simpler expression

$$\frac{d\left[Ca^{2+}\right]}{dt} = k_p K_{soc}\left[1 - 10^{pH - pH_s}\right] \tag{12.22}$$

Equation 12.22 yields estimates of the rate of calcite precipitation or dissolution based on the Langlier index. To illustrate this relationship, consider Ca^{2+} and HCO_3^- activities based on the chemical speciation of the soil solution described in Table 12.1: $(Ca^{2+}) \approx 2.59 \times 10^{-3}$, $HCO_3^- \approx 2.38 \times 10^{-3}$, and $pH_s = 7.02$. It follows from Eq. 12.22 that, at pH 8, the Langlier index equals 0.98 and the soil solution is supersaturated with respect to calcite (i.e., the right side of Eq. 12.22 is negative). The same conclusion is reached by calculating Ω directly for the soil solution:

$$\Omega = \frac{\left(Ca^{2+}\right)\left(CO_3^{2-}\right)}{K_{so}} = \frac{2.59 \times 10^{-3} \times 1.19 \times 10^{-5}}{3.3 \times 10^{-9}} = 9.3$$

Because $\Omega > 1$, the soil solution is supersaturated with respect to well-crystallized calcite.

The value of Ω just calculated leads to an IAP $\approx 10^{-8}$ for calcite in the soil solution described in Table 12.1. This high value in fact has been observed consistently in a large number of investigations of calcite solubility in arid-zone soils. The cause of ubiquitous supersaturation is not analytical error, crystalline disorder (Section 5.5), or Mg substitution for Ca (Section 2.5), but most likely is a kinetics-based mechanism relating to Eq. 12.16 (see Problem 6 in Chapter 5). One possibility is a reduction in k_p produced by the adsorption of soluble humus on the surfaces of calcite particles. Laboratory research has shown that calcite precipitation is inhibited greatly by adsorbed fulvic acid, which can reduce k_p by two orders of magnitude. Another possibility is sustained production of bicarbonate alkalinity through the oxidation of soil humus. Arid-zone soils incubated with plant litter at ambient P_{CO_2} readily produce HCO_3^- that increases in concentration as the plant materials decompose. High bicarbonate concentrations may be sustained under steady-state conditions, leading to a persistently high IAP for calcite.

12.4 Boron Chemistry

Boron is a trace element in soils (Table 1.1) that occurs typically as a coprecipitated element in secondary metal oxides, clay minerals, and mica, or as a substituent in humus (tables 1.4 and 1.5). Besides its occurrence as a separate solid phase in tourmaline, a number of Na, K, Ca, and Mg borates have been identified in saline geological environments. Among them are borax [$Na_2B_4O_5(OH)_4 \cdot 8H_2O$], nobleite [$CaB_6O_9(OH)_2 \cdot 3H_2O$], inyoite [$CaB_3O_3(OH)_5 \cdot 4H_2O$], colemanite [$CaB_3O_4(OH)_3 \cdot H_2O$], and inderite [$MgB_3O_3(OH)_5 \cdot 5H_2O$]. Dissolution equilibrium constants and a representative value for the activity of Na^+, Ca^{2+}, or Mg^{2+} in arid-zone soil solutions (Table 12.1) lead to the conclusion that borate minerals would support very high B solubilities in the soil solution. For example, to achieve the concentration of $B(OH)_3^0$ indicated in Table 12.1 (35 mmol m^{-3}), only very small amounts of these minerals would have to be present in soil—amounts that should be easily lost by normal leaching. The dissociation reaction

$$B(OH)_3^0 + H_2O(\ell) = B(OH)_4^- + H^+ \tag{12.23}$$

has an equilibrium constant equal to $5.8 \times 10^{-10} (\log K = 9.23)$. Therefore, the $B(OH)_4^-$ species will not be significant in soil solutions [i.e., will not be equal to the concentration of $B(OH)_3^0$] until the pH value approaches 9. This is true also for complexes like $CaB(OH)_4^+$ and $MgB(OH)_4^+$, which account for less than 0.5% of the boron species in the soil solution described in Table 12.1.

Boron concentrations in arid-zone soil solutions can range up to 2 mol m^{-3}, depending on the mineralogy of soil parent material or the composition of groundwater. Sensitive crop plants are affected by concentrations greater than 0.046 mol B m^{-3}, and almost all crops will be affected at concentrations greater than 0.5 mol m^{-3}. These threshold concentrations translate to irrigation water concentrations of 65 and 277 mmol m^{-3} respectively. Leaching experiments indicate that high B concentrations cannot be reduced easily by percolating fresh water (>3–5 years required). The rate of B removal is much less than that for chloride, and a resurgence of B concentration can occur after it has been reduced by extensive leaching. This behavior suggests not only that soil B is released slowly from minerals in which it is a trace component (Tables 1.4 and 1.5), but also that it adsorbs strongly onto soil particle surfaces.

An adsorption envelope for $B(OH)_4^-$ on a calcareous Entisol is shown in Figure 8.5 (see also Fig. 7.9). The resonance feature in it results from an interplay between adsorptive and adsorbent charge, as discussed in Section 8.4. The similarity in adsorption envelopes between F^- and $B(OH)_4^-$ in soils (cf. figs. 8.4 and 8.5), as well as studies of B adsorption by specimen minerals, suggest that the principal adsorption mechanism is ligand exchange with surface hydroxyls (Eq. 8.25):

$$\equiv SOH(s) + B(OH)_3^0 = \equiv SOB(OH)_3^- + H^+ \tag{12.24}$$

Support for this mechanism has come from infrared spectroscopy and from modeling studies wherein the *constant capacitance model* has been applied to describe adsorption envelopes like that in Figure 8.5. These studies and experimental investigations with specimen minerals indicate that surface OH groups are the main reactive sites for B adsorption. The low leachability of adsorbed boron then derives from the strong inner-sphere surface complex formed in conjunction with the reaction in Eq. 12.24.

The constant capacitance model describes specific borate adsorption based on Eq. 12.24 and two surface acid–base reactions like that depicted in Eq. 8.25a and the corresponding proton dissociation reaction:

$$\equiv SOH\,(s) + H^+ = \equiv SOH_2^+\,(s) \tag{12.25a}$$

$$\equiv SOH\,(s) = \equiv SO^-\,(s) + H^+ \tag{12.25b}$$

These reactions govern adsorbent surface charge, whereas Eq. 12.23 governs adsorptive charge, with the connection between them mediated by the reaction in Eq. 12.24. According to the model, the equilibrium constants for the reactions in eqs. 12.24 and 12.25 are

$$K_+ = \frac{[\equiv SOH_2^+]}{[\equiv SOH]\,[H^+]} \exp\left(F\sigma_p/a_sC/RT\right) \tag{12.26a}$$

$$K_- = \frac{[\equiv SO^-]\,[H^+]}{[\equiv SOH]} \exp\left(-F\sigma_p/a_sC/RT\right) \tag{12.26b}$$

$$K_B = \frac{[\equiv SOB(OH)_3^-]\,[H^+]}{[\equiv SOH]\,[B(OH)_3^0]} \exp\left(-F\sigma_p/a_sC/RT\right) \tag{12.26c}$$

where F is the Faraday constant (coulombs per mole of charge, C mol_c^{-1}), R is the molar gas constant (joules per mole per kelvin, J mol^{-1} K^{-1}), and T is the absolute temperature (kelvin, K).

The exponential factors in Eq. 12.26 contain the net particle charge (Section 7.3)

$$\sigma_p = \left\{ [\equiv SOH_2^+] - [\equiv SO^-] - [\equiv SB(OH)_4^-] \right\} / c_s \qquad (12.27)$$

where c_s is a solids concentration (e.g., Eq. 3.7), which is then divided by the product of specific surface area and a capacitance density (C) with a default value of 1.06 F m^{-2} (the same as coulomb-squared per joule per square meter, $C^2\ J^{-1}\ m^{-2}$), as determined through a broad variety of applications of the model to specific adsorption data. These factors are model expressions for the *activity coefficients* of the surface species that appear in the numerator in each equation for an equilibrium constant. In this sense, the model is a generalization of approaches like that in the biotic ligand model (Section 9.4), for which equilibrium constants contain only concentrations (Eq. 9.22).

Besides the universal parameter F/RT (= 38.917 C J^{-1} at 298 K), the valence of the surface species (e.g., +1 for $\equiv SOH_2^+$) and the particle charge enter into an activity coefficient. When these two parameters have the same sign, the exponential factor is more than one and the surface species concentration is reduced relative to its value at the p.z.c. [i.e., when $\sigma_p = 0$ (Section 10.4)]. When the two parameters are of opposite sign, the exponential factor is less than one and the reverse situation occurs. Thus, the constant capacitance model activity coefficient represents the effect of coulomb interactions between the adsorptive and adsorbent analogously to the way a single-ion activity coefficient does for aqueous species (Section 4.5). The capacitance density C modulates this effect, but, because of its unit value, acts mainly as a conversion factor between the units of σ_p/a_s and those of F/RT so as to render the exponent in the activity coefficient expression dimensionless. Application of the model to a large number of soils has led to regression equations that can be used to estimate the three equilibrium constants in Eq. 12.26 from just four soil properties: specific surface area, Al oxide content, and the content of both organic and inorganic C. The empirical coefficients in these equations are

$$\log K_+ = 7.85 - 0.102 \ln(f_{oc}) - 0.198 \ln(f_{ioc}) - 0.622 \ln(Al) \qquad (12.28a)$$

$$\log K_- = -11.97 + 0.302 \ln(f_{oc}) + 0.584 \ln(f_{ioc}) + 0.302 \ln(Al) \qquad (12.28b)$$

$$\log K_B = -9.14 - 0.375 \ln(a_s) + 0.167 \ln(f_{oc}) + 0.111 \ln(f_{ioc}) + 0.466 \ln(Al) \qquad (12.28c)$$

where a_s is in square meters per gram, whereas f_{oc}, f_{ioc}, and Al are in grams per kilogram. Once the model equilibrium constants have been estimated, chemical speciation calculations as described in Section 4.4 can be performed with the exponential factors in Eq. 12.26 treated like aqueous species activity coefficients and σ_p in Eq. 12.27 playing a role analogous to ionic strength. The typical ranges of values of the log equilibrium constants are approximately: $7.3 < \log K_+ < 9.4$, $-12.6 < \log K_- < -10.5$, and $-8.9 < \log K_B < -7.3$.

12.5 Irrigation Water Quality

The sustainable use of a water resource for the irrigation of agricultural land requires that there be no adverse effects of the applied water in the soil environment. From the perspective of soil chemistry, all irrigation waters are mixed electrolyte solutions. Their chemical composition, which reflects their source and postwithdrawal treatment, may not be compatible with the suite of compounds and weathering processes that exist in the soils to which they are applied. Adding to this the salt-concentrating effects of evaporation, crop extraction of water, and fertilizer amendments, one readily sees the possibility that irrigated soils can become saline or sodic without careful management.

The chemical *properties* of irrigation water that must be identified and controlled to maintain the water suitable for agricultural use are termed *irrigation water quality criteria.* The *numerical interpretation* of the water quality criteria to achieve goals in irrigation water quality management leads to *water quality standards.* These two distinct aspects of irrigation water quality are determined in the first case by the results of field and laboratory research and in the second by research data combined with the collective experience of extension scientists, farm advisers, and growers.

The three principal water quality-related problems in irrigated agriculture are *salinity hazard, sodicity hazard,* and *toxicity hazard.* Irrigation water quality

Table 12.2
Irrigation water quality standards to control soil salinity and sodicity hazards.[a]

	Restriction on water use		
	None	Slight to moderate	Severe
Salinity hazard EC_w ($dS\ m^{-1}$)	< 0.75	0.75–3.0	> 3.0
Sodicity hazard SAR_{pw} range[b] ($mol^{1/2}\ m^{-3/2}$)		EC_w ($dS\ m^{-1}$)	
0–3	>0.7	0.7–0.2	<0.2
3–6	>1.2	1.2–0.3	<0.3
6–12	>1.9	1.9–0.5	<0.5
12–20	>2.9	2.9–1.3	<1.3

[a]Adapted from Ayers, R. S., and D. W. Wescot. (1985) *Water quality for agriculture.* FAO irrigation and drainage paper no. 29, rev. 1. FAO, Rome.
[b]SAR_{pw} defined in Eq. 12.12 for total concentrations in *irrigation water.* The first three ranges correspond to sodicity hazard criteria of "none", "moderate", and "severe" respectively.

Table 12.3
The factor X(LF) in Eq. 12.30.[a]

LF	X(LF)	LF	X(LF)
0.05	3.2	0.30	1.0
0.10	2.1	0.40	0.9
0.15	1.6	0.50	0.8
0.20	1.3	0.60	0.7
0.25	1.2	0.70	0.6

[a]Adapted from Ayers, R. S., and D. W. Wescot. (1985) *Water quality for agriculture.* FAO irrigation and drainage paper no. 29, rev. 1. FAO, Rome.

standards to control salinity hazard are listed in Table 12.2. They are designated preferentially by three classes of conductivity (EC_W), measured in decisiemens per meter. These classes correspond approximately to groupings of agricultural crops into sensitive, relatively sensitive, and relatively tolerant categories respectively. Thus, for example, sensitive crops require $EC_W < 0.75$ dS m^{-1}, and only relatively tolerant crops can withstand $EC_W > 3$ dS m^{-1} without significant yield reduction. According to Eq. 4.22, the three EC_W ranges in Table 12.3 are equivalent to the ionic strength ranges: $I < 11$ mol m^{-3}, $11 < I < 44$ mol m^{-3}, and $I > 44$ mol m^{-3}.

The definition of a saline soil refers to the conductivity of the soil saturation extract (EC_e), not to that of applied water. Even though EC_W is recommended to be < 3 dS m^{-1}, the validity of this restriction depends on knowing the relationship between EC_W and EC_e in the root zone. This relationship continues to be the subject of much research in the chemistry of soil salinity, because many complicated factors enter into it, even in the absence of external effects from rainwater and shallow groundwater. As a rule of thumb, the steady-state value of EC_e that results from irrigation with water of conductivity EC_W is estimated from a knowledge of the *leaching fraction* (LF) of the applied water. The leaching fraction is defined by the equation

$$\text{LF} = \frac{\text{volume of water leached below root zone}}{\text{volume of water applied}} \tag{12.29}$$

Typically, LF is in the range 0.05 to 0.20, meaning that 5% to 20% of the water applied leaches below the root zone whereas 80% to 95% is used in evapotranspiration. With the value of LF known, the average value of EC_e in the root zone is estimated as

$$EC_e = X(\text{LF})EC_W \tag{12.30}$$

where X(LF) is a function with a dependence on LF that has been worked out on the basis of experience with typical irrigated, cropped soils. The function

X(LF) is given in numerical form in Table 12.3. As an example of its use, if water with $EC_w = 1.2$ dS m^{-1} is applied and LF = 0.25, then EC_e is predicted to be 1.44 dS m^{-1}, on average, in the root zone. Note that LF > 0.3 results in $EC_e < EC_w$, and that LF $\leq$ 0.1 will produce a saline soil if water with $EC_w >$ 2 dS m^{-1} is applied.

Irrigation water quality standards to control sodicity hazard are also listed in Table 12.2. They reflect the interplay between electrolyte concentration and exchangeable cation composition discussed in Section 12.2. Thus, for example, if SAR_{pw} is in the range of 3 to 6 $mol^{1/2}$ $m^{-3/2}$ and EC_w is > 1.2 dS m^{-1}, the development of poor soil structure from exchangeable sodium is unlikely because the electrolyte concentration in the applied water is large enough to maintain the integrity of soil aggregates. It is instructive to compare Table 12.2 with figures 12.1 and 12.2.

The cation exchange relationship on which the use of SAR_{pw} is based refers to SAR in the soil solution, not in applied irrigation water. Like EC_w and EC_e, the relationship between SAR_{pw} and the soil solution SAR is the subject of current research. The conversion of SAR_{pw} to an SAR_w value involving *free*-cation concentrations can be made with the help of the "12% rule of thumb" mentioned in Section 12.2. More serious, usually, is the need to account for calcite precipitation or dissolution as the irrigation water percolates into soil under the influence of ambient $CO_2(g)$ pressures. For this purpose, the *Suarez adjusted sodium adsorption ratio* may be used to estimate ESP with Eq. 12.8. This parameter, denoted *adj RNa,* is defined by the equation

$$\text{adj RNa} \equiv c_{\text{Naw}} \big/ \left[c_{\text{Mgw}} + c_{\text{Ca}}^{\text{eq}}\right]^{1/2} \tag{12.31}$$

where c is a free-cation concentration in moles per cubic meter, and c_{Ca}^{eq} is the concentration of Ca^{2+} in a soil solution having the same activity ratio (HCO_3^-) / (Ca^{2+}) as the irrigation water when it is in equilibrium with calcite at a soil value of P_{CO_2} (in atmospheres).

The relationship between c_{Ca}^{eq} and the $[HCO_3^-] / [Ca^{2+}]$ ratio in irrigation water, necessary to apply Eq. 12.31, can be derived from the relationship between the relative saturation and the Langlier index implicit in Eq. 12.22:

$$\Omega_w = 10^{pH_w - pH_s} \tag{12.32}$$

where pH_w is now the pH value of the irrigation water. Equation 12.2 can be combined with Eq. 12.21 to derive from Eq. 12.32 the alternative expression:

$$\Omega_w = \left(Ca^{2+}\right)_w \left(HCO_3^-\right)_w^2 / 10^{2.5} K_{so} P_{CO_2} \tag{12.33}$$

where the numerical factor is equal to $K_2/10^{7.8}$. The denominator in Eq. 12.33 is equal to $\left(Ca^{2+}\right)_{eq} \left(HCO_3^-\right)_{eq}^2$, as can be seen by setting $\Omega_w = 1.0$ in the equation and changing "w" to "eq" for that case. It follows that

$$(Ca^{2+})_{eq} = \frac{(Ca^{2+})_w (HCO_3^-)_w^2}{(HCO_3^-)_{eq}^2 \, \Omega_w}$$

which can be transformed to the equation

$$(Ca^{2+})_{eq}^3 = (Ca^{2+})_w^3 / \Omega_w \tag{12.34}$$

on multiplying by $(Ca^{2+})_{eq}^2$ on both sides, then multiplying by $[(Ca^+)_w/(Ca^{2+})_w]^2$ on the right side only using the condition $(HCO_3^-)_{eq}/(Ca^{2+})_{eq} = (HCO_3^-)_w/(Ca^{2+})_w$ assumed by hypothesis. The substitution of Eq. 12.33 into Eq. 12.34 yields the expression desired:

$$(Ca^{2+})_{eq} = \left\{ \frac{10^{2.5} K_{so}}{[(HCO_3^-)_w / (Ca^{2+})_w]^2} \right\}^{\frac{1}{3}} P_{CO_2^{1/3}} \tag{12.35}$$

Equation 12.35 can be used to calculate c_{Ca}^{eq} (which differs by a factor of 10^3 from $[Ca^{2+}]_{eq}$) after values of P_{CO_2}, $[HCO_3^-]/[Ca^{2+}]$, the activity coefficients of Ca^{2+} and HCO_3^- and K_{so} have been chosen. The value of K_{so} at 25 °C ranges from 3.3×10^{-9} to 4.1×10^{-7}, depending on the crystallinity of calcite (Section 5.5). Alternatively, the IAP value of 10^{-8} can be used as a surrogate for K_{so} in soils (Section 12.3). The single-ion activity coefficients can be estimated using a selected EC_w along with eqs. 4.23 and 4.24. For example, if $EC_w = 1$ dS m^{-1}, then $\gamma_{HCO_3} = 0.886$ L mol^{-1} and $\gamma_{Ca} = 0.616$ L mol^{-1}. Suppose that $[HCO_3^-]/[Ca^{2+}] = 2$. Then, if $K_{so} = 10^{-8}$ and $P_{CO_2} = 10^{-3}$ (Section 5.5), Eq. 12.35 yields $c_{Ca}^{eq} = 1.1$ mol m^{-3}. This prediction can be introduced into Eqs. 12.31 and 12.8 to estimate a soil ESP value.

For Further Reading

Goldberg, S. (1993) Chemistry and mineralogy of boron in soils, pp. 3–44. In: U. C. Gupta (ed.), *Boron and its role in crop production.* CRC Press, Boca Raton, FL. An advanced exposition on the soil chemistry of boron that amplifies the discussion in Section 12.4.

Karen, R. (2000) Salinity. Levy, G. J. (2000) Sodicity, pp G-3 to G-63. In: M. E. Sumner (ed.), *Handbook of soil science.* CRC Press, Boca Raton, FL. These two chapters provide a sound introduction to the chemistry of arid-zone soils.

Levy, R. (1984) *Chemistry of irrigated soils.* Van Nostrand, New York. Collected classic articles on a classic soil chemistry problem.

Mays, D. C. (2007) Using the Quirk–Schofield diagram to explain environmental colloid dispersion phenomena. *J. Nat. Resour. Life Sci. Educ.*

36:45. A useful introduction to the construction and application of Quirk–Schofield diagrams that include pH effects.

Quirk, J. P. (1986) Soil permeability in relation to sodicity and salinity. *Phil. Trans. R. Soc. (London)* **A316**:297, and Quirk, J. P. (2001) The significance of the threshold and turbidity concentrations in relation to sodicity and microstructure. *Aust. J. Soil Res.* **39**:1185. Two fine salty essays on the chemistry involved in the reclamation of sodic soils.

Shainberg, I., and J. Letey. (1984) Response of soils to sodic and saline conditions. *Hilgardia* **52**:1. A classic monograph on the physical chemistry and physics of soil permeability.

Sparks, D. L. (ed.). (1996) *Methods of soil analysis: Part 3. Chemical methods.* Soil Science Society of America, Madison, WI. Chapters 14, 15, and 40 of this standard reference describe methods of measuring EC_e, SAR, and calcite solubility.

Sumner, M. E., and R. Naidu (eds.). (1998) *Sodic soils.* Oxford University Press, New York. A comprehensive treatise on the causes and management of sodicity hazard.

Problems

The more difficult problems are indicated by an asterisk.

1. In the table presented here are data pertaining to the saturation extract of an Aquic Natrusalf. Use these data to calculate the corresponding equilibrium CO_2 pressures, in atmospheres.

pH	EC_e ($dS\ m^{-1}$)	Alkalinity ($mol\ m^{-3}$)	pH	EC_e ($dS\ m^{-1}$)	Alkalinity ($mol\ m^{-3}$)
8.05	0.709	1.13	8.25	0.930	1.75
8.10	0.849	1.50	8.30	1.279	1.63
8.20	0.954	1.25	8.35	1.012	1.88

2. In the table presented here are ESP values measured in the upper 0.3 m of an Alfisol irrigated for 8 years with waters of varying SAR. Use these data to calculate an average value of $K_{ex}^{Na/Ca}$ for the soil. Take $\Gamma \approx 1.3$.

	Irrigation water			
	Gage Canal	Colorado	Sulfate	Chloride
ESP	2.1	3.4	4.4	2.7
SAR ($mol^{1/2}\ m^{-3/2}$)	1.30	2.92	4.85	3.31

3. Explain conceptually, using Figure 9.2 and Eq. 12.8, why soil structure may become adversely affected as the Mg^{2+} concentration increases in a soil solution at the expense of Ca^{2+}.

4. In a study of soil permeability, it was found that the relationship between and the ionic strength above which good aggregate soil structure existed was related to SAR_p by

$$I = 0.6 + 0.56\ SAR_p \quad (0 \leq SAR_p \leq 32)$$

where I is in moles per cubic meter, and SAR_p is in units of square-root of moles per cubic meter ($mol^{1/2}\ m^{-3/2}$). Prepare a Quirk–Schofield plot based on this empirical relationship and Eq. 4.23. Compare your result with the data plotted in Figure 12.2.

*5. Derive a relationship between $\left[SO_4^{2-}\right]$ and the ESP of a soil containing gypsum. Calculate the ESP resulting from $c_{Na} = 5\ mol\ m^{-3}$ and $\left[SO_4^{2-}\right] = 0.0032\ mol\ dm^{-3}$ using Eq. 12.11. Ignore Mg^{2+} in these calculations, but consider ionic strength.

6. The kinetics of dissolution of well-crystallized calcite was observed to follow the empirical rate law

$$\text{rate}(\text{mol kg}^{-1}\text{s}^{-1}) = 4.14 \pm 0.46 \times 10^{-7}(1 - \Omega)^{1.25\pm0.16}$$

for an initial calcite solids concentration equal to $0.006\ kg\ dm^{-3}$. Estimate the value of the rate coefficient for calcite precipitation.

7. Derive Eq. 12.34 from Eq. 12.33. Indicate precisely where the assumption $\left(HCO_3^-\right)_{eq} / \left(Ca^{2+}\right)_{eq} = \left(HCO_3^-\right)_w / \left(Ca^{2+}\right)_w$ is involved in the derivation.

*8. a. Show that, in the absence of $B\,(OH)_3^0$, the p.z.c. for a soil adsorbent described by the constant capacitance model is given by the equation

$$\text{p.z.c.} = \frac{1}{2}\left(\log\ K_+ - \log\ K_-\right)$$

[*Hint*: In the constant capacitance model, the only adsorbed species are those described by Eq. 12.24 (or Eq. 8.25) and Eq. 12.25.]

b. Application of the constant capacitance model to B adsorption by a broad group of soils led to average values of K_+ and K_- given by

$$\log K_+ = 7.29 \pm 1.62 \quad \log K_- = -10.77 \pm 1.26$$

Estimate the average p.z.c. of the B-adsorbing soil constituents. What is the likely composition of the soil adsorbent that is interacting with $B(OH)_3^0$?

c. What effect will B adsorption have on the p.z.c. of the soils considered in (b)? (*Hint*: Apply the third PZC Theorem to Eq. 12.27.)

d. Use Eq. 12.26 to develop a rationale for the resonance feature near $pH = \log K$ for $B(OH)_4^-$ protonation ($\log K = 9.23$). (*Hint*: Combine eqs. 12.24 and 12.25b, then consider the effect of increasing pH on each of the two reactants in the resulting B adsorption reaction.)

*9. Goldberg et al. (Goldberg, S., H. S. Forster, and E. L. Heick. (1993) Temperature effects on boron adsorption by reference minerals and soils. *Soil Science* **156**:316.] have investigated the temperature dependence of the B adsorption envelope on specimen minerals and arid-zone soils. A variety of experimental studies indicates that the equilibrium constant for the overall acid–base reaction obtained by combining Eq. 12.25a with the reverse of Eq. 12.25b always decreases with increasing temperature.

a. What is the expected temperature dependence of p.z.c.?
b. What is the expected temperature dependence of the resonance feature in the B adsorption envelope? Does this expectation agree with the observations of Goldberg et al. (1993)?

*10. Goldberg et al. [Goldberg, S., S. M. Lesch, and D. L. Suarez (2000) Predicting boron adsorption by soils using soil chemical parameters in the constant capacitance model. *Soil Sci. Soc. Am. J.* **64**:1356.] derived Eq. 12.28 for a variety of soils representing six different soil orders. Calculate K_B for Diablo clay (a Vertisol), given $f_{oc} = 19.8\ g\ kg^{-1}, f_{ioc} = 0.26\ g\ kg^{-1}, Al = 1.02\ g\ kg^{-1}$, and $a_s = 0.19\ m^2 g^{-1}$. If $[SOH] = 3.1 \times 10^{-4} mol\ dm^{-3}$, calculate the concentration of $\equiv SOB(OH)_3^-$ that is in equilibrium with a $B(OH)_3^0$ concentration at the maximum permitted for unrestricted use of irrigation water with respect to boron toxicity hazard. Take pH = p.z.c. = 9.47 for this soil. Convert your result to an amount adsorbed in micromoles per gram given $c_s = 200\ kg\ m^{-3}$ as the solids concentration.

11. Given the following table of EC_w values, indicate which irrigation waters are likely to result in a saline root zone if a leaching fraction of 0.2 is used. What maximum SAR_{pw} values would be acceptable for these waters?

River water EC_w ($dS\ m^{-1}$)		
Salt	Colorado	Sevier
1.56	1.27	2.03

12. Give a rationale for why the equation

$$SAR_{dw} = (c_{Naw}/LF) / \left[(c_{Mgw}/LF) + c_{Ca}^{eq} \right]^{\frac{1}{2}}$$

should provide a reasonably accurate estimate of the SAR value for water draining from the root zone. Explain carefully why LF appears in the equation and why c_{Ca}^{eq} is used.

13. Gypsum is applied to a soil irrigated with water in which $EC_w = 1.3\ dS\ m^{-1}$. Given that $c_{Na} = 12\ mol\ m^{-3}, c_{Mg} = 5.2\ mol\ m^{-3}$, and $[SO_4^{2-}] = 0.014\ mol\ dm^{-3}$ in the soil solution at steady state, calculate the steady-state ESP in the soil if the leaching fraction is 0.20.

*14. Evaluate the factor within curly brackets in Eq. 12.35 for $K_{so} = 10^{-8}, I = 20\ mol\ m^{-3}$, and $[HCO_3^-]_w/[Ca^{2+}]_w = 1.0$, after converting the equation to an expression for c_{Ca}^{eq} using γ_{HCO_3} and γ_{Ca}. Compare your result with the appropriate entry in Table 1 of Suarez [Suarez, D. (1981) Relation between pH_c and sodium adsorption ratio (SAR) and an alternative method of estimating SAR of soil or drainage waters. *Soil Sci. Soc. Am. J.* 45:469.]

15. The Colorado River irrigation water referred to in Problem 11 has $EC_w = 1.3\ dS\ m^{-1}, \left[HCO_3^-\right]/\left[Ca^{2+}\right] = 1.12, c_{Naw} = 5\ mol\ m^{-3}$, and $c_{Mgw} = 1.3\ mol\ m^{-3}$. Calculate the value of adj RNa for this water using Eq. 12.35 using $K_{so} \approx 10^{-8}$ and $P_{CO_2} = 10^{-3.15}$ atm. Compare your result with the value of SAR based on $c_{Caw} = 2.6\ mol\ m^{-3}$.

Appendix: Units and Physical Constants in Soil Chemistry

The chemical properties of soils are measured in units related to *le Système International d'Unités,* abbreviated SI. This system of units is organized around seven *base physical quantities,* six of which are listed in Table A.1. (The seventh base physical quantity, luminous intensity, is seldom used in soil chemistry.) The definitions of the SI units of the base physical quantities have been established by international agreement.

One *meter* is a length equal to 1,650,763.73 wavelengths in vacuum of the radiation corresponding to the transition between the levels $2p_{10}$ and $3d_5$ in ^{86}Kr.

One *kilogram* is the mass of the international metal prototype mass reference.

Table A.1
Base units in the *Système International.*

Property	SI unit	Symbol
Length	meter	m
Mass	kilogram	kg
Time	second	s
Electric current	ampere	A
Temperature	kelvin	K
Amount of substance	mole	mol

One *second* is the duration of 9,192,631,770 periods of the radiation corresponding to the transition between two hyperfine levels of the ground state in ^{133}Cs.

One *ampere* is the electric current that, if maintained constant in two straight, parallel conductors, of infinite length and negligible cross-section, and placed 1 mm apart in vacuum, would produce between them a force of 0.2 μN per meter of length.

One *kelvin* is 1/263.16 of the absolute temperature at which water vapor, liquid water, and ice coexist at equilibrium (*the triple point*).

One *mole* is the amount of any substance that contains as many elementary particles as there are atoms in 0.012 kg of ^{12}C.

Fractions and multiples of the SI base units are assigned conventional prefixes, as indicated in Table A.2. Thus, for example, 0.1 m = 1 dm, 0.01 m = 1 cm, 10^{-3} m = 1 mm, 10^{-6} m = 1 μm (not 1 μ!), and 10^{-9} m = 1 nm. An exception to this procedure is made for the unit of mass, because it already contains the prefix *kilo.* Fractions and multiples of the kilogram are denoted by adding the appropriate prefix to the mass in units of grams. For example, 10^{-6} kg = 1 mg, *not* 1 μkg, and 10^3 kg = 1 Mg, *not* 1 kkg.

Several important units of measure are defined directly in terms of the SI base units. The time units, minute (1 min = 60 s), hour (1 h = 3600 s), and day (1 d = 86,400 s), are examples, as are the liter (1 L = 1 dm^3), the coulomb (the quantity of electric charge transferred by a current of 1 A during 1 s), and degrees Celsius (°C), which is equal to the temperature in kelvins minus 273.15. The pressure units, atmosphere (1 atm = 101.325 kPa) and bar (1 bar = 10^5 Pa), are common alternatives in the laboratory to the small SI unit, pascal (Pa). Other important units related to the SI base units are listed in Table A.3.

The unit mole is closely related to the concept of *relative molecular mass,* M_r. The relative molecular mass of a substance of definite composition is the ratio of the mass of 1 mol of the substance to the mass of 1/12 mol of ^{12}C (i.e., 0.001 kg). Although M_r is a dimensionless ratio, it is conventionally designated in daltons (Da). For example, the relative molecular mass of H_2O

Table A.2
Prefixes for units in the *Système International.*

Fraction	Prefix	Symbol	Multiple	Prefix	Symbol
10^{-1}	deci	d	10	deca	da
10^{-2}	centi	c	10^2	hecto	h
10^{-3}	milli	m	10^3	kilo	k
10^{-6}	micro	μ	10^6	mega	M
10^{-9}	nano	n	10^9	giga	G
10^{-12}	pico	p	10^{12}	tera	T

(ℓ) is 18.015 Da, which means that the absolute mass of 1 mol water is 0.018015 kg. The relative molecular mass of the smectite montmorillonite, with the chemical formula $Na_{0.9}[Si_{7.6}Al_{0.4}]Al_{3.5}Mg_{0.5}O_{20}(OH)_4$, is the weighted sum of M_r for each element in the solid: 0.9 × 22.990 (Na) + 7.6 × 28.086 (Si) + 3.9 × 26.982 (Al) + 0.5 × 24.305 (Mg) + 24 × 15.999 (O) + 4 × 1.0079 (H) = 739.54 Da. The same method of calculation applies to any other substance of known composition.

The SI unit of concentration is moles per cubic meter, which is equal numerically to millimoles per liter. The unit *molality* is preferred for measurements made at several temperatures, because it is a ratio of the amount of substance to the mass of solvent, neither of which is affected by changes in temperature. The concentration of adsorbed charge in a soil is measured in moles of charge per kilogram soil (Table A.3). For example, if a soil contains 49 mmol adsorbed Ca kg^{-1}, then it also contains 0.098 mol_c kg^{-1} contributed

Table A.3
Units related to SI base units.

Property	Unit	Symbol	SI relation
Area	hectare	ha	10^4 m^2
Charge concentration	moles of charge per cubic meter	mol_c m^{-3}	
Concentration	moles per cubic meter	mol m^{-3}	
Electric capacitance	farad	F	m^{-2} kg^{-1} s^4 A^2
Electric charge	coulomb	C	A s
Electric potential difference	volt	V	m^2 kg s^{-3} A^{-1}
Electric conductivity	siemens per meter	S m^{-1}	m^{-3} kg^{-1} s^3 A^2
Energy	joule	J	m^2 kg s^{-2}
Force	newton	N	m kg s^{-2}
Mass density	kilogram per cubic meter	kg m^{-3}	
Molality	moles per kilogram of solvent	mol kg^{-1}	
Pressure	pascal	Pa	m^{-1} kg s^{-2}
Relative molecular mass	dalton	Da	
Specific adsorbed charge[a]	moles of charge per kilogram of adsorbent	mol_c kg^{-1}	
Specific surface area[a]	hectare per kilogram	ha kg^{-1}	10^4 m^2 kg^{-1}
Viscosity	newton-second per square meter	N s m^{-2}	
Volume	liter	L	10^{-3} m^3

[a] *Specific* means "divided by mass."

Table A.4
Values of selected physical constants.

Name	Symbol	Value
Atmospheric pressure	P_0	101.325 kPa (exactly)
Atomic mass unit[a]	u	1.6605×10^{-27} kg
Avogadro constant	N_A, L	6.0221×10^{23} mol^{-1}
Boltzmann constant	k_B, k	1.3807×10^{-23} J K^{-1}
Faraday constant	F	9.6485×10^4 C mol^{-1}
Molar gas constant	R	8.3145 J K^{-1} mol^{-1}
Permittivity of vacuum	ε_0	8.8542×10^{-12} C^2 J^{-1} m^{-1}
Zero of the Celsius temperature scale	T_0	273.15 K (exactly)

[a] 1 u is defined numerically by the ratio 0.001 kg/N_A.

by adsorbed Ca. In general, as discussed in Chapter 9, the moles of adsorbed charge equal the absolute value of the valence of the adsorbed ion times the number of moles of adsorbed ion per kilogram soil. The cation exchange capacity (or CEC) of a soil is expressed in the units of specific adsorbed charge. In a similar manner, the concentration of ion charge in a soil solution ($\tilde{Q}$) is measured in moles of charge per cubic meter (mol$_c$ m^{-3}) and is equal numerically to the absolute value of the ion valence times the ion concentration in moles per cubic meter.

The values of the most important physical constants used in soil chemistry are listed in Table A.4. These fundamental constants appear in theories of molecular behavior in soils. Note that $R = N_A k_B$ and that $F = N_A e$, where e is the elementary charge.

For Further Reading

Cohen, E.R. et al. (2007) *Quantities, units and symbols in physical chemistry.* 3rd edition. RSC Publishing, Cambridge, UK. The standard reference for units of measure and definitions based on the *Système International.* It is available in an online version at http://goldbook.iupac.org.

Problems

1. Show that a pascal is the same as a force of 1 N acting on 1 m^2. (*Hint:* Use Table A.3 to express pascals in terms of newtons.)

2. Using the information in Table A.3, show that a volt is the same as a joule per coulomb. Calculate the electrode potential scale factor, RT/F (Eq. 6.18), in volts at 298.15 K. (*Answer:* 0.025693 V)

3. Use the data in Table A.4 to calculate the mass of N_A atoms of ^{12}C. (*Answer:* 0.012 kg)

4. Calculate the relative molecular mass of a Ca-vermiculite having the chemical formula $Ca_{0.7}[Si_{6.6}Al_{1.4}]Al_4O_{20}(OH)_4$. [*Answer:* $M_r = 0.7\ (40.078) + 6.6\ (28.086) + 5.4\ (26.982) + 24\ (15.999) + 4\ (1.0079) = 747$ Da]

5. Calculate the relative molecular mass of a fulvic acid "molecule" with the chemical formula $C_{186}H_{245}O_{142}N_9S_2$. [*Answer:* $M_r = 186\ (12.011) + 245\ (1.0079) + 142\ (15.999) + 9\ (14.007) + 2\ (32.066) = 4943$ Da]

6. Calculate the mass of 1 mol humic acid with the chemical formula $C_{185}H_{191}O_{90}N_{10}S$. (*Answer:* 4.027 kg)

7. Use the result of Problem 6 to calculate the concentration of humic acid in a solution containing 0.5 g humic acid L^{-1}. (*Answer:* 0.1242 mol m^{-3} = 124.2 μM)

8. Ten milliliters of soil solution contain 5.5 mg $CaCl_2$. Calculate the concentration of $CaCl_2$ and the charge concentration of Cl^- in the solution assuming complete dissociation. (*Answer:* The concentration of $CaCl_2$ is 4.96 mol m^{-3}, and the charge concentration of Cl^- is 9.92 $mol_c\ m^{-3}$. Note that 49.6 μmol $CaCl_2$ is dissolved in the 10 mL water.)

9. Given that the mass density of liquid water is 997 kg m^{-3}, calculate the molality of $CaCl_2$ in the solution described in Problem 8. (*Hint:* Derive the following relation: concentration = molality × mass density of solvent.)

10. Calculate the adsorbed charge of Ca on the vermiculite with the chemical formula given in Problem 4. (*Hint:* What is the mass of 1 mol Ca-vermiculite? How many moles of Ca charge does 1 mol of the clay mineral contain? *Answer:* 1.87 $mol_c\ kg^{-1}$.)

Index